D0165993

LEGAL ASPECTS
OF HOSPITALITY
MANAGEMENT

Second Edition

John E. H. Sherry, J.D.

Professor of Law

School of Hotel Administration

Cornell University

An Educational Foundation Textbook

National Restaurant Association
EDUCATIONAL FOUNDATION

This publication is designed to provide accurate and authoritative information in regard to the subject matter covered. It is sold with the understanding that the publisher is not engaged in rendering legal, accounting, or other professional service. If legal advice or other expert assistance is required, the services of a competent professional person should be sought. FROM A DECLARATION OF PRINCIPLES JOINTLY ADOPTED BY A COMMITTEE OF THE AMERICAN BAR ASSOCIATION AND A COMMITTEE OF PUBLISHERS.

Director of Product Development: Marianne Gajewski
Project Manager: Lisa Parker Gates
Production Manager: Virginia Christopher
Assistant Editor: Kate E. Sislin
Cover Design: Julie Streicher
Art Director: Ed Wantuch

Copyright © 1984, 1994 by the National Restaurant Association Educational Foundation.

Library of Congress Catalog Card Number: 93-80734

ISBN: 0-915452-74-X

All rights reserved. No part of this publication may be reproduced, stored in a retrieval system, or transmitted, in any form or by any means, electronic, mechanical, photocopying, recording, or otherwise, without the prior written permission of the copyright owner.

Printed in the United States of America

10   9

National Restaurant Association
EDUCATIONAL FOUNDATION

A Message from The Educational Foundation

The Educational Foundation of the National Restaurant Association is proud to present the second edition of *Legal Aspects of Hospitality Management*, a law book for hospitality operators. This book, written by Professor John E. H. Sherry, is designed to be read and used as a management tool by supervisors and managers of hospitality businesses and by students of hospitality management.

Managing a hospitality operation is a many-faceted job, requiring skills in different fields such as human resources, marketing, accounting and cost control, and areas particular to operations, such as menu management and quantity food production. These skills are essential to the process of maintaining a business and making it grow. Too often, operators are not aware of the legal aspects of managment until it is too late.

Employee relations, food liability, liquor liability, patron civil rights, and federal regulations are subjects that concern hospitality operators, who know that preventing legal problems is the best way to keep profits from being siphoned into expensive legal hassles. Since the publication of the first edition of this text, the industry has experienced an incredible increase in such regulation and litigation. Industry experts predict that this trend will continue to increase as we move toward the year 2000 and beyond, which makes education in this subject even more crucial. This book is designed to give readers an opportunity to look at hospitality operations from a legal standpoint and to develop management strategies to prevent legal problems.

Professor Sherry attacks head-on the common reaction of businesspersons that 1) the law is too complex to understand, and 2) ignorance of the law will somehow protect operators from legal pitfalls. He debunks both reactions in the first chapter and carries this theme throughout the book.

Legal Aspects of Hospitality Management, Second Edition, is a *practical* book, designed for those seeking a down-to-earth explanation of legal subjects relevant to hospitality. It is a *readable* book. For students and teachers, it contains outlines, chapter objectives, cases, review questions, and footnotes for further study. For operators and students alike, explanations are thorough and concepts are illustrated with examples.

Once read, this book should be kept as part of the hospitality manager's personal library. Its chapters, the glossary, and the appendixes can be referred to as specific questions arise.

Legal Aspects of Hospitality Management is part of the National Restaurant Association Educational Foundation's Professional Management Development (ProMgmt.[SM]) program. As such, it is combined with a student manual and final examination and is offered in schools. Students who successfully complete the final examination are awarded a certificate of completion from the Foundation. Students who fulfill the requirements of the program are awarded a certificate of program completion.

Special recognition is due the Foundation staff who assisted Professor Sherry in the development of this book. Project Manager Lisa Parker Gates and Assistant Editor Kate E. Sislin coordinated the reviews of each chapter, conducted research, and skillfully edited the text. Production Manager Virginia Christopher coordinated the design of the cover and text as well as the actual production of the book.

The Foundation is dedicated to the advancement of professionalism in the restaurant and foodservice industry through education and training. Our objective is to provide the resources managers need to reach the highest possible level of achievement in a very competitive environment.

It is our hope that *Legal Aspects of Hospitality Management* will contribute to your professional growth.

Acknowledgements

Special thanks are owed to the following people.

David A. Dittman, Dean of the School of Hotel Administration, Cornell University, for his interest and encouragement in support of this undertaking.

Tom Atkinson, Professor at Columbus Community College; Steven Barth, Assistant Professor at the Conrad N. Hilton College of Hotel and Restaurant Management at the University of Houston; Robert Bennett, Professor at Delaware County Community College; Bruce Lazarus, FMP, Associate Professor at Purdue University; Anthony Marshall, Dean of the School of Hospitality Management at Florida International University; and Andrew Schwarz, FMP, Professor at Sullivan County Community College, who contributed their technical expertise in reviewing the revised manuscript.

Bertha Hubbell, my secretary, for assisting me immeasurably in the preparation of both editions.

Finally, my wife, Eleanor Fullerton Sherry, for providing editorial guidance and moral support throughout my period of authorship.

John E. H. Sherry, J.D.

Contents

Preface to the Second Edition

"Don't bother me with a legal headache, I have a business to run. Let my lawyer handle it." This often-heard plea illustrates two common fallacies people believe about the law: 1) the law is the enemy; and 2) all legal problems are a lawyer's problems.

The objective of this book is to put to rest the notion that law is a necessary evil intended to undermine hospitality operators in the conduct of their businesses.

First, the book's premise is that *hospitality operators need help in preventing avoidable legal situations*. Second, this book takes the position that *the law works best when it is used as a management tool,* and not as a last resort. Those in the hospitality industry, as well as students about to enter it, must know the legal basics of how to run a business. The law does not become important only when something goes wrong; it is an ongoing factor in running a hospitality operation.

The traditional limitation of current business law texts is their stress on legal concepts without practical examples of law in action. Moreover, such texts are so general in scope that they lack relevance to the everyday operational needs of the hospitality manager. The pressurized demands of running a hospitality operation do not afford managers the luxury of extensive business law training. This book is designed to teach problem solving, not legal knowledge for its own sake. The hospitality industry is representative of the small businessperson in general, as well as a specialized industry, some of whose members cannot afford continuous legal services or costly legal hassles.

This is not a theoretical text, although students and operators should be able to determine what legal pitfalls are possible, through both comprehensive explanation and the legal cases presented. To anticipate and prevent problems, and to obtain the best use of this book, the reader must bring a thinking framework into play. The objective is not to have the reader think like a lawyer, but like a businessperson who is aware of the law regarding major foodservice issues, knows what is needed to prevent serious legal problems, and knows how to make the law work in his or her favor, whether on the job or in a courtroom.

Legal Aspects of Hospitality Management is organized into 16 chapters, covering the areas of law most likely to confront hospitality managers. A new chapter has been developed for the second edition dealing with current issues concerning innkeeper-guest liability.

Specific learning aids are included to provide students, managers, and instructors with a ready means of understanding hospitality law in action. Chapter outlines highlight the major topics. A "Case in Point" appears at the beginning of each chapter to structure the subject matter into a practical management context. Each chapter also contains objectives, exhibits, excerpts from relevant court decisions, summaries, questions, and end-of-chapter notes. Various state-by-state charts are contained in the appendixes to provide state provisions on key issues. A new appendix outlines some of the guidelines of the Americans with Disabilities Act, which promises to affect hospitality managers and operators greatly in the coming years. Finally, although any critical legal terms are explained in the text at first mention, a glossary is added to provide a complete reference.

Notes to Students and Teachers: Use of End-of-Chapter Notes

Notes are included at the end of each chapter to enable you to read actual statutes and opinions, since the reasoning of the court is as important as the principle of law the court states or adopts. Under the American common-law legal system, the reasoning of the court may serve as a guide in determining the outcome of future cases. In legal language, cases which establish a governing principle of law are called *precedents*. Teachers and students of foodservice law may find these notes helpful for further research.

Very few trial court decisions are published. Rather, they are filed with the clerk of the court, where copies can be made available for inspection. Appellate court decisions are published in book form. These decisions contain principles of law and reflect trends that may affect foodservice business decisions. They are published in volumes, called reports, that are numbered in order, starting with volume 1. The West Publishing Company publishes reports for all states in a National Reporter System: Atlantic (A. or A.2d), Northeastern (N.E. or N.E.2d), Pacific (P. or P.2d), Southeastern (S.E. or S.E.2d), Southern (So. or So.2d), and Southwestern (S.W. or S.W.2d). For each case, the title of a decision is *cited*, meaning reported.

For example, let us use the case of *Kauffman v. Royal Orleans, Inc.*, cited as 216 So.2d 394 (La. 1968), to illustrate how to find such a report. After the name of the parties, normally *plaintiff* (the party suing) *v.* (meaning versus or against) the *defendant* (the party being sued), the opinion is listed in a way similar to the following: Volume 216 of the Southern Reporter, Second Series (the most recent series) on page 394; the state court is abbreviated (La., for Louisiana, which, standing alone, means the Supreme Court of Louisiana), and the year 1968 (the year the decision was filed).

The majority of federal court decisions are reported in the Federal Supplement (F. Supp. for District Courts) and the Federal Reporter (F. or F.2d, for Circuit Courts of Appeal). The United States Supreme Court Reports (U.S.), Supreme Court Reporter (S. Ct.), and the Lawyer's Edition (L.Ed.) each contains all United States Supreme Court opinions. The same procedure is followed in finding a particular federal opinion. First, the name; then volume number; page number; District, Circuit, or Supreme Court designation (S.D.N.Y., or U.S. District of New York, for example); and the date.

In some cases, the names of the parties are switched to reflect the fact that the *appellant* (party appealing) is bringing an appeal against the *respondent* (the party against whom the appeal is taken). Therefore, to avoid confusion, you must sort out the parties by reading the opinion carefully. By doing so, you will understand which of the parties ultimately prevailed.

1 Introduction to Hospitality Law

OBJECTIVES

After reading this chapter, you should be able to:

1. Explain how laws apply to hospitality operators and operations, and why you need to know these laws.
2. Describe the kinds of foodservice and hotel establishments the law governs.
3. Outline the rights and responsibilities the law creates to protect both operators and customers.

Case in Point

Paul Rourke knew he would have a bad day when he arrived to open his restaurant. First, he had all weekend to think about the shortage in the cash register drawer last week. Then, when he arrived, he found that his order for ten cases of tomatoes was one case short and the weekend inventory clerk had already signed for the order. Also, the safety inspector was coming, and Paul had been cited for a safety violation the week before by the fire inspector. And the day was just beginning.

When he went through his mail, Paul noticed that a credit card company had refused to pay an overcharge on a credit card account after Paul's cashier failed to phone the company regarding the credit limit.

The seafood restaurant down the street was being picketed by local representatives of a national restaurant workers' union. Paul was worried that he was next on their list, and wondered what his rights were if he was. Paul had bought the empty lot next door for parking, but the city said the lot, only a few yards from his restaurant, was not zoned for business.

Later that night, while Paul was away making a deposit at the bank, a patron became unruly and began verbally attacking other patrons. Not only did the bartender fail to oust the patron, but he leaped into the resulting fray, breaking a bottle over the head of another customer who had only been trying to leave the premises. The next day, the injured patron's lawyer called Paul, asking him if he wanted to settle out of court or in.

At this point, Paul needs a lawyer, but a lot of his problems could have been prevented with only rudimentary knowledge about how the law affects his foodservice operation. Such knowledge would allow him to supervise his employees better, stand up for his rights as a businessperson, and prevent other problems. However, even if Paul gets a lawyer, odds are he won't know how to work with one to obtain the most effective help. Finally, going to an attorney and saying, "Get me out of this," is not much different from closing the barn door after the horse has gone—and is twice as costly.

LAW FOR FOODSERVICE AND HOTEL OPERATORS

The law is a body of rules that applies to given situations, exists to maintain order in society and in business and personal dealings, and balances personal rights with the rights of those with whom you deal. These rules are backed by an enforcement system to interpret them, hand out punishments for violations, and settle disputes.

Hospitality law is the area of consumer law that governs the legal rights of owners and operators and their responsibilities to consumers of their products or services. It includes the following duties and corresponding rights: 1) the duty to admit and the legal right to refuse to serve customers; 2) the duty to construct, maintain, and supervise the premises in a proper manner so as to protect customers, employees, and their property against harm; and the right to remove a customer who threatens the person or property of others; and 3) the duty to sell products and services in a manner fit for the intended purpose, and the right to impose lawful restrictions on sales of the product or service.

Types of Laws that Apply to Hospitality Operators

Laws vary in the way they come into being, how they are administered, and how they apply to certain segments of the population. Several kinds of law make up the body of hospitality law.

1. *Common law* or *case law* refers to rules of law that originate in court decisions. Common law is judge-made or jury-made law. American common law is inherited from the English common law, and in many states remains in effect unless altered or repealed by the state legislature.[1] In practice, common law is usually the law formulated by the highest reviewing court. These *appellate courts* have the authority to make legal rules binding on all other courts within that state or federal system.
2. *Statutory law* originates in the lawmaking bodies of the United States at either the local, state, or federal level. Criminal law is purely statutory and involves government prosecution of those who violate these laws. Punishment as a deterrent to protect society is the goal, rather than compensation for the victim.
3. *Administrative laws* include rules created by various federal and state regulatory agencies. They have the force of law and are subject to court interpretation. For foodservice operators, administrative law appears in the form of consumer laws and employment laws. In some cases, the system passing out punishment is not the court system, but a regulatory agency.
4. Historically, a person's oral promise was binding. The community enforced that promise by refusing to deal with anyone who failed or refused to keep their word. In today's impersonal world, courts and administrative agencies perform this function.

Expansion of Consumer Law[2]

Consumers, in increasing numbers, are using the law to recover compensation for wrongs inflicted by manufacturers or sellers of a product or service that has caused them personal injury or economic harm.[3] The foodservice industry is not exempt from this trend. Food services deal in foods and beverages, and while the potential for economic harm to a customer is not as serious as with some other products, the potential to cause injury or illness to patrons is always there, and with it the accompanying potential for legal snarls.

Patrons may also seek legal compensation for injuries that occur to themselves or their property on your premises that are directly or indirectly caused by foodservice managers or employees, directly or indirectly.[4]

The traditional approach is for the victim to sue and obtain a monetary judgment. However, when the amount involved is inadequate to justify hiring an attorney, or the wrong is widespread within a particular industry, consumers may lobby for a broad legislative solution. The result of *consumer lobbying* may be a regulatory agency created to license industry members, set standards of conduct, and punish violators by fine or license revocation. Whether through a regulatory agency or by statute, the objective of consumer law is to protect the legal rights of consumers.[5]

How the Law Affects Hospitality Operators

Many types of foodservice operations exist; all are affected, to varying degrees, by the law.

Commercial foodservice operations are units that compete with each other in the private marketplace for profits from selling food and beverages. They can be owned by individual proprietors or corporate chains. Commercial operations include full-service restaurants, quick-service establishments, cafeterias, drive-ins, and vending operations. *Contract feeding* is also commercial in that a contract feeder will contract with an institution, such as a hospital, to feed patients or with a recreational facility, such as a racetrack, to feed the patrons. *Business and industry food services* contract with companies to feed employees. Another type of food service that falls into the commercial category is *transportation food service*, found on planes, cruise ships, and trains. Although the food service may be incidental to the main service (transportation), it is nonetheless a commercial, and competitive, part of the industry.

Noncommercial operations make up the rest of the foodservice industry. A profit motive is *not* a legal requirement for entry into the foodservice field, and every establishment that provides food and beverages prepared, served, sold, or provided for consumption, on or off the premises, is considered a foodservice operation.

Noncommercial operations include *institutional facilities*, such as hospitals and nursing homes, military clubs and messes, schools and universities, and correctional facilities. Noncommercial operations normally earn enough in sales to recover the costs of labor and supplies, or *break even*.

All foodservice operations have in common at least two premises. First, they are all affected, in varying degrees, by the law. They are affected in the areas of labor relations, liability, and security, and by regulations affecting food services in particular and small businesses in general. Second, whether or not they operate for profit, many foodservice operations operate close to the bottom line, particularly in the beginning, and cannot afford legal snarls that are easily preventable.

Why You Need to Know the Law

Because of the complexity of modern law and the day-to-day struggle of trying to stay competitive in an unpredictable economy, it should come as no surprise that most businesspeople instinctively shy away from the subject of law and justify this attitude with the excuse: "Legal problems? Call my lawyer." However, shifting the burden to a lawyer, with the expectation that he or she can work miracles while we continue to mind the store, is dangerous because it rests on the premise that ignorance of the law is a defense to a civil lawsuit or criminal prosecution. The contrary is the rule. Ignorance of the law is usually *not* a defense;[6] nor are good intentions that fail to satisfy the requirements of the law. The absence of intent to inflict harm does not excuse responsibility.[7] It may serve merely to lessen the damages or punishment.

Ultimately, failure to prevent or minimize legal risks will jeopardize your ability to operate successfully. For commercial operations, success means profitability and customer goodwill. For noncommercial establishments, it means operating within a budget. Your business depends on good legal management—yours and your lawyer's.

Overview of Hospitality Law

For purposes of the law, hospitality managers who do not own the business occupying the premises under their jurisdiction, are considered responsible if they are in direct, continuous contact with patrons. As such, managers have duties imposed by the law regarding the protection of patrons. Managers cannot ignore these responsibilities by pleading that they are solely the owner's; they are *directly responsible* to patrons, as well as to regulatory agencies concerned with public safety and welfare.[8] Additional duties may also be imposed by the operating agreement with the owner. However, managers cannot ignore statutory responsibilities, such as those governing alcohol sales to minors, just because they were not included in the owner-manager agreement.[9] The law is not something that can be ignored when it is convenient to do so. Any agreements operators make with others must be made within a legal framework.

Managers not only are responsible for their own legal wrongs, *but may also be held responsible for the acts or omissions of employees who injure someone or who damage a patron's property while on the job.*[10] This general rule, referred to in legal terms as *respondeat superior*, makes the employer liable for an employee's conduct when that conduct occurs in the course of the employment and is intended to benefit the employer.[11] This means that managers not only have a responsibility to know the laws that affect their own business relations with patrons, but also to *train employees* to avoid legal pitfalls that might result in costly litigation with patrons or regulatory agencies.

In other ways, the law can be a great benefit to you. *For every legal duty you are required to perform, a corresponding legal duty is required of persons or parties with whom you do business.* Anyone on whom the law imposes a legal duty or for whom it creates a legal right is responsible for the performance of that duty or the exercise of that right. An individual claiming to have been injured, or a regulatory agency claiming a violation, must initiate legal action. By the same token, you are responsible for initiating legal action if you feel you have been wronged.

Foodservice and hotel law, like the law in general, attempts to balance two interests. One is your business interest of maximizing your profitability and minimizing your liability to your patrons, employees, competitors, and the government. The other is the interests of all these groups, or governmental interest in protecting their legal rights, health, welfare, and safety. You are responsible to all these groups for physical and economic injuries and losses caused by your failure or refusal to obey the law. Your responsibility to individuals is usually to compensate them for the cost of such injuries or losses. Your responsibility to government is to halt illegal acts, to pay fines for violations, or to pay damages either to the regulatory body established to police your activities or, in some cases, to aid the victims.

Since the law usually attempts to strike a reasonable balance between the economic needs of the foodservice and hotel industry and the welfare of the industry's patrons, it is important to recognize that awareness and compliance with the law is not a necessary evil, but a positive thing. Protection of patrons enhances your credibility, goodwill, and profits.

The law affects you in almost every phase of your business. The following is a capsule view of some of the important legal implications in major areas of your hospitality business.

1. As the seller of products to the public, you are obligated by the government to provide a warranty, and to see to it that your food and beverage products are in fit condition for consumption.
2. As an operator of a business, you occupy premises and are required by federal, state, and local statutes to maintain them in a safe condition. The degree to which you are obligated for this varies from place to place.

3. You are obligated to conform to laws and ordinances enforced for businesses.
4. You are obligated both to yourself and to the law to keep your premises and customers reasonably secure from crime by patrons, third parties, and employees.
5. You are obligated to obey the consumer protection laws and civil rights laws of the federal government, and, increasingly, those of state and local governments.
6. As an employer, you are obligated to obey all employment laws.

SUMMARY

It is a major premise of this book that prevention of legal pitfalls, consistently practiced, not only protects the public but also enhances your credibility as a hospitality manager, and ultimately the credibility of the entire industry. Foodservice owners and managers must recognize those aspects of food service that have legal implications, train themselves and their staffs to comply with regulations, and, most important, prevent needless exposure to liability.

A hospitality operator cannot usually prevent patrons from filing a lawsuit or complaining to a regulatory agency once they are determined to do so. However, obedience to statutory requirements and the use of reasonable care will, in many cases, allow the operator to prevail in any litigation, contain legal costs, and remain competitive.

QUESTIONS

1. How is the law an aid to hospitality owners and operators?
2. Describe the differences between statutory law, common law, and administrative law.
3. What are some of the specific ways that consumer laws affect hospitality operators?

4. What is the common negative attitude of businesspersons regarding the law? Why is this attitude faulty?
5. A patron is injured in a restaurant by a careless employee and decides to sue. Who may be held responsible and why?

NOTES

1. The 1895 Revised Statutes, originally enacted in 1840 by the Texas Republic, contain the following representative language: "The common law of England so far as it is not inconsistent with the Constitution and laws of this State shall, together with such Constitution and laws, be the rule of decision and shall continue in force until altered or repealed by the Legislature."
2. See generally J. McCall, *Consumer Protection: Cases, Notes and Materials* (1977); D. Epstein and S. Nickles, *Consumer Law in a Nutshell* (2d. ed. 1981).
3. See generally D. Noel and J. Phillips, *Cases on Products Liability* (1977); D. Noel and J. Phillips, *Products Liability in a Nutshell* (2d. ed. 1981).
4. See J. Sherry, *The Laws of Innkeepers* 3d. ed., sec. 11:13 (Restaurant Keeper's Duty to Protect Patrons), secs. 15:1–15:7 (Responsibility Arising from the Sale of Food [and] Beverages), secs. 19:1–19:8 (Responsibility of Restaurant Keepers for Patrons' Property) (rev. ed. 1981).

5. See R. Anderson, *Social Forces and the Law* 1–31 (2d. ed. 1981), E. Tucker, *Adjudication of Social Issues* Chapters 19–20, 296–331 (2d. ed. 1977).

6. See W. LaFave and A. Scott, *Criminal Law* 356 *et seq.* (1972).

7. This rule is illustrated by the offense of selling liquor to minors by those licensed to sell alcoholic beverages. This crime does not recognize the defense that the licensed vendor believed the minor was of age; or that the minor appeared to be of age; or that the minor represented himself or herself as having reached the required age; or that the minor produced false credentials showing that he or she was of age. See LaFave and Scott, *supra*, note 6, at 359 nn. 23–27.

8. See *Fitzgerald v. 667 Hotel Corp.*, 103 Misc. 2d 80, 426 N.Y.S.2d 368 (Sup. Ct. 1980). (Both owner and occupier held liable to injured residential and commercial tenants. Owner's statutory duty to keep building in repair held *nondelegable* to occupier of building.)

9. Statutory duties such as these operate independently of the agreement between owner and occupier of the premises. This is because these duties were enacted to protect the public, whose rights cannot be diminished by a contract to which the public was not a part.

10. *Block v. Sherry*, 43 Misc. 342, 87 (N.Y.S. 160 1904). (Restaurant owner held liable for waiter's negligence in spilling water on patron's clothing.)

11. *Fruit v. Schreiner*, 502 P.2d. 133 (Sup. Ct. Alaska 1972). (Employer responsible for intoxication of employee causing harm to third party.) See also *Riviello v. Waldron*, 47 N.Y.2d 297, 391 N.E.2d 1278, 418 N.Y.S.2d 300 (1980). (Operator of tavern liable for injuries inflicted by employee upon patron.)

Hospitality Operators and Government Regulations

OUTLINE

Other Areas of State and Local Regulation
    Sanitation
    Criminal Trespass and Disorderly Conduct
    Blue Laws
    Truth-in-Menu Regulations
    Kosher Food
    Smoking Regulations
    Zoning Laws
    Building and Safety Codes
Living with Regulations
Summary
Questions

OBJECTIVES

After reading this chapter, you should be able to:

1. List and describe the federal agencies regulating the sale of food and beverages, employment practices, and safety.
2. Identify federal laws that act as models for state laws.
3. Discuss state and local regulations, including alcohol sales, sanitation, criminal trespass, disorderly conduct, blue laws, truth-in-menu regulations, kosher food, smoking regulations, zoning laws, and building and safety codes.

Case in Point

You open a restaurant in a retail district in your community and secure a liquor license. The community then enacts a zoning ordinance prohibiting the sale of alcoholic beverages by restaurants in that district. Can you be forced to surrender your liquor license because your district has been zoned to ban liquor sales in restaurants?

*   No. Only the State Liquor Authority has the right to regulate traffic of alcoholic beverages. Having granted you a liquor license, the Authority's decision overrides the community zoning ordinance.[1]*

The foodservice industry is highly regulated, with regulations appearing at the federal, state, and local levels. A foodservice operation is more immediately affected by the state and local regulations; however, federal regulatory agencies are important, as a few of them have broad powers that can affect businesses.

FEDERAL REGULATIONS

Federal regulations affect the preparation and sale of food and beverage products, as well as such nonfood areas as advertising, labor, and even the organization of a business. Federal regulations often result in the formation of federal agencies to enforce or amend rules. Enforcement may range from an agency with tight controls and harsh punishments to a laissez-faire agency that serves simply as a federal "watchdog."

Federal regulations tend to result from abuses in certain areas of business over a period of time. For example, turn-of-the-century conditions in meat packing resulted in much of our modern-day federal regulation of meat processing and packing. Abuses of immigrants and children by employers in some textile industries resulted in stiff labor laws for all employers. Some federal regulations historically have fallen by the wayside as the abuses that prompted them have ceased. Others continue and are even expanded or amended as new situations arise.

Federal Agencies Regulating Sale of Food and Beverages

From the time of the Industrial Revolution and the related development of a complex, highly diversified food and beverage production and distribution system, the federal government has attempted to protect the public against unwholesome products.[2] Since food products are necessary for human survival, and the risk of contagious diseases caused by unfit food and beverages is so great, a number of regulatory bodies have been created to eliminate or minimize risks to consumers.

The role of federal regulatory agencies in the food products field is twofold: 1) to prevent unsafe food products from reaching the marketplace; and 2) to maintain a minimum level of food quality, to furnish adequate and reliable descriptions of foods, and to provide standard measures and descriptions of food quantities.[3] The first role aims to regulate the food production industry and has little direct effect on the foodservice business. The second is broader. It covers all foods produced for human consumption. The following federal agencies regulate the sale of food and beverages.

Food and Drug Administration (FDA)   Created in 1906 by the Food, Drug, and Cosmetic Act, and established in its present form in 1938, the FDA is empowered to: 1) require truthful and informative labeling and proper packaging of foods and beverages; 2) protect against commercial misrepre-

Exhibit 2.1   Nutrition Label

	NUTRITION FACTS
Standardized	Serving size: 1/2 cup (114g)
	Servings per container: 4

New requirement

Calories 260	Calories from fat 120

New requirement
The daily values allow an evaluation of food in terms of a total daily diet of 2,000 calories. The suggested daily intake for these items is shown in the table below.

AMOUNT PER SERVING	% DAILY VALUE*
Total Fat 13g	20%
Saturated fat 5g	25%
Cholesterol 30mg	10%
Sodium 660mg	28%
Total Carbohydrate 31g	11%
Sugar 5g	
Dietary fiber 25g	0%
Protein 5mg	

New requirement
Nutritional values for these items were not required on the previous label.

Vitamin A 4% • Vitamin C 2% • Calcium 15% • Iron 4%

*Percentage (%) of a Daily Value are based on a 2,000 calorie diet. Your Daily Values may vary depending on your calorie needs:

New requirement
This table shows recommended daily intake for two levels of calorie consumption.

Nutrient		2,000 calories	2,500 calories
Total Fat	Less than	65g	80g
Sat. Fat	Less than	20g	25g
Cholesterol	Less than	300mg	300mg
Sodium	Less than	2,400mg	2,400mg
Total Carbohydrate		300g	375g
Fiber		25g	30g

New requirement

1g Fat = 9 calories
1g Carbohydrate = 4 calories
1g Protein = 4 calories

Copyright © 1992 by The New York Times Company. Reprinted by permission.

sentation by enforcement of food and beverage standards; and 3) provide for food safety. It enforces these objectives in the federal district or trial courts by seizure of unfit products, criminal prosecution of violators, and obtaining of court orders barring further distribution or sale.

FDA labeling guidelines are still being developed, but within the next several years the guidelines are expected to affect the kind and amount of information foodservice operations must provide. Currently, when food labels are used, they must give several basic pieces of information: product name; net weight; name and address of the manufacturer, packer or distributor; and ingredients in order of decreasing weight. **Exhibit 2.1** shows a typical nutrition label.

FDA regulations concerning commercial misrepresentations are aimed mainly at food growers, producers, and processors rather than retailers, and apply only to products prepared for *interstate commerce*, in which a food product prepared or grown in one state is shipped to another state for sale. Since most foodservice operators have a single place of business within one state and obtain most of their food items from local markets (*intrastate commerce*), they are usually not subject to FDA regulations. However, where Maine lobsters, for instance, are ordered by a New York restaurant for service to New York consumers, the commerce is interstate rather than intrastate, and FDA regulations apply.

Local commerce is often covered by state food and drug acts patterned on the federal model. These laws may be stricter, but cannot be more lenient, than the federal law.

Nonetheless, foodservice operators are indirectly subject to these FDA regulations since the FDA has the power to step in and seize harmful products, regardless of whether the buyer caused or contributed to the seller's conduct.[4] An operator may be completely innocent of any violation of the Food, Drug, and Cosmetic Act, yet be subject to the drastic remedies the FDA can enforce.

Many state and local ordinances concerning food safety and sanitation are patterned after the Model Food Service Sanitation Ordinance recommended by the FDA. **Exhibit 2.2** is a list of the main provisions of that ordinance.

The Federal Food, Drug, and Cosmetic Act provides for severe criminal penalties for violations of the Act.[5] Normally, the person or business against whom the enforcement action is taken has the right to contest the action in a court hearing, conducted by the federal district court where the action was taken by the FDA.[6]

Exhibit 2.2 Main Provisions of the FDA's Model Food Service Sanitation Ordinance

1. Food care: supplies, protection, storage, preparation, display and service, transportation
2. Personnel: health, personal cleanliness, clothing, hygiene practices
3. Equipment and utensils: materials, design and fabrication, equipment installation and location
4. Equipment and utensil maintenance: cleaning, sanitizing, storage
5. Sanitary facilities and controls: water supply, sewage, plumbing, toilet facilities, lavatory facilities, garbage disposal and solid waste management, integrated pest management (IPM)
6. Construction and maintenance of physical facilities: floors, walls, ceilings, cleaning facilities,lighting, ventilation, dressing rooms and locker areas, poisonous materials
7. Mobile foodservice units
8. Temporary foodservice units
9. Compliance procedures: issuance and suspension of permits
10. Examination and condemnation of food

Reprinted with permission from *Applied Foodservice Sanitation Certification Coursebook, Fourth Edition.* Copyright © 1992 by The Educational Foundation of the National Restaurant Association.

Federal Trade Commission (FTC)   Another area of federal regulation concerns deceptive or unfair advertising of food and beverages. The Federal Trade Commission (FTC), established in 1914, has authority over this area in cooperation with the FDA. This authority applies only to interstate commerce, or false and misleading advertising claims for food products originating in one state for distribution in another. Such advertising claims include those found on labels on food and beverage products, in television and radio commercials, and in advertisements in newspapers and magazines.[7] It may also regulate the wording on menus shipped by a foodservice chain from its home office to out-of-state retail outlets.[8]

The FTC has the power to stop false food advertising, including ads that fail to reveal that products are dangerous or that their use, under certain conditions, can cause harm.[9]

A false advertisement for food, when the use of the product may injure health or where the ad is intended to defraud or mislead the buyer, is a *misdemeanor* (a lesser crime than a felony, but more serious than a traffic violation).[10] The FTC requires that advertisers make no statements that are *materially misleading*, meaning that consumers might not have purchased the food product if they had known the facts.[11] The deception may be *implied*; that is, it need not be expressly stated.

As a foodservice operator, you could violate this federal law if you share in or underwrite the cost of out-of-state advertising by your supplier, and the advertising contains materially false and misleading statements. The fact that you did not cause or contribute to the false statements is no defense to an FTC action. If you own a chain of restaurants in different states and prepare the advertising for all of them, FTC rules apply.

United States Department of Agriculture (USDA)   The Department of Agriculture establishes and enforces standards, and inspects and grades a variety of food products, including meat and poultry. This agency generally works in favor of foodservice operations by enforcing standards of wholesomeness that affect your buying of food products. In the case of fish, inspection is voluntary. Inspection services are provided by the Commerce Department. Meat and poultry inspection is required by law; quality grading is voluntary. These functions enable foodservice buyers to evaluate and select meat and poultry products with reasonable assurances of quality and safety, and to obtain the right cuts of meat to suit their operations.

Recently, consumer rights groups have called for inspection of fish, presently exempt from regulation, and for improved standards governing meats, particularly in regard to laboratory testing to detect disease.

Federal Control in Practice

Federal law directly governs multistate chain and franchise operators who obtain food ingredients from central stores or commissaries outside their own state. It also applies to any out-of-state food and beverage purchases operators make. State and local laws apply to commerce within a single state or city.

In cases of conflict between federal, state, and local regulations, *the federal regulations are controlling*. When federal acts do not prohibit state and local legislation on the same subject, the *state and local acts may strengthen, but may not weaken, the federal laws*. Some cities are authorized to enact their own laws, independently of the state, but they may not weaken the state's regulations. They are also free to adopt state codes.

The threat to public health and welfare that lack of uniformity would cause in the area of food and beverages explains why federal law governs interstate sales. States are free to adopt their own standards within their own borders. State standards vary, as does the amount of money provided for enforcement. The penalties for violations also vary. However, a degree of cooperation and unity may yet emerge.

Nonfood Areas of Federal Regulation

Among the nonfood areas regulated by the federal government are those related to employment practices and safety. The following federal agencies affect foodservice operators.

Department of Labor   The Department of Labor affects all employers. This federal department sets hourly guidelines and labor standards, defines work weeks, and regulates child labor. The Department of Labor also sets the minimum wage. Each state is free to set its own wage, which may be higher than, but not lower than, the federal minimum. The Department of Labor also sets tip credits administered by the Internal Revenue Service.

Equal Employment Opportunity Commission (EEOC)   The Equal Employment Opportunity Commission was set up as the enforcement agency for the employment section of the Civil Rights Act of 1964. This agency's authority was further broadened by the Equal Employment Opportunity Act of 1972.

The EEOC enforces equal opportunity laws regarding race, color, religion, national origin, sex, and pregnancy. All employers with 15 or more employees or who engage in interstate commerce are regulated by the EEOC.

This agency differs from other federal agencies in that it may initiate lawsuits on behalf of employees against employers. It also investigates discrimination complaints by employees, and has the power to request company records as part of an investigation.

The EEOC may directly affect employers who fail to comply with federal civil rights laws. In addition, many states have their own civil rights laws, which may require even stricter compliance. The chapter on staff selection discusses employment civil rights laws in more detail. States and localities have expanded the classes of wrongful discrimination, to include such topics as marital status and sexual orientation.

Federal Occupational Safety and Health Administration (OSHA)[12]
OSHA is responsible for enforcing measures protecting the job-related health and safety of employees.[13] The goal of this agency is to prevent industrial health hazards and injuries, the costs of which are ultimately paid by operators and their customers.

Department of Justice   The hospitality industry must comply with federal and state laws that forbid certain types of activities the laws broadly define as *antitrust*.[14] The Justice Department is the arm of the executive branch of government that, among other things, enforces the federal antitrust laws. These laws are designed to prevent trade monopolies. For example, foodservice operators may not cooperate or agree to cooperate with competitors to fix the prices of the products and services provided to customers. Antitrust laws also prevent groups of competitors from monopolizing the industry to prevent others from entering.[15] Monopolies hurt two groups: the competitor, who is frozen out of the market; and the consumer, who must pay higher prices not justified by the normal forces of supply and demand.

Severe criminal and civil fines, as well as other forms of enforcement, are dealt by the Department of Justice.[16]

Federal Immigration and Naturalization Service (INS)   Historically the INS prosecuted illegal aliens, not their employers. Today this exemption is removed, which requires strict compliance with INS regulations.

Federal Trade Commission (FTC)   The FTC regulates some aspects of franchising agreements for hotels, motels, and restaurants.[17] The chapter on franchising discusses this type of regulation in more detail.

Fire Safety Regulations   One of the most severe problems that confronts operators of multistory hotels, and to a lesser degree, operators of large on-site commercial foodservice facilities, is the threat of fire and smoke inhalation. State and local standards that deal with the issue have lacked uniformity, making compliance as well as enforcement difficult, expensive, and frustrating.

In 1990 Congress, reacting to several devastating hotel fires in the United States and abroad, and to the unfavorable media response they generated, enacted the Federal Hotel-Motel Fire Safety Act. This is the first federal law aimed specifically at fire prevention rather than fire response within the lodging industry. In addition to existing laws mandating installation of sprinkler and smoke detection systems in all properties, the act prohibits federal employees from obtaining reimbursement for food and lodging expenses incurred in hotels and motels not providing these devices. An exception governs hotels and motels that are no higher than three stories and meet other requirements. The law impacts heavily on those properties not in compliance, since major private convention hosts would hesitate to book conventions into such properties and face potential legal liability that might arise from a hotel fire that injures convention guests. The federal law sets a uniform minimum fire safety standard and avoids the problem of *grandfathering* existing properties: state and local laws require such installations in new, but not existing properties. Grandfathering refers to exempting existing premises from the requirements of laws passed after their construction.

Federal Laws as Models

Federal laws and agencies often serve as models for similar state laws and agencies. One such model is the FDA Food Service Sanitation Ordinance, which has been adopted by a number of states.

Many state and local ordinances are patterned on the Model Food Service Sanitation ordinance recommended by the Food and Drug Administration. The following outline of the main provisions of that ordinance will illustrate the range of interests of all levels of government.

- Protection, storage, preparation, display, service, and transportation of food
- Health, personal cleanliness, clothing, and hygiene practices of employees
- Materials, design, fabrication, installation, and location of equipment and utensils
- Cleaning, sanitizing, and storage of equipment and utensils
- Sanitary facilities and controls: water supply, sewage, plumbing, toilet facilities, lavatory facilities, garbage and refuse, insect and rodent control
- Construction and maintenance of physical facilities: floors, walls and ceilings, cleaning facilities, lighting, ventilation, dressing rooms and locker areas, poisonous materials
- Mobile units
- Temporary units
- Compliance procedures: issuance and suspension of permits
- Examination of food

Some states go beyond the FDA model and enforce stricter compliance of sanitation standards. State governments are encouraged, but not required, to use federal guidelines in lawmaking. This results in state health and sanitary codes, agriculture and marketing laws, state or city truth-in-menu laws, indoor clean air laws, and other consumer protection measures that affect your business.

STATE AND LOCAL REGULATIONS

Foodservice operators are most likely to feel the regulatory pinch not at the federal level, but at the state and local levels. Here is where the cost of regulation, in time and money, is most often felt by owners and operators of foodservice businesses. State and local governments regulate your business in a number of areas and with varying degrees of enforcement, so that you may not always know by whom and to what extent you are being regulated. Some states regulate more than others. However, one area where states are consistently strict and have broad enforcement powers is in the regulation of the sale of alcohol.

Regulation of Alcohol Sales

The sale of alcoholic beverages is heavily regulated. Many food services offer alcohol to patrons in order to increase profits. The production of alcohol is closely regulated by the federal government, *but its retail sale for on or off premise consumption is regulated primarily by the states.*[18]

History explains this heavy regulation. Liquor has often been viewed as a moral evil to be stamped out. The temperance movement, promoted politically by the Temperance Party that flourished in the 19th century, eventually brought about Prohibition. This was accomplished by ratification of the federal Volstead Act, outlawing commercial traffic in liquor, beer, and related beverages, as the Eighteenth Amendment to the Constitution on January 16, 1919.

Because of the demand for alcohol by many Americans, and the denial of consumption through legal channels, criminal gang members fought rival gang members to gain control over the lucrative black market in alcohol. When Prohibition ended in 1933 with the Twenty-First Amendment, there were strong fears that notorious criminal syndicates, collectively called "the mob," would retain their grip over the liquor traffic to further their other criminal enterprises. Such fears, together with renewed public clamor for temperance, caused most states to create regulatory agencies to license and police the commercial sale of alcohol at all levels, including retail sales by hotels, restaurants, taverns, clubs, and other authorized outlets.[19]

State Licensing Regulations   Today, you have the option to offer or not offer your patrons alcohol, as you choose, but once you opt to sell, your state regulates your entry into the field and your ongoing conduct for as long as you choose to market such beverages.

Each state has a liquor authority or liquor commission to regulate the sale of alcoholic beverages. States set up the regulations and the enforcement apparatus. The authority may be very broad, overlapping into gray areas such as regulation or restriction of prostitution and loitering, or it may be limited simply to regulating liquor sales to adults and minors. The states have the exclusive right to regulate liquor sales to consumers by creating an authority or commission. All states have created such bodies.

Although every state has the authority to regulate traffic in liquor, some states have a "dry option," meaning that some communities may opt to prohibit the sale of liquor completely. Such areas are often found in rural communities with a strong religious bias against alcohol consumption. In these cases, operators would fall under local jurisdiction and would be expressly prohibited from selling alcohol. Once the city or community opts to allow the sale of alcohol, however, the state standards take precedence over any local laws regarding liquor. Some communities allow only the sale of beer and wine. Even so, these limited sales are still controlled by the state.

Operators in a "wet" area must usually apply for a liquor license from the state, which will enable them to sell alcohol. (Foodservice businesses usually come under slightly different regulations than package liquor stores. In some areas, the standards for sales of packaged liquor are less stringent.)

License requirements vary from state to state, with some states more lenient than others. The following factors are used in New York, which is considered a strict state:

- Operating experience
- Financial assets and source of financing
- Good character and freedom from organized criminal activity
- Compliance of premises with all building code requirements
- No interest in any wholesale liquor business, brewery, winery, or distillery
- Physical locations of premises within prescribed distance from church or school
- Number of liquor licenses within predetermined area in which premises are located
- Environmental factors, such as potential noise and traffic that might be created by your operation[20]

Of these requirements, the most sensitive and most open to interpretation is the "good character and freedom from organized criminal activity" condition. New York and most other states frown on any criminal record but are usually concerned only with criminal convictions involving *moral turpitude*, meaning immoral criminal conduct, unsavory business dealings, or more serious crimes. New York does not permit any association with organized criminals, either directly or indirectly through financial support. This is true in most other states. Any past dealings with crime syndicates in whatever form will automatically disqualify an applicant.

The courts have the power to review the denial of any liquor license application, either when an application is first made or at the time of license renewal. The test used is whether the liquor authority or board acted arbitrarily in denying the license. This means that the reviewing court must find in the record no reasonable basis on which the authority could have acted. Otherwise, the authority's or board's actions are presumed valid and such bodies are given wide discretion in denying a new license or renewal of a current license.[21]

The second major regulatory power of liquor control bodies is license revocation or suspension. Regulatory agencies have the power to regulate the conduct of license holders once their licenses are granted. All of the grounds for disciplinary action against violators apply equally to all license holders, regardless of the type of license held.[22] The kinds of liquor licenses vary from state to state, but the grounds for suspension or revocation are fairly uniform.

Disorderly Conduct—Most states interpret this term to include allowing fights or assaults by either patrons or employees, creating a disturbance or nuisance, or otherwise violating public order. Threats to the peace and safety of the surrounding neighborhood would fit the definition of a nuisance. Disorderly conduct also may include public displays of and attempts to satisfy sexual urges. One single or isolated incident will not establish such a violation.

Gambling—Gambling is any game of chance where people pay for the opportunity to receive any property offered in return for their payment. It is illegal in many states. Some states make exceptions for charitable raffles, bingo games, pari-mutuel (mutual stake) betting, and sales of lottery tickets in licensed operations. Gambling is typically prohibited on licensed premises except in states that allow casino gambling.

As in the case of disorderly conduct, there must be proof that the license holder allowed gambling. This means proof that the license holder knew or should have known that gambling was in progress. An isolated, onetime instance of gambling is normally insufficient for revocation of a liquor license.

Prostitution—Prostitution and related forms of sexual misconduct, such as soliciting and inducing minors to commit sexual acts, are uniformly prohibited. Proof that the solicitation of customers took place only once is usually not sufficient reason for revocation of a liquor license. More than one isolated incident is usually required to justify license revocation. The license holder must have consistently *tolerated* the misconduct.

Adult Entertainment—Traditionally the courts made a distinction between partially or completely nude entertainment and lewd, indecent, or obscene entertainment. However, the Supreme Court of the United States has ruled that the states are free to ban all forms of nude dancing, including topless entertainment, and that the First Amendment right to free speech and expression must bow to the Twenty-First Amendment, which gives states very broad powers to regulate conduct on premises licensed to serve alcohol.[23]

Narcotics Traffic—The sale or distribution of narcotic drugs by employees or patrons is disorderly conduct under most liquor control laws and is forbidden. A license holder will not be held responsible for the criminal sale of narcotics by employees unless the sale forms a pattern or practice for a long enough time to put the proprietor on *constructive notice*—meaning that he or she should have known what was happening had he or she exercised reasonable diligence to supervise the premises.

Show Boat of New Lebanon, Inc. v. State Liquor Authority
Supreme Court of New York, Appellate Division
33 A.D.2d 954, 306 N.Y.S.2d 850 (1970)

The following case is an example of how poor management almost cost an operator a liquor license; moreover, it is a rare example of a successful appeal of a state liquor authority decision.

Facts: The State Liquor Authority canceled the restaurant operator's liquor license after it was proved that drug trafficking was taking place on the premises, thus causing the operation to become disorderly and in violation of the license agreement. The operator appealed and the court upheld the Authority's determination, but decided that the cancellation was excessive punishment and that a three-month suspension would be more appropriate.

Reasoning: The court said that: 1) the decision as to whether a licensed operation has become disorderly and is supported by substantial evidence is up to the State Liquor Authority; 2) a single act of disorder with no proof showing the licensee knew or should have known of the disorder is insufficient for cancellation; 3) the alleged disorder need not have been known to the licensee, but it is sufficient if he or she should have known (the court noted that the operator "had knowledge or had the opportunity through the exercise of reasonable diligence to acquire knowledge" about trafficking in narcotics and drugs occurring and allowed it to take place); 4) police officers obtained narcotics from two of the licensee's employees on separate occasions, and this was enough to determine that there was drug traffic on the premises that the licensee should have been aware of; but, nonetheless, 5) the cancellation was excessive punishment and suspension of three months was more appropriate.

Conclusion: In noting that the operator could or should have been aware of the trafficking through "reasonable diligence," the court made an interesting observation. The operator must make sure that a manager is aware of what the employees are up to on the job—and during on-premises break time—and guard against liquor law violations. This also holds true for patrons who may carry on a little illegal (or illicit) business on poorly managed premises—at the operator's expense.

Sales to Minors—In virtually all states, sale of alcohol by a license holder to any person under the state's legal drinking age is both a crime and a violation of the liquor control statutes. Each liquor authority has the right to suspend or revoke the license, regardless of whether the minor concealed or lied about his or her age. In most localities it is a misdemeanor to serve a minor alcohol.[24]

The law does not require proof that the operator intended to serve a minor alcohol. Moreover, the fact that he or she is found not guilty of the crime does not prevent the liquor authority from suspending or revoking the liquor license. The two types of cases are not the same. One is a criminal trial, before a court of law; the other is an administrative hearing before an officer employed by the state authority. A finding of not guilty by the criminal court does not require the administrative officer to dismiss the authority's charges at the hearing. A single finding of a sale to a minor is normally insufficient to justify revocation. A pattern or practice of such sales would justify it.

Consistently serving minors is a certain, proven way to lose a liquor license, or at least have it suspended.

The practical problem for the foodservice operator and employees is to know how to verify age. Liability may also be imposed for failure to verify age or for inadequate verification of age, if a minor injures someone. It is not enough just to ask for the required forms of identification. They may be phony and must be verified by careful observation of the person to whom they allegedly belong. As the following case illustrates, being in a hurry or otherwise ignoring the statutory requirements regarding service to minors carries a heavy penalty, as the following case shows.

5501 Hollywood v. Department of Alcohol Beverage Control
Court of Appeals of California
155 Cal. App. 2d 48, 318 P.2d 820 (1957)

Facts: After being found guilty of selling liquor to a minor in violation of California statute, the defendant, a tavern owner, had his liquor license suspended. He appealed the Alcohol Beverage Control Board's decision. A superior court reversed the Board's decision. Then the Board appealed, the defendant arguing that he had met the requirements of a companion statute authorizing acceptance upon demand of a valid driver's license, Selective Service certificate, or similar form of identification. His defense was that he acted in good faith to obtain proper identification. The court of appeals rejected the argument, on the grounds that the identification must be checked against the physical characteristics of the holder, and that her appearance in this case did not match the age stated on the driver's license she presented to the defendant.

Reasoning: The court noted that: 1) the defendant had failed to adhere to the present demands of the statute requiring bona fide evidence of identity, and merely producing a driver's license was not enough because the person bearing the card may not be the person described on the card; 2) the girl, Peggy, was "too young in appearance to be 21; she weighed 19 pounds more than the person to whom the license was issued; she was three and one-half years younger, and had blue eyes instead of hazel." The referee who had acted for the Board in making the initial decision had said that the discrepancies between the license and Peggy's appearance, " . . . were such that a reasonable, prudent licensee or employee in premises licensed for the sale of alcoholic beverages would not in good faith accept said driver's license as a bona fide documentary evidence of the identity of Peggy. The apparent discrepancies between the driver's license presented and the minor presenting it were sufficient to put appellant's employee on guard and to indicate that further inquiry was in order."

Conclusion: This case destroys the fallacy of assuming that mere demand of identification is enough to satisfy liquor control laws. Identification must be verified by checking the appearance of the presenter. Virtually all states authorize suspension or revocation of a liquor license, independently of criminal penalties, for sales of alcohol to minors. Any statutory defense, such as good faith reliance upon a false driver's license or other authorized proof of age, is strictly interpreted in favor of the minor, even though the minor may have sought purposely to mislead the operator. Operators must act reasonably in accepting any otherwise valid proof of age. If the proof of age accepted deviates in any degree from the appearance of the patron, it is a risk to serve that person. The law imposes an affirmative non-delegable duty of reasonable care to check proof of age on licenses. The carelessness of employees is treated as the licensees' carelessness, whether or not he or she is present at the time.

The following are guidelines for verifying drinking age.

- Know what your state law requires as acceptable proof of age and demand to see that proof.
- Do not accept any proof of age other than items specifically mentioned in your state statute.
- Ask the owner of an identification to verify information, such as an address or social security number.
- Hold and feel the identification to catch forgeries.
- Do not serve anyone who cannot provide acceptable proof of age, no matter how above age he or she appears nor how authentic the proof may seem.
- Always check the required proof of age against the physical appearance and characteristics of the person presenting it. If any discrepancies exist, either demand other acceptable proof or do not serve that person.

Train and supervise your bar staff and table servers to follow the above rules.

Sales to Intoxicated Persons (or to Persons Actually or Apparently Under the Influence of Alcohol)—Sale of liquor to intoxicated patrons or persons believed to be under the influence of alcohol are forbidden by most states. Such sales also are crimes in many states, as well as violations of the beverage control laws. However, unlike the prohibited sales to minors, the license holder is given some leeway in determining whether or not the patron is actually or apparently intoxicated. If the patron, on the basis of speech, walk, and manner, appears sober, he or she may be served without running afoul of the law. This is called the *objective appearance test*. A few states refuse to apply the objective appearance test. This means that you may jeopardize your license even if you honestly believe a patron to be sober and serve him or her.

Sales to Known or Habitual Alcoholics—The same factors that govern responsibility for sales to intoxicated persons apply to sales to known habitual drunkards. The critical factor is the word *known*. If the person serving alcohol does not know that an otherwise sober person is an alcoholic, the server does not violate the law by serving that patron.

Dramshop Liability   The other area of concern to foodservice operators is *dramshop liability*. Dramshop is the term used for the state laws that govern operator liability for liquor service to patrons that results in injuries or death to third persons. In many states, operators may be directly responsible to third parties or their families for injuries or death caused by liquor service to patrons. This and other areas of liquor liability are covered in the chapter on sales of food and beverages.

Judicial Review

Judicial review is usually available to an applicant whose license has been denied, suspended, or revoked. In the case of initial denial, the courts are generally more liberal in reviewing the agency's grounds, especially if they feel the agency has overstepped its powers.[25]

In cases of suspension or revocation, it is generally more difficult to have the agency's decisions overturned.[26] An exception would be a court finding inadequate evidence to support an agency's decision.[27]

Since the profitability or very existence of an operation is at stake in many cases, adverse decisions by alcohol control agencies are appealed more frequently than are those for health violations.

The scope and regulation of liquor laws are fairly uniform, but each state is free to add or drop provisions or to interpret its liquor laws strictly or liberally. Usually the courts, in reviewing liquor law violations, are especially concerned with sales to minors, and this is the area of greatest risk to a liquor license.

OTHER AREAS OF STATE AND LOCAL REGULATION

Sanitation

The regulation of the foodservice field in the area of sanitation begins at the federal level, where uniform standards and enforcement mechanisms are set. These regulations, however, are generally aimed at the source or the originator of the food product, not necessarily foodservice operators. These relatively few, but highly complex, laws are supplemented by state and local health sanitary codes, with their own enforcement schemes, and usually less uniformity. State health and sanitary codes are enforced by municipal health departments.[28] Larger cities usually set up their own agencies for health enforcement. Smaller communities and rural areas often come under the state umbrella, with regulation and enforcement handled by the county. It is these regulatory agencies that supervise the day-to-day operations, investigate violations, and prosecute violators. Local regulation is the cutting edge of the regulatory apparatus regarding sanitation.

Food and beverage operators must comply with state and local health codes. These codes usually regulate all aspects of the storage, preparation, and handling of food products from receiving through preparation. Sanitation regulations are often strict, since poor sanitation and unhealthy conditions could lead to the outbreak of disease in the local community as well as injury to individual patrons.

Regulation begins with the licensing of the operation. A health or foodservice permit is issued after an inspection and approval by the local regulatory body. A new permit must be obtained whether an operator is opening a new operation or purchasing or leasing your business from someone else.

Inspections may be held without prior warning to ensure that health and sanitary standards are maintained throughout ownership or occupancy period. Many local agencies pattern their inspections and their inspection forms on the federal model. **Exhibit 2.3** shows a typical inspection form. Violations are treated either as *offenses* (minor) or *crimes* (major), and if proven in court, may result in a fine, imprisonment, or both. Regulatory bodies are not bound by any court decision as to whether an offense was criminal, and may still suspend or revoke the operating permit.

A violation of a health or sanitary code may be proof of *negligence* (a legal wrong where the victim is exposed to an unreasonable risk of harm), which can give an injured customer the right to sue and recover damages. In some states, such a violation is in itself conclusive proof of a wrongdoing. In other states, the violation creates a *prima facie* (literally, "on the face of it", or held to be true unless disputed) case of wrongdoing, meaning that the victim will recover unless the foodservice operator can prove that he or she

Exhibit 2.3　Inspection Form

HAZARD ANALYSIS CRITICAL CONTROL POINT MONITORING PROCEDURE REPORT

County	Dist.	Est. No.	Month	Day	Year

Establishment Name _____　Operator's Name _____

Address _____　County _____

(T)(C)(V) _____　Zip Code _____

Food _____

Critical Control Point	Priority	Process (Step)	Monitoring Procedure	Name or Title of Responsible Person
Yes ☐　No ☐		CONDITION AT DELIVERY	☐ Approved source (inspected) ☐ Temperature check less than 45°F ☐ Not spoiled ☐ Shellfish tag complete	
Yes ☐　No ☐		STORAGE	☐ Refrigeration: product temperature less than 45°F ☐ Raw/cooked/separated	
Yes ☐　No ☐		THAW	Methods ☐ Under refrigeration ☐ Running water less than 70°F ☐ Microwave ☐ Less than 3 lbs, cooked frozen	
Yes ☐　No ☐		COOK	☐ Greater than or equal to 165°F ☐ Greater than or equal to 150°F ☐ Greater than or equal to 140°F ☐ Greater than or equal to 130°F How determined _____	
Yes ☐　No ☐		HOT HOLD	☐ Product greater than or equal to 140°F Temperature checks every _____ minutes	
Yes ☐　No ☐		COOL	120° to 70°F in 2 hours; 70° to 45°F in 4 additional hours by the following methods (check all that apply). ☐ Shallow pans ☐ Ice water bath and stirring ☐ Reduce volume by _____ ☐ Rapid chill refrigeration Temperature checks every _____ minutes	
Yes ☐　No ☐		PREPARE AND SERVE	Maximum total time between preparation and service _____ hours _____ minutes	
Yes ☐　No ☐		SLICE, DEBONE, MIX, ETC.	☐ Wash hands ☐ Use gloves, utensils ☐ Workers' health ☐ Wash and sanitize equipment and utensils ☐ Minimize quantity of food at room temperature ☐ Use pre-chilled ingredients	
Yes ☐　No ☐		REHEAT	☐ Rapidly heated to 165°F How determined _____	
Yes ☐　No ☐		HOLD FOR SERVICE	☐ Hot product greater than or equal to 140°F ☐ Cold product less than or equal to 45°F Temperature checks every _____ minutes	

Actions that will be taken when the monitoring procedures are not met: _____

_____

I have read the above food preparation procedures and agree to follow and monitor the critical control points and to take appropriate corrective action when needed. If I want to change any monitoring procedure, I will notify the Health Department prior to such a change.

_____　　　　_____
Signature of person in charge　　　　　　Signature of inspector

SOURCE: New York State Department of Health, Bureau of Community Sanitation and Food Protection, Albany, NY 12203

did not cause or contribute to the violation.[29] Violation of a fire regulation or regulation of a swimming pool ordinance are illustrative of this concept.

There is one more weapon in the arsenal of the regulators. Many operators naturally try to avoid bad publicity, especially regarding sanitation. This concern is understandable. The consumer complaint or claim may or may not be justified, yet the attendant publicity may injure an otherwise good reputation. Some large metropolitan newspapers publish the sanitary and health code violations of foodservice establishments and a list of forced closings.[30] Although not printed in banner headlines on the front page, such publicity has the effect of warning customers away, which provides an additional reason to comply with local sanitation guidelines.

Each state may establish its own sanitation standards, as long as they do not weaken or dilute the FDA model. In other words, states may improve upon the federal model for sanitation, but may not limit or curtail its scope and effectiveness. The states are not required to adopt any standards, since the federal Food, Drug, and Cosmetic Act will serve in the absence of any state legislation.[31] The same is true of agricultural and marketing requirements governed by federal law, such as the Federal Meat Inspection Act of 1906 and the Poultry Products Inspection Act, enacted in 1957, administered by the U.S. Department of Agriculture.[32]

Criminal Trespass and Disorderly Conduct

Independently of liquor law violations, most local legislatures have written laws to protect private property from trespassers and intruders who have no legitimate interest in being on your property and whose conduct infringes on your rights, and on the rights of your patrons and of others on the premises. The Supreme Court, in a landmark decision, ruled that the United States Constitution protects government action, or private action on public property, not action on private property.[33]

You have a right to keep intruders out of your private business premises, or remove them, as long as you use no more force than absolutely necessary under the circumstances.[34]

Blue Laws

Under common law, business activity was legal on Sunday. Many states, however, have statutes called *blue laws*, which modify the "business as usual on Sunday" common law rule. They prohibit any nonreligious business transactions on Sunday. Some states modify the rule to allow communities to choose which day businesses must close. Other states prohibit the sale of merchandise other than for "necessity" and "charity" on Sunday. The most typical sales prohibited by blue laws are sales of alcoholic beverages. Each state is permitted to define sales of "necessity" and "charity" and to include as well as exclude alcohol from the scope of such laws. Blue laws are so named because they were "as wide as the sky is blue" in their original form. This no longer holds true. They are currently called "Sunday closing laws."

Works of charity are those acts that involve religious worship or helping persons in distress. Works of necessity include acts that must be done to save life, health, or property. Some courts have interpreted the "works of necessity" exception very broadly to permit sales where their prohibition would cause serious economic loss to the seller or inconvenience to the buyer.

Some communities do not enforce their blue laws at all, others do so strictly, and in many states such laws have been declared unconstitutional.

Truth-in-Menu Regulations

Truth-in-menu guidelines used to be the domain of state laws, which were based on federal regulations regarding food and beverage advertising and packaging. These state guidelines were replaced by the federal Nutrition Labeling and Education Act (NLEA) of 1990. This act is meant to protect consumers against economic losses suffered when they do not get the product they pay for—a loss of the benefit of the bargain a consumer makes with the foodservice operator.[35] The NLEA act affects foodservice operators in that any nutritional or health-related claims made on a menu must be substantiated and adhere to proper use of terms (light, fat-free, low-cholesterol). (See Appendix B for more detailed menu guidelines.)

Some valid customer complaints deal with product substitutions, where the product served differs materially from the one advertised. For example, if a restaurant advertises prime rib of beef on the menu and the waiter brings a beef shoulder cut instead without telling the customer about the substitution, then that patron is justified in making a complaint. The substitution is often inferior in quality and less costly, and may make the patron believe that he or she is being cheated.

Kosher Food

Some local governments go a step further than state truth-in-menu regulations and make it illegal for restaurants to advertise falsely that a food is prepared in the Jewish orthodox kosher manner when it is not.[36] These regulations make the misrepresentation a crime, regardless of whether there was any intent to defraud.

Smoking Regulations

Another form of regulation involves the issue of segregation of smokers in dining facilities. With increasing medical evidence to support the connection between smoking and the risk of lung cancer as well as other health hazards to smokers and nearby nonsmokers, indoor clean air acts have been adopted in a few states.[37] The first state to do so was Minnesota, which adopted its Indoor Clean Air Act in 1975. Twelve other states have followed this pattern (See **Exhibit 2.4**.)

Exhibit 2.4   States with Indoor Clean Air Acts

California	Nevada
Colorado	New York
Connecticut	North Dakota
Michigan	Rhode Island
Minnesota	Utah
Montana	Vermont
Nebraska	

Minnesota's Clean Air Act, the most comprehensive of its type, prohibits smoking in eating establishments, except in designated smoking sections. At least 25 percent of restaurant tables must be set aside for nonsmokers, who must be advised that they have the right to be seated and served in such a section. Bars and taverns—that is, establishments whose primary purpose is serving alcoholic beverages rather than food—are exempt, but signs to that effect must be posted. Violators are usually subject to a civil fine.[38]

Clean air acts represent a legislative compromise between the competing interests of smokers and nonsmokers. In practice, it may be good business for you to provide separate seating for nonsmokers, who make up the majority of the population. Voluntary action may serve to forestall legal regulations of yet another aspect of foodservice management.

In addition to segregating smokers and nonsmokers in foodservice establishments, a growing number of hotels have voluntarily advertised the availability of rooms for nonsmokers. These measures follow the trend to segregate hotel foodservice smoking patrons, which is required by the above state laws. In some instances, local communities have passed their own laws imposing an outright ban on smoking in such establishments. This was the case in Beverly Hills, California. After a dramatic drop in restaurant patronage, the city ordinance was repealed. The question of whether local communities may ban smoking in the face of less stringent state requirements on smoking in public places is a growing issue.

Zoning Laws

Local governing bodies have the power to regulate land use within their boundaries. Almost every local government exercises this power to some extent.

Usually the objective is to allow for both business and residential needs, without one land use conflicting with another. A secondary goal of land-use regulation is to protect certain institutions, especially churches and schools, from being exposed to business activities, such as the sale of alcohol or adult books, that might be perfectly acceptable elsewhere.

Most local governments regulate land use through *zoning* (dividing up land for specific uses). One area may be a residential zone, another a commercial zone, and another mixed residential and commercial. These laws have the effect of regulating *both* private and public land. In other words, if you buy a piece of property in a residential area, although you have full ownership rights, the local government may still prevent you from opening a restaurant there, or even from conducting a catering business in your home. This topic will be explored in detail in the chapter on property rights.

Building and Safety Codes

In addition to regulating land use, most local governments regulate what operators may do to a restaurant building to improve or expand it. Many local governments have building and safety codes to guide restaurant owners. Business owners encounter more building and safety regulations than owners of private homes, and foodservice operators have even more regulations to deal with.

The purpose of building and safety codes is to protect the surrounding area—and, in the case of a restaurant, the patrons—from poor building or safety practices. Building codes usually specify requirements for building a new structure or for adding to an existing one. At some point in your business, you may want to add to your restaurant building. Always, before you spend any money on building materials and advice, consult the local building authorities.

Safety codes are designed to prevent accidents to the public and your employees. These codes usually concern fire regulations. Along with a visit from the health inspector, you will probably be inspected by the local fire department, which usually is looking for faulty wiring, unsafe lighting, and other fire hazards. Get a copy of the local codes so you can be sure you stay within the requirements. Violating safety codes is not only illegal and may cost you a fine, but if a patron is injured due to a code violation, he or she is one step closer to holding you liable.

LIVING WITH REGULATIONS

Each state has the power to create regulatory agencies independently of the federal government, as long as the state does not override federal authority to regulate interstate commerce, in which federal law is supreme. State and local commerce are within the power of state and local bodies to regulate. However, federal regulations often provide models for state and local regulations, and cooperative state and local guidelines are often written into federal laws.

Normally, the regulatory maze does not create conflicting rules for the single-operation foodservice operator. However, the multiunit operator within a single state may find that regulations vary from locality to locality, thus adding to the overall cost of doing business. This problem is substantially greater for the multistate chain operator, who must comply with more federal regulations as well as state and local rules.

Regulation increases the individual operator's cost of doing business. The operator usually passes these costs on to consumers, and may suffer the consequences of lost patronage because of more expensive products. To the extent that costs associated with regulation are uniform, and affect each operator equally, consumers are not directly affected, since the costs of regulation are equalized for operators in that market. However, when the regulations overlap and are inconsistent, then the costs of doing business also vary, making the operator suffer or benefit, depending on the locality. If regulations are made much tougher, the manager/owner may find it difficult, if not impossible, to continue operating at a profit.

When regulations involve the preservation of patron's lives and health, they must be accepted as a cost of entering into, or continuing in, the hospitality business, since the benefit to the public clearly outweighs any economic detriment to the foodservice operator.

When a regulation does not involve any threat to life or health, the industry must examine it to see if it meets reasonable public expectations at a reasonable cost. When the cost of regulations unreasonably exceeds the benefits to the public, you have the power to help eliminate such regulations through the political process.

The hospitality industry as a whole should educate the public to promote useful regulation and to do away with regulation that is costly but of little benefit. The form, content, and scope of regulations prepared by legislatures are based upon input from consumers *and* industry representatives, so that a proper compromise may be reached when those interests differ. The role of the courts is to interpret statutes to suit the collective values of the individual judges. Ultimately, the legislatures control the process of adoption and change, and so it is to the lawmakers that you must address your concerns.

Wherever possible, a voluntary plan of action is always superior to legally compelled compliance. This holds true for everyone in the industry. The more operators can do to prevent consumer abuses, the less likely will be the need for legislation. Cooperative efforts are not only good business, but reflect pride in the hospitality profession and a resolve to weed out those few operators who would tarnish the reputation of the industry as a whole. On the other hand, indifference to a consumer desire or attempts to cut corners in the proper operation of business can only invite governmental regulation.

SUMMARY

A number of federal agencies regulate the process and distribution of food products. Most of those will not affect the foodservice operator directly, but will provide standards for sources of food products, and hence buying guides.

Foodservice operators will encounter regulations in the area of employment. These federal regulations are supplemented by state guidelines as well.

Operators will really feel the regulatory pinch at the state and local levels. The states usually have some apparatus to regulate employment practices in the area of civil rights, and local metropolitan governments may, also.

State regulation is heaviest with regard to liquor sales. Operators who wish to sell alcohol are regulated from the time they apply for a license through the time they sell liquor on their premises. The severity of alcohol regulations and enforcement will vary from state to state, but all states are uniformly strict in the area of liquor sales to minors, and such sales pose the greatest threat to a liquor license.

Local regulations usually deal with building codes, sanitation regulations, and fire and safety inspections. Preferably, operators should know these regulations before building or opening a hospitality establishment. Generally, preventive medicine is the best way to avoid problems with local regulations.

QUESTIONS

1. Why are federal inspection and grading of meat products important to the foodservice operator?
2. What effects do federal regulations have on lawmaking bodies at the state and local levels?
3. A state sanitation law prohibits employees from smoking in food preparation and service areas. A city law prohibits employees from smoking in food preparation areas, but not in service areas. Which law will control city restaurants and why?
4. What are some on-premise activities that may threaten a liquor license? Which activity is most closely regulated by every state?
5. How may a violation of a sanitation code leave an operator open to liability?

NOTES

1. *Tad's Franchises, Inc. v. Incorporated Village of Pelham Manor*, 42 A.D.2d 616, 345 N.Y.S.2d 136 (2d Dep't 1973), *aff'd* 35 N.Y.2d 672, 319 N.E.2d 202, 360 N.Y.S.2d 886 (1974).
2. H. Schultz, *Food Law Handbook* 1–30 (1981); Regier, *The Struggle for Federal Food and Drugs Legislation* in Law and Contemp. Probs. 3 (1933).
3. Schultz, *supra*, note 2, at 196.
4. 21 U.S.C. secs. 302, 304(a), (b), (d), 306, 704, 705 (1976).
5. 21 U.S.C. sec. 303(a) and (b). Each violation of the Act is a *misdemeanor*, a criminal offense punishable by a maximum $1,000 fine or up to a year in jail, or both. Repeat offenders are subject to a maximum three-year prison term and a maximum $10,000 fine or both for *each* violation. *Willful* violations of prohibited acts (including adulteration, misbranding of products, refusal to permit inspection of premises, and introduction or receipt of any adulterated or misbranded product) with intent to defraud or mislead can result in a maximum three-year prison term or a maximum fine of $10,000, or both, for each violation.
6. 21 U.S.C. secs. 304(b), 333, 335. A right to be heard is afforded any person subject to criminal prosecution after the FDA has notified the federal Attorney General to prosecute. The burden of proof is on the FDA, but will be satisfied by the weight of credible evidence. See also Christopher, *Cases and Materials on Food and Drug Law* (1966), Chapter 8, sec. 9 (Burden of Proof), pp. 577–78 citing food cases. Either side may appeal to a higher reviewing court, and with that court's permission, to the U.S. Supreme Court. Administrative Procedure Act, 5

U.S.C. secs. 701(a), 702. Either side may file for a *writ of certiorari*, a procedure giving the Supreme Court the right to grant review as it sees fit.

7. The Federal Trade Commission Act, 15 U.S.C. secs. 12(a), 12(b), 13, 14, 15, 16, 41–58 (1976).

8. To date, the FTC has not exercised its authority to regulate chain restaurant menu advertising in interstate commerce.

9. 15 U.S.C. secs. 12(a), 13(a), 14(a) (1976). The "affirmative failure to reveal" rule has been most frequently applied to drug advertisements by the FTC. FTC cease and desist orders that made the "failure to reveal" itself a deceptive act, have been affirmed in *Wybrant System Products Corp. v. Federal Trade Commission*, 266 F.2d 571 (2d Cir.), *cert. denied* 361 U.S. 883 (1959) and in *Erickson and Scalp Specialists v. Federal Commission*, 272 F.2d 318 (7th Cir. 1959), *cert. denied* 362 U.S. 940 (1960). Contra *Alberty v. Federal Trade Commission*, 182 F.2d 36 (D.C. Cir.), *cert. denied* 340 U.S. 818 (1950). *Federal Trade Commission v. Merck & Co.*, 69 FTC 526 (1966) (drugs); *in re* Porter and Deitsch, Inc., noted in *Legal Developments in Marketing*, 42 J. Mkting. 90 (Oct. 1978) (weight-loss pills).

10. 15 U.S.C. sec. 54(a) (1976).

11. *Id*. sec. 55(a)(1). Section 55(b) defines "food" to mean "(1) articles used for food or drink for man or other animal, (2) chewing gum, and (3) articles used for components of any such articles."

12. 29 U.S.C. secs. 651–78 (1976).

13. Section 651 of the Act sets forth a Congressional statement of findings and declaration of purposes and policy. See 29 U.S.C. sec. 651 (1976).

14. See J. Sherry, *The Laws of Innkeepers*, 3d. ed. secs. 12:1–12:19 (rev. ed. 1981).

15. Sherry, *supra*, note 14, sec. 12:2.

16. Sherry, *supra*, note 14, secs. 12:12–12:14.

17. The FTC issued a trade regulation, effective October 21, 1979, having the force and effect of law, entitled Disclosure Requirements and Prohibitions Concerning Franchising and Business Opportunity Ventures, 16 C.F.R. sec. 436 (1982).

18. The Federal Alcohol Administration Act, 27 U.S.C. secs. 201–12 (1976) controls traffic in distilled spirits, wine, and malt beverages in interstate (between-the-states) commerce and between states and foreign nations. The United States Constitution, Twenty-First Amendment, grants the states power over the sale and distribution of liquor within their borders.

19. The powers and duties of alcohol beverage control boards or commissions vary from state to state. Essentially, they license, police, and discipline distillers, wholesalers, and retailers who are subject to their respective rules and regulations.

20. N.Y. Alcoholic Beverage Control Law, sec. 1 *et seq*. (McKinney 1970 & Supp. 1982); Rules and Regulations Promulgated by The New York State Liquor Authority. See *Show Boat of New Lebanon, Inc. v. State Liquor Authority*, 33 A.D.2d 954, 306 N.Y.S.2d 859 (3d Dep't), *aff'd*, 27 N.Y.2d 676, 262 N.E.2d 211, 314 N.Y.S.2d 2 (1970).

21. *Show Boat of New Lebanon, supra*, note 19; *Circus Disco Ltd. v. New York State Liquor Auth.*, 73 A.D.2d 354, 426 N.Y.S.2d 495 (App. Div. 1st Dep't) *rev'd*, 431 N.Y.S.2d 491, 409 N.E.2d 963 (N.Y. 1980).

22. Section 65, N.Y. Alcoholic Beverage Control Law, states: No person shall sell, deliver, or give away, or cause or permit or procure to be sold, delivered or given away any alcoholic beverage to: (1) Any minor, actually or apparently, under the age of twenty-one years; (2) any intoxicated person or to any person, actually or apparently, under the influence of liquor; and (3) any habitual drunkard known to be such to the person authorized to dispense any alcoholic beverages. Neither such person so refusing to sell or deliver under this section nor his employer shall be liable in any civil or criminal action or for any fine or penalty based upon such refusal, except that such sale or delivery not be refused, withheld from or denied to any person on account of race, creed, color or national origin.

23. In New York, operators may provide topless dancing without jeopardizing their license. *Bellanca v. State Liquor Auth.*, 54 N.Y.2d 228, 429 N.E.2d 765, 445 N.Y.S.2d 87 (1981) *cert. denied*, 456 U.S. 1006 (1982). Other states, however, are free to outlaw such entertainment. *Cianci v. Division of Liquor Control, Department of Business Regulation*, 50 U.S.L.W. 3798 (April 4, 1982), *cert. denied*, 102 S. Ct. 1769 (1982).

24. See sec. 65 (1)–(3) of the N.Y. Alcoholic Beverage Control Law, *supra*, note 20.

25. See *Circus Disco, supra*, note 21.

26. *Show Boat of New Lebanon, supra*, note 20.

27. See *Circus Disco, supra*, note 21.

28. See Schultz, *supra*, note 2.

29. See Chapter 4, *infra*.

30. The *New York Times* does so regularly.

31. Schultz, *supra*, note 2, at 4–5, 1–30.

32. 21 U.S.C. secs. 541–570 (1976). Section 661 of the Federal Meat Inspection Act and sec. 451 of the Poultry Products Inspection Act provide federal and state cooperative guidelines.

33. *Lloyd Corp. v. Tanner*, 407 U.S. 551 (1972). See also *Hudgens v. National Labor Relations Board*, 424 U.S. 507 (1976) (peaceful picketing of retail store).

34. Trespassing members of the news media are liable in damages to foodservice operators. *Le Mistral, Inc., v. Columbia Broadcasting*, 61 A.D.2d 491, 402 N.Y.S.2d 815 (1st Dep't 1978).

35. See Sherry, *supra*, note 14, sec. 14:10.

36. Florida and New York make it a crime to represent falsely that food sold is kosher when, in fact, it is nonkosher.

37. See Sherry, *supra*, note 14, sec. 14:11.

38. *Id.*

Civil Rights Implications of Admission Policies

OBJECTIVES

After reading this chapter, you should be able to:

1. Outline the rights of operators to admit or refuse guests or patrons and their duties to avoid unlawful discrimination against guests and patrons.

2. Discuss the scope and effect of civil rights laws as related to operator admission policies.

3. Define house rule and provide guidelines for formulating such a rule.

Case in Point 1

The Golden Nugget Saloon is located in a small community in the Pacific Northwest. Twenty percent of the total population is Mexican-American, most of whom are native-born U.S. citizens. The owner of the saloon had issued a rule that directed his bartenders not to allow a foreign language to be used at the bar, "if it interfered with the regular trade." If a problem arose, employees were to turn the jukebox up (using house money) and then ask the "problem" people to move to a table.

One night, three Mexican-Americans went to the saloon where the bartender served them beer. While drinking, the three men began to speak Spanish. Other customers, also sitting at the bar, became irritated and complained to the bartender. The bartender told the three that if they persisted in speaking Spanish, they would have to sit at a table or leave. An argument broke out, the police were called, and the three Mexican-Americans left peacefully. Two days later the scene was reenacted with different Mexican-Americans. The bartender "pulled" the beers of these customers, and they left. This time, however, they were followed out of the tavern and assaulted by three English-speaking customers.

The Mexican-Americans sued, charging a violation of their civil rights not to be discriminated against because of their race and national origin. Were their civil rights violated?

The court found that the tavern management had violated the civil rights of the plaintiffs—namely, a federal statute that allows all persons to make and enforce contracts and to enjoy the full and equal benefits of all laws and proceedings for the security of person and property as enjoyed by white English-speaking citizens.[1] The rule, the court said, deprived Spanish-speaking Mexican-Americans of their right to drink their purchases and enjoy equal hospitality with English-speaking customers.[2]

The tavernkeeper's defense that the policy was adopted simply to avoid trouble and not to arouse the anger of the "preferred" English-speaking customers was rejected. The equal protection of the laws guaranteed by the Fourteenth Amendment of the Constitution cannot be ignored when patrons enter in a fit and proper condition, conduct themselves in a peaceful manner, and are provoked by other patrons.[3]

Case in Point 2

Fred Fineburg and Gloria Goodfrank operate a foodservice establishment that caters to families with preteen children who are served a cafeteria-style selection of quickly prepared foods and soft drinks. No alcoholic beverages are sold. Fred and Gloria make every effort to accommodate families in clean, pleasant surroundings at a low cost, and the operation's profitability depends on high turnover.

Lately, Fred and Gloria have experienced a loss of family patronage due to the presence of a large number of town toughs who smell of alcohol, are boisterous, and generally disrupt normal business. They usually occupy tables and chairs for a much longer time than needed to eat their food, verbally harass other customers, and threaten to "cause trouble" if anyone complains about their aggressive behavior. Fred and Gloria wish either to deny admission to these customers or to remove them. Do they have the legal right to do so?

Yes. In contrast to innkeepers, who are generally required by law to admit anyone who appears in a fit condition and able to pay, the foodservice operation has more freedom to pick and choose customers. The only restrictions the law imposes are civil rights laws that prohibit discrimination in all places of public accommodation, including restaurants, on the basis of race, creed, color, or national origin, and in some states, on the basis of sex or physical disability.

In this case, Fred and Gloria can legally deny access to the disruptive patrons or remove those who overstay their welcome and annoy or verbally abuse other customers.

ADMISSION OF GUESTS
AND PATRONS

The first contact you or your employees have with customers is at the front door. Your rights to admit or refuse patrons are limited by your own standard admission policies and by federal and state civil rights laws prohibiting discrimination.

Most people do not intend to violate others' civil rights, including hospitality operators. It is not easy for operators, subjected to other business pressures, to always recognize where their business rights end and patrons' civil rights begin. It is important that you be able to differentiate between those rights, so that you can develop admissions policies that are in keeping with the civil rights of patrons and your own rights as a businessperson.

COMMON LAW ADMISSION
POLICIES: BRITISH ROOTS

From the earliest recorded evidence, places of public hospitality have been a fact of life in the development of modern civilized societies. Our present-day common law reflects the early English recognition of hospitality businesses as places of public refuge, entertainment, and accommodation.

However, medieval English law made a sharp distinction between innkeepers and proprietors of restaurants or taverns as to their respective duties to admit patrons or guests. These distinctions were based on the different functions of each establishment.

The innkeeper was required to admit and receive all travelers, as long as a prospective guest was in a fit condition, and able and willing to pay for the lodging. This was so because a traveler was assumed to be a stranger to the locality, and thus a person in need of shelter and protection as well as food and lodging. Travel during the Middle Ages was dangerous. The roads were infested with outlaws who would prey on unwary travelers, especially at night. The only places of refuge were the wayside inns, but these were few in number and located sporadically along the way. As a result, admission was a serious matter because of the threat of loss of life as well as of property that confronted the traveler if shelter could not be found. It was for these reasons that common law required innkeepers to admit all travelers—with very few exceptions.[4]

Although tavernkeepers, like innkeepers, provided food and drink, they were not under the same duty to admit everyone. The English common law gave the proprietor of a tavern the right to discriminate as to customers, because the function of the tavern was to serve local inhabitants who knew the risks of venturing out at night and who did not require shelter and protection.[5]

MODERN ADMISSION POLICIES: HOUSE RULES AND CIVIL RIGHTS

The law does not require foodservice operators, tavernkeepers, and others who invite the public on their premises to admit everyone.[6] Only innkeepers and public transportation carriers must do so.[7] In all cases the potential patron, guest, or passenger must be in a fit condition to be received and be prepared and able to pay for food and services provided.[8] As a hospitality operator, you are limited only by civil rights laws and your own admissions policies.

The modern trend is to require all operators of public businesses to base any denial of access on a *house rule*, which is a policy adopted voluntarily and used consistently to protect the reputation and class of that establishment, as well as the safety and welfare of other patrons.[9]

In practice, most courts allow you a good deal of discretion in deciding whether or not to admit patrons. For you, the hospitality manager, the first test of your admissions policy that must stand up in court is whether or not a house rule exists, is reasonable and adequately communicated to patrons, and has been violated by a prospective patron.

The term "house rule" is relative, since there is no standard that uniformly fits every proprietor. Generally the courts view *reasonableness* in terms of your circumstances; that is, the size, class, and nature of your operation. For example, a coat and tie for a man and formal wear for a woman might be a reasonable requirement for admission to a first-class French restaurant. However, less formality would be expected in a quick-service establishment or a family restaurant. Likewise, the management of a family restaurant might reasonably refuse to admit or serve a person wearing a swimsuit and sandals because other customers might object to an appearance that would be totally acceptable in a beach resort dining area.

In some cases the force of statutory health codes will override any discretion you may wish to exercise. The admission of people in bare feet or who are accompanied by pets other than seeing eye dogs is often forbidden regardless of your own more liberal admission standards.

Under no circumstances are you required to admit anyone who is drunk, disorderly, filthy, or acting in a threatening manner. Here you and your employees must use relatively consistent powers of observation *at the door*, because you may have trouble ejecting these patrons after they have been admitted.

The following example illustrates the application of the reasonable house rule concept in a typical restaurant setting.[10]

A young man and woman, upon leaving a fraternity house party, went to a quick-service restaurant. Both went through the cafeteria line, selected and paid for their food, and sat at a table. When she started to eat, the woman was approached by the foodservice manager, who told her to leave because she was not wearing shoes. The manager informed her that it was company policy not to serve anyone not wearing shoes, although there was no sign to that effect posted. She replied that she would leave when she was through eating. An argument started, and she began verbally abusing the manager. The manager called a police officer, who informed the patron that she was violating the city's unlawful-entry criminal statute. When she persisted in refusing to leave, the officer arrested her. Later the criminal charge of unlawful entry was dropped, but not until she had been fingerprinted, photographed, and placed in a police lineup. She then sued the owner of the restaurant for false arrest. Was the arrest justified?

The court held that the arrest was lawful and dismissed her suit, noting that she unlawfully refused to leave after being properly ordered to do so by the manager.

The court reasoned that the restaurant manager had the common law right to refuse service to any patron, and that there was no violation of the woman's civil rights (a refusal based on race, creed, color, or national origin) that would establish an exception to this rule. The court said that while it is unlawful for a foodservice operator to refuse to serve any "quiet and orderly person," the unlawful entry statute did not prevent an operation from establishing a reasonable dress requirement. A rule that all patrons must wear shoes as a condition of admission and service is typically viewed as reasonable.

Today, as in the Middle Ages, a restaurant patron's legal status is not the same as that of a guest at an inn, who, once admitted, has a right to remain until proven guilty of misconduct. The restaurant patron merely has a permissive right to remain, and cannot lawfully remain once ordered to leave. In this example the patron's only legal remedy was to sue the foodservice owner for the breach or violation of her contract for the food she purchased. Obviously, she should not have been admitted at all, since she was in violation of the dress code from the beginning.

Dress codes have to be clear and consistently enforced, however. In the following California case, a house rule regarding dress codes was too vague.

Hales v. Ojai Valley Inn and Country Club
Court of Appeals of California, Second District
73 Cal. App. 3d 25, 140 Cal. Rptr. 555 (1977)

Facts: The Hales family entered an inn's restaurant to eat a meal. Mr. Hales had made room reservations for the facility, using information from a brochure he had obtained before his vacation. Referring to its three dining facilities, the brochure said:

> Sports and casual clothes are in order during the day. A warm sweater or wrap is suggested for the evenings, which are frequently cool. Gentlemen are requested to wear jackets and ties to dinner.

When Mr. Hales tried to order, however, he was told he could not be served because he was not wearing a tie. He alleged that at the time, female patrons who were attired in suits without ties were being served. On this basis, he sued the inn for sex discrimination and false advertising. The trial court dismissed his complaint and he appealed. The trial court order dismissing the complaint by the guest, which claimed sex discrimination and false advertising, was reversed.

Reasoning: According to the California Unruh Civil Rights Act, all members of the public of lawful age have a right to patronize a public restaurant as long as they are acting properly and are not committing unlawful or illegal acts. Further, proprietors have no right to eject patrons "without good cause" and may be liable otherwise. The appeals court found that, while the plaintiff should provide more information, there appeared to be enough facts to justify a *cause of action* (*suit*) for sex discrimination.

As for false advertising, the appeals court determined that there was enough information to uphold the complaint. The court said that the advertising indicated that a suit and tie were preferred, *but not required*. The defendant said that he intended the advertising to indicate that jackets and ties were required. The plaintiff said if he had known that to be the case, he would not have made reservations at the inn for himself and his family.

Conclusion: The following rules apply: 1) Dress codes must be specific and not arbitrary or open to confusion by either patrons or employees: in this case, the dress preference could have been interpreted either way by a potential patron. 2) The court cited a California Supreme Court decision that prohibits owners of public establishments from refusing to admit persons because they wear their hair long or dress unconventionally, in the absence of proof of illegal or immoral conduct. 3) This ruling was based on a violation of the California Civil Rights Act. A denial of admission on the grounds of violation of a reasonable dress code would be permissible in common law. The California Civil Rights Act, as interpreted, alters the common law rule. In a similar case, the New York Supreme Court ruled in favor of a patron who complained that he was denied admission because he had long hair, whereas women with even longer hair were served. 4) State civil rights laws vary from state to state, and these laws may be stricter than federal civil rights laws.[11]

Civil Rights Legislation

The history of access to places of public accommodation and entertainment in the United States reveals a clear-cut pattern of refusals to admit certain groups of people, particularly racial minorities.[12] Even innkeepers practiced such discrimination, in clear violation of the common law.

Some states, applying the common law duty to admit all, passed special legislation prohibiting whites and nonwhites from occupying space within the same hotel, restaurant, or other facilities open to the public, thereby legalizing discrimination.[13]

The U.S. Supreme Court adopted a "separate but equal" accommodations doctrine, which was ignored by all states, including those that did not have discriminatory public access statutes.[14] *Separate but equal* meant that the product or service offered to a member of a minority must be equal in all ways to that offered to a white person. Although more commonly applied to schools, this doctrine also included restaurants and inns. In other words, an inn could have separate rooms for minorities as long as they were equal in class, decor, and services provided to rooms offered to nonminorities. However, innkeepers in some states ignored the doctrine. In restaurants there could be a separate entrance and separate seating, as long as the food and service were equal. Restaurant owners and tavernkeepers, never under any common law duty to admit everyone, did as they pleased. In each case, as the Supreme Court decided regarding schools in 1954, separate was not usually equal.[15] For example, forcing minorities to obtain foodservice through a back window, and prohibiting them access to the main dining room defied the separate but equal doctrine.

Federal Laws   The federal Civil Rights Act of 1964 drastically changed these common law and statutory rules. Title II of the Act, called the *public accommodations section*, outlaws discrimination by all businesses serving the public, including foodservice operators, on the basis of race, creed, color, and national origin. No state statute to the contrary is enforceable.[16] Only gender-based discrimination in public accommodation remains outside the scope of the Civil Rights Act.

State Laws   The majority of states have passed their own civil rights acts governing public accommodations. Some have also outlawed sex-based discrimination in public places.[17]

State civil rights acts often cover areas not mentioned under the federal act. Not only may sex-based discrimination be prohibited, but discrimination based on such matters as sexual orientation, marital status, age, and physical or mental disability may also be prevented by state laws. (See **Appendix C**.)[18] In addition, some large cities have passed their own civil rights laws.[19]

You may refuse to admit potential patrons on neutral grounds, such as for abusive conduct or their use of your premises for drug sales or other illegal activities.[20] In the first situation, the misconduct is open and obvious. In the second situation, it is hard to detect the illegal act until the drug sale is attempted on the premises. Only if the patron was known by you to have sold or attempted to sell drugs in your place of business on previous occasions would you be within your legal rights to bar access to that person.

Penalties   The common law imposes penalties for unlawful discrimination. Usually a damage award is imposed to compensate the victim. Damages include humiliation, emotional harm, and mental suffering.[21]

The federal Civil Rights Act of 1964 permits federal courts to stop discriminatory practices against prospective and admitted patrons. However, the courts lack the power to award compensatory or punitive damages.[22] (*Compensatory* damages are designed to compensate a plaintiff for injuries or losses; *punitive* damages are meant to punish the defendant.) Only when a restaurant operator violates a court order to stop discriminating may that court impose a fine or jail sentence.

However, state civil rights laws provide penalties for violators. New York, for example, authorizes its courts to fine violators up to $500 and to imprison them for up to 90 days, as well as to award compensatory damages to the victims not to exceed $500 for each violation.[23] New York also empowers its State Commission on Human Rights to halt future violations and fine violators up to $500 and to impose a maximum one-year jail sentence.[24]

Exemptions to the Civil Rights Act   Clubs not open or advertised to the public are exempt from the 1964 Civil Rights Act. The general guidelines are that a club that calls itself private, maintains selectivity in the admission of members, does not advertise for members or does so selectively, does not operate strictly for profit, and has a history of membership selectivity is deemed private.

However, public clubs that advertise for members and are less arbitrary in their selection, are not usually exempt from either federal or state antidiscrimination laws.

Recently, Minnesota amended its human rights act to include men-only civic clubs within its scope. Litigation involving the United States Jaycees ultimately resulted in a U. S. Supreme Court decision holding that Minnesota could prohibit the Jaycees from excluding women from full membership rights in the local chapters. New York City has also adopted a law requiring private clubs of 400 members or more that authorize employer payment of member business expenses including dues, meals, and other house charges, all of which are deducted for income tax purposes, to admit qualified women members. That city law was also upheld as permissible by the U.S. Supreme Court under the First Amendment. In both cases, the Jaycees and the New York City Club Association argued that First Amendment protection in favor of private association was being violated by these laws. The Supreme Court held that the civic clubs were not sufficiently small, secluded, and selective to qualify for First Amendment protection. When challenged as violation of civil rights laws, private clubs continue to carry the burden of proving their private status.

In most states and localities, private clubs may discriminate. However, in the area of sex discrimination depriving women of full membership privileges, the door to private civic clubs has been opened. However, each case must be reviewed on its merits, since the U. S. Supreme Court has not created a standard definition of privacy that can be applied uniformly to all such establishments. New York City has amended its civil rights law to prohibit sex discrimination by private clubs in membership policies if the club has 400 or more members and permits employers of members to write off tax-related club dues and house accounts.

There is disagreement in federal courts as to whether clubs that refer to themselves as private but operate on public land, have unclear admissions policies, or advertise may be held liable for violations of civil rights or other discriminatory statutes.

There is one exception to this. *Guests of members are protected by civil rights laws during their stay on the premises.* The following case, based on a New York civil rights statute, illustrates this issue.

Batavia Lodge No. 196 v. Division of Human Rights
Court of Appeals of New York
35 N.Y.2d 143, 316 N.E.2d 318, 359 N.Y.S.2d 25 (1974)

Facts: Plaintiffs, African-Americans invited by white club members to the private Moose Lodge club premises to attend a fashion show, sued to recover damages for the club's refusal to serve them at the bar and for verbal abuse inflicted by club members. White nonmembers were served at the same bar without incident. The Commissioner of the Human Rights Division awarded each claimant $250 in compensatory damages. The appeals court affirmed the finding of discrimination but eliminated (struck out) the damage award as not permissible, since the claimants suffered no actual out-of-pocket expenses. The plaintiffs appealed to have the damages reinstated. The modification by the appeals court striking the damage award to minority guests of the private lodge was reversed and the award made by the Commissioner of the Human Rights Division was reinstated.

Reasoning: The appeals court upheld the damage award to the guests, and based that decision on the weight New York State gave to its antidiscrimination policy as proof that an individual did not need to produce large amounts of evidence to prove discrimination.

The court said that the evidence was ample to support the Commissioner's determination of damages.

Conclusion: Private clubs are not totally exempt from civil rights laws, especially when it comes to members' guests. Some state and city civil rights commissions may prescribe a punitive damage award to the wronged party in addition to a compensatory one.

Customer and Employee Discrimination

Gender Discrimination   Many states have laws prohibiting sex discrimination, particularly in employment. (See Appendix F.) Some of these laws are also aimed at protecting people from being denied admission to public places or being subjected to discriminatory action once they are admitted. For states that have sex discrimination laws, their general guidelines regarding public accommodations apply to foodservice operations open to the public. In states without sex discrimination laws, the high court is left to rule on each case.

Age Discrimination   According to common law, a minor, usually a person 18 years of age or younger, who is otherwise fit and able to pay is entitled to admission and service by an *innkeeper*.[25] *Foodservice* operators are free to admit or reject minors.[26]

What effect do current civil rights laws prohibiting age discrimination have on your right to refuse a minor access or service? You may not discriminate against minors solely because of their age. Only in the case of liquor service to minors, a crime in all states, does an exception exist.

Once a minor is admitted, however, he or she is entitled to the same consideration and protection as an adult patron, with the caution that the minor may be entitled to the benefit of the doubt as to the ability to assume risks that might lessen your legal liability to an adult.[27]

Most age discrimination laws are applied to the area of employment, specifically regarding people over 40. However, age discrimination laws can affect public businesses, and are usually meant to protect both minors and the elderly.

Guests or Patrons with Disabilities   The rights of people with mental or physical disabilities in public places are covered by federal law. You are prohibited from discriminating against people with disabilities as long as they are otherwise in a fit condition to be served and able to pay. Additionally, a number of states have laws protecting people with disabilities. The general rule is to accommodate all people with disabilities, unless their admission would cause an undue hardship in the normal conduct of your business.

Title III of the Americans with Disabilities Act of 1990 prohibits discrimination "on the basis of disability in the full and equal employment of the goods, services, facilities, advantages or accommodations of any place of public accommodation." It further requires that any new construction or alteration to existing buildings be made accessible to people with disabilities. The law protects people with disabilities, with a record of disabilities, and those regarded as having such an impairment. (See **Appendix G**.)

The Public Accommodations section is sweeping and includes every type of operation open for business to, or in contact with, the general public. Only sales or rentals of residential housing are exempt. All business property owners and lessors of such business are equally affected. Note that there is no exception for small business as there is in Title I of the Act governing employment. Private clubs are not required to comply, except that guests of members attending a public function (such as a golf tournament open to nonmembers) are covered.

The law governing provisions of goods and services imposed upon public accommodation prohibits exclusion of inferior goods or services, segregation on the basis of disability, standards, criteria, and methods of administration that have the effect of discriminating, whether or not that was the intent. An exception is made in serving a person who poses a "direct threat" to the safety of others.

Public accommodations must modify the way they operate to provide goods and services to individuals with disabilities unless they can prove that to do so would fundamentally alter the nature of the business, or that they would assume an "undue burden" in doing so. Such accommodations must also take steps necessary to communicate effectively with their customers who have disabilities affecting hearing, vision, or speech. The auxiliary aids do not include personal devices (such as wheelchairs) or personal services (such as eating or dressing).

All architectural and communication barriers must be removed where removal is "readily achievable" or by the use of alternative methods.

Newly constructed facilities must be approached, entered, and used by people with disabilities easily and conveniently. New construction is exempted from compliance only in cases of structural impracticality; that is, only by means that threaten the structural integrity of the building. Any alteration to existing facilities must comply with the alteration standards unless compliance would be "virtually impossible" under the circumstances. Paths of travel to primary function areas must also be provided, including restrooms and telephones. These requirements are waived only if the cost of doing so is disproportionate to the cost of the overall alteration.

These laws are enforced in one of two ways: by private action to obtain an injunction halting the discrimination, so that a court may order alteration to make the facilities accessible; or by Justice Department enforcement. In such a case, the Justice Department may seek a civil penalty not available to private parties to indicate the public interests, not exceeding $50,000 for the first violation, and not exceeding $100,000 for any subsequent violation. Punitive damages are not available.

Sexual Orientation   Most states do not have laws specifically protecting homosexuals from discrimination. The common law is normally the guide. The general common law rule is that unless these patrons are engaging in openly offensive acts or otherwise acting unruly, they should be admitted and served along with other patrons.[28]

Discriminatory Action Once on Premises   Once patrons have been admitted, they should not be subjected to discriminatory action as long as they behave in a proper manner and are prepared to pay. Patrons of a minority race, color, or religion should be served and treated with the same respect as all patrons. Abusive language, rude treatment, failure to serve, or extremely slow service to patrons who are not your idea of "preferred" customers is discriminatory.

GENERAL GUIDELINES FOR PREVENTING DISCRIMINATION

The laws regarding admission to places of public accommodation are generally not as strict as those regarding employment discrimination; however, it is to your own advantage to standardize your admissions procedure, *not to prevent* admission of certain groups of people, but only to exercise your right to maintain the smooth operation of your business, whether it is a fine-dining establishment or a quick-service restaurant. You should develop an admissions policy, including dress codes, that will best suit the economic needs and patrons of your operation.

Keep in mind that some dress codes shift from midday to the evening. You should be specific about your dress code, and place signs specifying what it is at the outer entrance to your operation, or in the lobby, or both. Relying on employees to enforce your dress code by observation alone often leads to arbitrary decisions by them and a loss of customer patronage, or even legal action.

Managers must be more than aware of the potential for legal pitfalls regarding discrimination. You must train your employees to enforce house rules fairly and professionally. Embarrassing patrons in the process of enforcing house rules will only aggravate them further. Have a sign to which you or your employees can point. Some restaurants even keep jackets and ties on hand, or make polite suggestions to return properly dressed. This can appease an otherwise disgruntled patron, *and* maintain your good business reputation.

SUMMARY

Foodservice managers are given more room to deny admission to prospective patrons under the common law than are innkeepers to prospective guests. These rights are not unlimited, though, and the best practice is to establish reasonable house rules and enforce them in a consistent way.

The common law has been supplemented by federal and state civil rights statutes prohibiting discrimination in public restaurants, bars, taverns, and other food and beverage establishments, but are limited to forbidding discrimination based on race, creed, color, national origin, age, and disability. Refusals to admit on the basis of gender or sexual orientation may be prohibited at the state and local levels, if at all. Neither the federal nor state acts prevent you from excluding or removing from your operation people who are drunk, disorderly, abusive, or in violation of reasonable dress and appearance codes. This law recognizes that your management policies are your own, and that the courts were not created to supervise your daily business responsibilities.

QUESTIONS

1. What criteria do the courts normally use in reviewing a house rule?

2. How may private clubs be exempt from the Civil Rights Act of 1964? What requirements must they meet to be exempt? Describe a situation in which such clubs would not be exempt.

3. The following is an example of a sign posted in a restaurant lobby: *We prefer our gentlemen patrons to wear dress suits. Ladies must wear dresses.* What is wrong with this sign?

4. Aside from a house rule, give an example in which a person might legally be denied admission.

5. What is the general common law rule regarding the admission of homosexuals?

NOTES

1. Civil Rights Act of 1866, 42 U.S.C.A. sec. 1981 (1976).

2. *Hernandez v. Erlenbusch,* 368 F. Supp. 752 (D. Or. 1973).

3. There was, however, no violation of the common law duty of tavern owners to protect their patrons from assaults within the premises, since the defendant had no reason to anticipate the assault that took place outside the tavern.

4. Wyman, *The Law of the Public Callings as a Solution of the Trust Problem,* 17 Harv. L. Rev. 156, 159 (1903).

5. *Id.* See also J. Sherry, *The Laws of Innkeepers,* 3d. ed., secs. 1:1–1:6 (rev. ed. 1981).

6. *Jacobson v. New York Racing Ass'n, Inc.,* 33 N.Y.2d 144, 305 N.E.2d 765, 350 N.Y.S.2d 639 (1973); *Rockwell v. Pennsylvania State Horse Racing Comm'n,* 15 Pa. Commw. Ct. 348, 327 A.2d 211 (1974), and authorities cited therein.

7. The original English common law duty to admit is noted in 21 *Halsbury's Laws of England* 445–46 (3d. ed. 1957). The New Jersey Supreme Court cited and followed this rule, which is uniform throughout the United States, in *Doe v. Bridgeton Hospital Ass'n, Inc.,* 71 N.J. 478, 483, 366 A.2d 641, 646 (1976), *cert. denied,* 433 U.S. 914 (1977). See Sherry, *supra,* note 5 at sec. 3:3.

8. *Doe v. Bridgeton Hospital Ass'n, Inc., supra,* note 7.

9. Sherry, *supra,* note 5, at sec. 16:1.

10. *Feldt v. Marriott Corp.,* 322 A.2d 913 (D.C. App. 1974) (Junior Hot Shoppe).

11. *Braun v. Swiston,* 72 Misc. 2d 661, 340 N.Y.S.2d 468 (1972).

12. See J. Sherry, *The Laws of Innkeepers,* 3d. ed., sec. 6:5 (1972). The public accommodations section of the Civil Rights Act of 1875 was struck down by the Supreme Court in the *Civil Rights Cases,* 109 U.S. 3 (1883).

13. *Peterson v. City of Greenville,* 373 U.S. 244 (1963).

14. The initial doctrine applicable to innkeepers and common carriers was enunciated by

Judge Dick in his charge to the grand jury in North Carolina. *Charge to the Grand Jury—The Civil Rights Act*, 30 F. Cas. 999, 1001 (W.D.N.C. 1875). Also see Sherry, *supra*, note 5.

15. *Brown v. Board of Education*, 347 U.S. 483 (1954).

16. 42 U.S.C. secs. 2000(a), (d), 2002 (1976).

17. Sherry, *supra*, note 5, at sec. 4:13.

18. N.Y. Exec. Law sec. 291 *et seq.* (McKinney Supp. 1982) is a representative state civil rights law extending prohibited discrimination beyond the federal Act, to include sex, disability, marital status, and extension of credit.

19. New York City and Chicago have adopted such legislation.

20. Sherry, *supra*, note 5, at sec. 4:5.

21. *Id.* at sec. 3:7. *Cornell v. Huber*, 102 A.D. 293, 92 N.Y.S. 434 (2d Dep't 1905).

22. Sherry, *supra*, note 5, at sec. 4:7.

23. N.Y. Civ. Rights Law, sec. 41 (McKinney 1976).

24. N.Y. Exec. Law secs. 297(9), 299 (McKinney Supp. 1982).

25. Sherry, *supra*, note 5, at sec. 3:5.

26. See note 7, *supra*.

27. See, as representative legal authorities, *Young v. Caribbean Assoc. Inc.*, 358 F. Supp. 1220 (D.V.I. 1973) (resort dining room); *Peterson v. Haule*, 304 Minn. 160, 230 N.W.2d 51 (1975) (fast-food establishment); *Baker v. Dallas Hotel Co.*, 73 F. 2d 825 (5th Cir. 1934) (hotel).

28. The state of California is the only state thus far to outlaw discrimination in public places based on sexual preference. *Stoumen v. Reilly*, 37 Cal. 2d 713, 234 P.2d 969 (1951).

4 Liability for the Sale of Food and Beverages

OBJECTIVES

After reading this chapter, you should be able to:

1. Define areas of liability created by sales of foods and beverages.
2. Explain the role of consumer laws and the Uniform Commercial Code in the development of food liability law.
3. Contrast the theories for food liability claims.
4. Discuss liabilities to patrons and third parties created by alcohol service.
5. Discuss the various defenses to liability claims for the service of foods and beverages.
6. Identify management action to avoid liability.

Case in Point

Nancy Merrill ordered a Super Sub in Gwynn's Gourmet, a sit-down delicatessen. The sandwich included bologna, sausage, turkey, olive loaf, ham, and tomato, with either mustard or mayonnaise, served on a buttered French roll. When she bit into the sandwich, she broke her tooth on what was discovered to be a section of the spatula used to butter the bread. Is the owner of Gwynn's liable to Nancy?

Most likely, Gwynn's will be liable. A section of spatula is not naturally expected to be found in a sandwich or any other food product. However, if the object Nancy broke her tooth on was found to be a bit of olive pit, the court might apply a different interpretation, saying that the pit, which is natural to the olive loaf, might occur in a sandwich despite the best efforts of the preparer to keep it out.[1]

LIABILITY

According to the law, *liability* is an enforceable responsibility one person has to another. Liability is not just there; it is *created*. Foodservice operators create liability by careless preparation and service of food, poor judgment in the service of alcohol, or unsafe conditions in the operation. But since liability is created, it can also be *prevented*.

FOOD LIABILITY: THREE THEORIES

There are three approaches a patron can take to recover for injuries suffered as a result of eating or drinking unfit food or beverages. These three legal theories form the backbone of liability law for the service of foods and beverages.

The first is the *negligence theory*, a theory used under common law. In its most basic form, this theory requires that the injured patron be able to prove that negligence on the part of the foodservice operator resulted in the food being unfit to eat, which, in turn resulted in illness.

The second theory, and the most commonly used, is the *breach of warranty theory* applied under the Uniform Commercial Code. Here, the patron is not required to prove the foodservice operator was negligent in the storage, preparation, or service of the product, but only that the product caused the illness and that the seller breached the contract of sale by serving unfit food. Depending on which alternative of the Uniform Commercial Code a particular state has adopted, the operator may be held liable not only to the buyer of the food, but also to third parties who ate the food after the buyer brought it home. An advantage of the breach of warranty theory for foodservice operators is that they have recourse against the grower, producer, or packer of the food item if it was contaminated or unwholesome and the operator lost profits because it had to be thrown out.

Finally, *strict liability* is the most recent common law theory. This theory has evolved out of a number of court decisions and is most favorable to the consumer, if not the seller. Strict liability does not require that negligence or breach of warranty be established, but simply that the unfit food caused the person to become ill. This theory rests on the assumption that the consequences of eating unfit food are so threatening to human health and life that the law requires that someone be held responsible for injury to a patron. That someone could be the grower, processor, the foodservice operator, or all three.

COMMON LAW ROOTS:
THE NEGLIGENCE THEORY

From its early development in England, the common law has been criticized for its lack of uniformity. In the United States, each local judge or court was given wide discretion to mold the common law to fit the particular needs of the locality.[2] These officials had control over the content of the law, subject only to the state appellate courts. Under our multi-state legal system, this meant that each state's high court was free to formulate its own legal principles, subject only to the state's constitution and statutes.[3] Since the high courts of individual states were not required by law to confer with each other, the law in one state could, and often did, differ from the laws of other states.[4]

Nowhere was the lack of uniformity less desirable than in the area of commercial transactions (sales of foods between merchants), and sales by merchants to consumers. In early England, transactions were dealt with through the *law merchant*, a system of rules and customs used by sellers of goods. The law merchant was the businessperson's law for handling business transactions and for the resolution of controversies. English common law ultimately absorbed the law merchant, and the common law was transferred to the American colonies in the 17th century.[5]

In its formative period, the common law honored the widest degree of freedom of contract between sellers and buyers of products and services. Products and services were exchanged on a one-to-one basis, and the law adopted the maximum *caveat emptor* (literally, "let the buyer beware") because every product was available for inspection.

The common law did not suffer the foolish or careless buyer lightly.[6] Each state applying the common law could treat the sale of foods as a *service* or as a *sale*. If a transaction was classified as a sale, there was an *implied warranty*, or unspoken guarantee that the product was fit for consumption. However, if it was a service, no warranty was implied.

Most often the transaction was a service. In the absence of deceit or fraud on the part of the seller, the buyer assumed all the risks of a purchase, and had legal recourse only if the product was defective or deficient and the defect or deficiency caused the buyer physical harm.[7] The *negligence theory* of common law required that the injured person prove fault or personal harm on the part of the foodservice operator for the injury.[8] No warranties of product fitness were implied because the courts, applying the common law, treated the transaction as a service and not as a sale.[9] The buyer was at the mercy of the seller unless the latter expressly (orally or in writing) guaranteed the fitness of the product.[10] For obvious reasons, very few express warranties were offered, since no one wished to guarantee product performance against all risks.

The law of the state where the foodservice establishment was situated governed the rights of the patron and the foodservice operator.[11] This meant that in each state the highest court was free to adopt the theory of service, for which no contract liability existed, or to treat the transaction as a sale with an implied warranty or guarantee that the food sold was reasonably fit for human consumption.[12] Under this *warranty theory*, the liability of the foodservice operator was extended to include members of the immediate family, and not just the patron who purchased the food.[13]

In terms of protection, the negligence theory left the consumer somewhere out in left field. Obviously it was difficult for the patron to prove that the foodservice operator was negligent in the storage, preparation, or service of the food. If the food had been contaminated at the source—by the processor or grower—the patron was out of luck, because the common law required that *privity of contract*, a direct contractual relationship, exist between the parties to a sale for the buyer to recover for injuries.[14] Also, the common law made the seller responsible only to the buyer of unfit foods and beverages, and not to any other injured person, such as family members who ate the food later after receiving it from the buyer.[15]

These two common law elements of the negligence theory made it difficult for consumers to pursue justifiable complaints. First, they only had recourse to the seller of the product—the foodservice operator—even though he or she might not have been the source of the problem. Second, negligence, even if it existed, was difficult to prove. As for the foodservice operator, the unhappy consumer was another lost customer.

The negligence theory may still be used by anyone bringing suit as a result of eating unfit food, especially if there is strong proof of negligence. However, the Uniform Commercial Code breach of warranty theory eliminated most of the problems consumers had with the negligence theory. The strict liability theory, which is applied by some courts, goes even further.

THE UNIFORM COMMERCIAL CODE: BREACH OF WARRANTY THEORY

After the Civil War, the industrial revolution took the United States into a new economic era. The transformation of an agricultural society into a vast industrial and commercial society added to the problems consumers already had with the common law. This was true because previously the producer and the seller of food products were often the same person. The seller dealt directly with the consumer, and the consumer who had a complaint knew where to take it—directly back to the seller. Now, foods and beverages are sold through a complex system, often with numerous stages from manufacturer to retailer. With the advent of a more advanced food distribution system, the common law proved inadequate to protect consumers.

During the 19th century, the need for uniformity in the commercial area was finally recognized. Legislation was enacted to overcome its limitations. With the coming of the Industrial Revolution, many states adopted more uniform laws, including sales acts. However, these laws were not collectively adopted by every state.[16]

Finally, in 1942 the Uniform Commercial Code project was set in motion by the American Law Institute and the National Conference of Commissioners on Uniform State Laws. Various official texts were adopted, the latest being the 1972 Official Text.[17] The Uniform Commercial Code is the most integrated and uniform law governing commercial transactions, and has been adopted by all states except Louisiana.[18]

Under the Uniform Commercial Code, every food sale automatically comes with an implied warranty that the food is fit for consumption. It also extends the scope of liability to protect others in addition to the buyer. The importance of the Uniform Commercial Code is that *it imposes a stricter standard of foodservice operator responsibility than the common law negligence theory*. Patrons no longer need to prove negligence or intentional misconduct to recover in unwholesome food cases. Rather, they need only to prove that the unfit food *caused* the injury or illness, and that by selling the unfit food, the operator breached the warranty of the sale. Patrons can sue either the foodservice operator, the grower, or the producer. Privity of contract is not required.

The Code also offers some protection to the foodservice operator by recognizing the right of the retailer to sue growers and producers to recover any losses sustained because of their negligence.[19] This principle is called *equitable risk distribution*.

APPLICABLE SECTIONS OF THE CODE

Section 2-314 of the Uniform Commercial Code (UCC), uniformly enacted into law in *all* states except Louisiana, says specifically that the serving for value of food or drink to be consumed either on the premises or elsewhere is a sale. This section 2-314 also says that a warranty that the goods are "merchantable" is implied in a contract for their sale if the seller is a merchant with respect to goods of that kind. "Merchantable goods" must be fit for the ordinary purposes for which they are meant. Under this provision, the service of food or drink consumed either on the foodservice premises or elsewhere creates a warranty or guarantee that the items are fit. The service itself is a sale.

The UCC imposes an *implied warranty*, a guarantee independent of any express guarantee made by the seller to the buyer or to third parties. An *express warranty* is an oral or written guarantee made by the seller to the buyer.

However, UCC section 2-318 provides three alternative provisions regarding who other than the purchaser (i.e., a third party) may obtain the protection of the implied or express warranty. The states are not required to adopt this section or its alternatives. The states may substitute their own language in place of this section, or they may do nothing, allowing the courts to decide the issue. Once the choice is made, however, the state courts are bound to apply the alternative selected in all appropriate cases. (See **Appendix D**.)

Alternative A says that a seller's warranty, whether expressed or implied, extends to any natural person who is in the family or household of his or her buyer, or who is a guest in the home, if it is reasonable to expect that such person may use, consume, or be affected by the goods and who is injured by breach of the warranty. A seller may not exclude or limit the operation of this section.

Under *alternative B*, a seller's warranty, whether expressed or implied, extends to any natural person who may be reasonably expected to use, consume, or be affected by the goods, and who is injured in person by breach of the warranty. A seller may not exclude or limit the operation of this section.

Alternative C states that a seller's warranty, whether expressed or implied, extends to any person who may be reasonably expected to use, consume, or be affected by the goods and who is injured by breach of the warranty. A seller may not exclude or limit the operation of this section with respect to injury to the person of an individual to whom the warranty extends.

Under alternative A, all guarantees by the seller, whether expressed or implied by law, extend to any person in the buyer's family or household, or to anyone who is a guest in the home, when it is reasonable for them to expect that they will use, consume, or be injured because of a breach of the guarantee that the product is fit. For example, say John buys hamburgers, fries, and soft drinks at Sam's Takeout and takes them home to his family. His daughter, Fran, eats one hamburger and gives another to her friend, Betty, who is having dinner with the family. If they become ill, both Fran and Betty can recover damages if the hamburgers turn out to be unfit for human consumption. However, Betty must be in the home of the purchaser, John, in order to recover. If she eats the hamburger outside the home (on her way to the home or after leaving), she may not recover. Alternative A is most favorable to foodservice operators.

Alternative B eliminates the restriction on liability that the guest must eat the product in the home of the purchaser to recover from the seller. All persons are included, not just family or household members. So, Betty could recover regardless of where she ate the hamburger; she would not have to prove she ate it in John's home. This alternative strikes a balance between A and C.

Neither alternative A nor B permits corporations or other business entities to recover. Nor do they permit individuals to recover property damages or lost profits resulting from a violation of a guarantee of food fitness. In other words, Sam's Takeout could not recover from the meat manufacturer for profit losses associated with meat purchased for preparation and service to patrons, if that meat were found to be unwholesome and subsequently seized by the local health board. Sam's could, however, recover for liability suffered when sued by John.

Alternative C provides virtually unlimited protection to third parties, and permits recovery for either personal injury or property damage. However, the seller may exclude or limit the operation of that liability for property damage or losses, but may not limit liability for individual personal injuries.

The reason for the three alternatives is to give each state a wide range in choosing the provision that best meets its commercial needs and policies with regard to the competing interests of growers and producers, retail vendors, and consumers.

Alternative A is most restrictive in terms of remedies available to a third party beneficiary who sues a retail vendor. Only *natural persons*, and not artificial legal entities, such as corporations, may sue, and the person must be a family member or guest who eats the food or beverage in the buyer's home. The majority of states have adopted alternative A since it appears to offer more protection to growers, producers, and foodservice vendors, in that restrictions are placed on who can recover in terms of buyer status, on the location where the product was consumed, and on damages.

Alternative B widens the scope of alternative A by permitting any person to sue for personal injuries regardless of where the unfit food or drink was consumed. Alternative B opens the door to recover further by eliminating the requirements of buyer status and location of consumption, but maintains the other restrictions contained in alternative A.

Alternative C offers the widest scope of recovery because it applies to all persons, and includes artificial entities such as corporations. It permits recovery for both personal injuries and property damage, and allows sellers, including growers, producers, and retail foodservice operators, to exclude or limit liability for economic or property losses, but not for personal injuries. For example, a corporation may recover for a loss of profits associated with unfit foods, unless the seller limits or expressly excludes his or her liability.

For these reasons, alternative C appears most favorable to all parties. Growers, producers, retail vendors, and foodservice operators can protect themselves against property losses. Consumers and beneficiaries can sue the growers, producers, vendors, or operators without being bound by any express or implied warranty restrictions. Every potential plaintiff and defendant gets some degree of protection.

The Code in Practice

The need for proof of negligence or intentional misconduct on the part of the seller was eliminated by the UCC. The seller is now required to guarantee the fitness of every food and beverage product.[20] This places a heavy burden on the retail foodservice operator, who must guarantee the performance of food growers and producers, even though not directly involved in these activities. Foodservice operators cannot be in control of what happens at the

Exhibit 4.1   Ten Rules of Safe Foodhandling

1. Require strict personal hygiene on the part of all employees, and relieve infected employees of foodhandling duties. Instruct employees not to touch cooked food or food that will not be cooked.
2. Identify all potentially hazardous foods on your menu and create a flowchart to follow these foods throughout the entire operation.
3. Obtain foods from approved sources.
4. Use extreme care in storing and handling food prepared in advance of service.
5. Keep raw products separate from ready-to-eat foods.
6. Avoid cross-contamination from raw to cooked and ready-to-serve foods via hands, equipment, and utensils.

7. Cook or heat-process food to recommended temperatures.
8. Store or hold foods at temperatures below 45°F (7.2°C) or above 140°F (60°C) during preparation and holding for service.
9. Heat leftovers quickly to an internal temperature of at least 165°F (73.9°C) within two hours.
10. Cooked food should be rapidly chilled in shallow pans in a refrigerator, in a quick-chilling unit, or in an ice water bath and stirred or agitated frequently during the chilling.

Reprinted with permission from *Applied Foodservice Sanitation, Fourth Edition.* Copyright © 1992 by The Educational Foundation of the National Restaurant Association.

source of the food distribution system. However, managers and employees must exercise great care in inspecting incoming foods and beverages. The operation's role should be to make sure the product is fit on delivery and that it stays that way up to service to customers. (See **Exhibit 4.1.**)

A few courts have carved out an exception to the Code by judicial interpretation; that is, the implied warranty of fitness does not extend to products sold in sealed containers for resale to consumers.[21] Since most retailers further process foods for resale to consumers, the exception, even if applicable to food growers and producers, is not usually available to retail foodservice vendors. The exception would apply only to someone who sold home-canned or bottled products directly to patrons.

Relief is not granted automatically to consumers under the UCC, but is determined by the courts on a case-by-case basis. Either a jury or a judge decides on the facts presented in court.

Fitness under the Uniform Commercial Code

What is "fit" for human consumption? Normally any ingredient or object which is natural to the food product is considered fit. Two court tests are used to determine food fitness under the breach of warranty theory: The foreign or natural test and the reasonable expectations test.

Foreign or Natural Test   The courts, in applying the applicable section of the UCC, usually apply the *foreign or natural test* to determine whether the foodservice operator is liable to the injured patron. What would establish liability under this test? A stone or piece of glass embedded in a dinner role would do so, since such objects are totally foreign to the food.[22] A cherry pit, however, *might* be expected in a cherry pie, and the restaurant owner might not be held liable if the court applied the *foreign or natural* test.[23]

The following case gives one example of a court interpretation of the foreign or natural test.

Webster v. Blue Ship Tea Room, Inc.
Supreme Judicial Court of Massachusetts
347 Mass. 421, 198 N.E.2d 309 (1964)

Facts: Webster, a patron of the Blue Ship Tea Room, swallowed a fish bone while eating a bowl of fish chowder. As a result, she required medical treatment for removal of the bone, and sued for injuries. She argued that the fish chowder was unfit for human consumption and violated the applicable warranty of food fit for consumption established under the Massachusetts version of the Uniform Commercial Code (alternative A). A jury found for Webster. The restaurant owner made a motion for a *directed verdict* (a verdict made by the judge), which was denied. The directed verdict, however, was approved by the Massachusetts Supreme Court, in favor of the restaurant owner.

Reasoning: The court ruled on the question of whether a fish bone should be considered a foreign substance that made the fish chowder unwholesome and not fit to be eaten. By applying the foreign or natural test, the Supreme Court reasoned that consumers expect the possibility of fish bones in chowder, and ruled in favor of the Blue Ship Tea Room.

Conclusion: This is a classic foodservice case involving an alleged breach of an implied warranty of fitness under the Uniform Commercial Code. This Massachusetts court decision (now called the "Massachusetts rule") established that *the unfit object must be foreign to the food product to impose liability.*

However, the court was careful to point out that a breach of implied warranty of fitness would exist had the food itself been found *unwholesome.* In other words, when a food product is tainted, the foreign or natural test has no application.

Reasonable Expectations Test   The Massachusetts rule is still followed by the majority of states, but a growing number now apply the *reasonable expectations test*. This test is applied to food only in its final, processed form.[24] Under this test, the condition of the product served must be reasonable in terms of the expectations of the patron.[25] This means that a patron might expect to find a chicken bone in fried chicken, but not in a chicken salad.

Under the foreign or natural test, the determination of fitness is made by the courts as a matter of law.[26] Under the reasonable expectations test, the determination is made by the jury as a question of fact.[27] Since juries are not consistent, the outcome of cases applying the second test is always less predictable.

Ex Parte Morrison's Cafeteria of Montgomery, Inc.
Supreme Court of Alabama
431 So.2d 975 (1983)

Facts: Rodney, a three-year-old boy, accompanied by his mother, visited Morrison's Cafeteria for lunch. The mother purchased a serving of fish amondine and served her son a portion from her plate. Rodney choked on the first bite of fish and was taken to the hospital. The cause of the problem was diagnosed as a one-centimeter long piece of fish bone lodged in Rodney's throat. The bone was removed and Rodney suffered no permanent injuries. No one at Morrison's claimed the fish was boneless nor was it advertised as boneless. Morrison's purchased the fish from Pinmella's Seafood Company. Pinmella's established that the fish was filleted by machine for sale to Morrison's, that this method could not prevent the presence of occasional small bones in the fillets, and that government regulations allow for the presence of small bones in fillets. To remove such bones would require the fillets to be cut up into small pieces, which would destroy the fillets.

Rodney and his mother sued Morrison's claiming a breach of warranty of fitness under Alabama law. The lower court rejected the foreign or natural test of fitness and applied the reasonable expectations test and the jury found in favor of Rodney and his mother. The Alabama Supreme Court reversed, holding as a matter of law that the presence of a one-centimeter bone did not render the fish in question unreasonably dangerous or unfit for human consumption.

Reasoning: Most courts that apply the reasonable expectations test of liability in foodservice cases will let the jury decide the issue of fitness of the food product for human consumption. However, in exceptional circumstances, a reviewing court will overturn a jury verdict for the injured parties. By doing so, the court is saying that on the evidence presented no reasonable jury could have found for the victims. The jury may have acted out of sympathy for the mother and child, an understandable but impermissible ground on which to render verdicts.

This case, one of the first of its kind decided in Alabama, must be limited to its unique facts. Had the fish bone been larger, or had Morrison's advertised the fish as boneless, the court might have affirmed the jury verdict.

Conclusion: This case demonstrates that a jury verdict can be overturned where it would be totally unreasonable to hold the foodservice establishment liable. This conclusion holds true under a foreign or natural test of fitness for human consumption or the more recent reasonable expectations test. However, under both tests, a jury is given reasonable discretion to find for or against the consumer. It is only in exceptional cases that a jury verdict is overturned on appeal. For example, if existing consumer health and safety regulations concerning the amount of fish bone in the Morrison case had been violated, the case might have been decided otherwise. Moreover, if a larger number of bones had been involved, the court might have affirmed the jury verdict.

Although it is unique, this case does indicate that foodservice operators are not necessarily insurers of the foods they serve and that under proper circumstances courts will deny liability based on unreasonable application of either test of liability.

STRICT LIABILITY

There is a growing trend for patrons seeking compensation for injuries from unfit food to ask the courts to apply *strict liability*. This judge-made legal theory, applied in some states, is generally beneficial to patrons because it does not require that they prove negligence or breach of warranty on the operator's part. They have only to prove that the food they ate made them ill. The law imposes a guarantee of food purity to the consumer without the need to determine whether an express or implied warranty between seller and buyer exists.

Where this theory differs slightly from the UCC breach of warranty theory is this: The UCC generally provides that patrons must prove breach of warranty and that the food purchased made them ill. Strict liability does not require patrons to prove a breach, only that the food made them ill. Once a victim has been established—that is, the injured patron has proven that the food was the source of the illness—the patron has legal grounds to recover.

The difference is one of law and not of fact. In other words, the law *imposes* liability, once the plaintiff has proven damages. The plaintiff does not have the burden of proving negligence or breach of warranty, using the two tests given for the latter theory. The thinking behind strict liability is that someone in a better position to pay should be held liable for injuries.

The Uniform Commercial Code is still the dominant product liability theory, with strict liability a growing one. The negligence theory is seldom used, but is available for patrons who can prove negligence.

THE THEORIES IN CONTRAST

Suppose an operator purchases chicken salad from a well-known national distributor. Unknown to him, the salad contains chicken bones, which he cannot detect in the course of normal preparation. The cook prepares menu items with the chicken salad, using all proper sanitary preparation procedures. A patron orders and eats a chicken salad sandwich and the bones get caught in her throat, requiring hospitalization. The patron sues for medical and hospital expenses, lost wages, pain, and suffering. Can the patron recover under either the negligence, breach of warranty, or strict liability theory?

Under the negligence theory, the patron cannot recover damages because the operator did not violate his legal duty to exercise reasonable care. He did not know, nor could he have known, the chicken salad contained tiny bones and would cause the patron injury. He complied with all reasonable sanitary standards and did not contribute in any way to the harm suffered.

According to the negligence theory, the operator is not liable for risks caused by the grower, producer, or packager of the product served in its original container, unless he caused or contributed to those risks.[28] An operator causes or contributes to a risk by knowing of the chicken bones and failing to warn patrons or remove the bones from the food. Since many food products purchased for restaurants are served "as is," except for heating and refrigeration, operators can escape liability for any unfitness in the original product that caused harm to a patron. The injured patron would have to prove that heating or refrigeration was negligently maintained, or that the operator allowed the chicken salad with bones to be served. It would not be enough to prove that the distributor was negligent in processing or packaging the product. In that case, the patron would have to sue the distributor and not the operator.

Under the breach of warranty theory, all the members of the distribution chain—the producer, processor, distributor, and operator are liable to the patron if she: 1) can prove that the food caused injury; and 2) can prove breach of warranty, using the reasonable expectations test. The law does not require the patron to prove negligence. None of the parties sued can defend on the grounds that they exercised all the care required of them under the circumstances, and were not at fault. Why? Because negligence is equated with a finding of fault. Breach of warranty does not depend on any finding of fault, only on the breach, since each party guarantees the fitness of the product for human consumption.[29]

The breach of warranty theory rests on a contract of sale between a patron and the foodservice operator. How then can the theory apply to the producer or distributor of the chicken salad, since neither party sold or served anything to the patron? It applies because the producer advertises the product for *ultimate* use by foodservice patrons, and is held to the same *implied* representation of fitness as the retail foodservice operator.

It is also important to note that service of a food or beverage is always deemed a sale, regardless of who pays the bill.

Under the strict liability theory, both the chicken salad distributor and the foodservice operator could be liable to the patron. The strict liability theory imposes a guarantee of product fitness, regardless of any express or implied guarantee imposed by the Uniform Commercial Code, and does not require proof of breach of warranty. If the food is found to be unfit to eat, someone will pay, either the operator, or the chicken salad distributor.[30]

DEFENSES FOR FOOD
LIABILITY CLAIMS

There are four general defenses against patron lawsuits; however, not all of these defenses can be used in every situation. A lawyer is best qualified to choose which defense is most appropriate. As with all legal areas, the best defense against lawsuits is to make every attempt to ensure the wholesomeness and safety of food being served.

Privity of Contract

The *privity of contract* defense limits operator liability. According to the negligence theory, only the direct patron (*proper plaintiff*) can take action against an operator, and not a family member who may eat the food after the buyer brings it home. In most states, however, third parties usually can take action. Yet, in some states there is still some confusion as to whether the UCC extends or limits who may seek compensation from the retailer.

The *proper defendant* also must be established through privity of contract. The plaintiff whose only contract was with you, the operator, can take action only against the operator and not the manufacturer. In addition, they must prove that the operator was the *source* of the negligence. Patrons trying to prove breach of warranty would have to prove that the operator breached the warranty of implied or express fitness. However, in some states the operator would then have recourse to sue the manufacturer or processor to recover losses.

Proximate Cause

Under all three liability theories, the patron must prove that the unfit (unwholesome) product caused the illness or injury. This means that the patron must eliminate alternative causes equally likely to be the cause of the illness.[31]

The burden rests on the plaintiff to demonstrate that: 1) the food or beverage was unfit to eat; and 2) the unfit product was the cause of the illness or injury.[32]

For example, if a patron became ill after having dinner in one restaurant and dessert in another, that patron must prove which restaurant's food caused the illness. Likewise, if a patron is taking a prescription or nonprescription drug, there is a legitimate question as to whether the drug and not food caused an illness. In all cases of liability, the patron must be able to prove that the illness or injury was the result of the food or beverage product, and not from some other cause.

Assumption of Risk

According to this defense, the courts expect that the patron assumes risks when eating certain types of foods. For example, in *Webster v. Blue Ship Steam*

Room, Inc., the plaintiff should have assumed that a fish bone might naturally be found in a bowl of chowder. However, the plaintiff would not expect that the dish contained a chemical contaminant. *This defense may not be used in strict liability cases.*

Contributory Negligence

This defense is good only against a negligence claim and is seldom used. Here the operator can try to prove that the patron contributed to the risk, by exposing the food to patron contamination, improper storage, or poor reheating.

Hoch v. Venture Enterprises, Inc.
United States District Court for the Virgin Islands
473 F. Supp. 541 (1979)

Facts: After consuming a native Caribbean fish at the defendant's food establishment, Hoch, a customer, suffered from ciguatera fish poisoning, conceded to be a latent natural condition in fish of that type. The poisoning was not fatal, but Hoch suffered stomach cramps, nausea, diarrhea, and extreme sensitivity to temperature changes. Hoch sued for breach of expressed and implied warranties that the fish served was wholesome and fit for human consumption. He moved for partial summary judgment based on the complaint. The defendant opposed the motion because of questions of fact as to proximate cause, unfitness of the fish, and whether the assumption of risk doctrine was an available defense.

The court denied Hoch's motion for several reasons. First, the court said that Hoch submitted insufficient evidence to prove proximate cause (that the fish caused the illness) to a jury.

On the issue of whether the fish product was unfit, the court adopted the reasonable expectations test of liability. This meant that the jury would have to decide whether the plaintiff could reasonably expect to find the poison natural to the fish in the product consumed. Only if the jury found that Hoch could not reasonably expect to find such a substance in the fish would the defendant be liable under an implied warranty theory of liability.

Lastly, and especially noteworthy, the court ruled that the assumption of risk defense was available in a breach of warranty case such as this, requiring that issue to be submitted to the jury for determination at trial. This conclusion also precludes granting the plaintiff's motion for summary judgment.

Reasoning: On the first issue, proximate cause is a jury question, and cannot be resolved by the plaintiff's motion for judgment without trial. In other words, the facts presented in this case are not so conclusive on this issue that judgment without trial would be appropriate.

Second, the facts also require the court to submit to the jury the question of whether the plaintiff could reasonably expect to find the poison in the fish. That legal standard, the reasonable expectations test, also must be submitted to the jury.

Lastly, the assumption of risk defense is available to the defendant in the U.S. Virgin Islands. This means that if the consumer is fully aware of the danger and nevertheless proceeds voluntarily to make use of the product and is injured by it, the consumer is barred from recovery. That defense can only be developed at trial, again requiring denial of the plaintiff's motion.

Conclusion: This case raises the possibility of asserting an assumption of risk defense in a foodservice breach of warranty claim. This is important because proof and acceptance of that defense by a jury will prevent recovery, at least in the Virgin Islands. This defense should be raised wherever it is relevant, since in the court's view a consumer who voluntarily consumes a product he knows to be harmful cannot claim reliance on any implied foodservice warranty of fitness. Rather, this consumer (Hoch) relied on his own willingness to risk injury, for which no recovery against the foodservice operator should be allowed.

Although there is a dispute among the states as to the availability of the assumption of risk defense in tort cases, the *Hoch* case reminds us that the defense may still have vitality in breach of warranty cases.

GENERAL MANAGER GUIDELINES: TRAINING AND FOOD SAFETY

The development of the law from English common law to the Uniform Commercial Code demonstrates a shift away from *caveat emptor* ("let the buyer beware") to *caveat vendor*, or "let the seller beware." This places a heavy responsibility on foodservice operators and their employees to minimize liability.

Almost every foodservice operator will encounter some form of local regulation concerning sanitation. These regulations vary. Some may be so complete that if the operation follows the standards, it may be less open to liability. Others may keep operators out of trouble with the government, but since they do not provide maximum standards, can still result in liability.

It is up to the manager to adopt the maximum sanitation standards justified by cost and training. At the bottom line, operators should enforce sanitary procedures concerning time and temperature and safe food handling, mentioned earlier. Food can be contaminated from the time it enters the restaurant until it is served to the patron. Sanitary procedures all along that route can prevent contamination.

Since foodservice operators must guarantee the fitness of their food products, they are under a continuing legal duty to exercise the standards of *reasonable care*. An operator must anticipate, warn, and/or remove unreasonable risks of harm. Regular inspection of the entire foodservice operation will help protect patrons against disease due to impurities, contamination, and spoilage.

LIABILITY FOR THE SALE OF ALCOHOLIC BEVERAGES

Alcoholic beverage sales have occupied a unique place in our social and economic development. Our puritanical social heritage made drinking alcohol a vice and temperance a virtue. The involvement of organized crime during the Prohibition era in the late 1920s reinforced this historic social concern by associating the commercial sale of alcohol with criminal profiteering.

Public attention has recently shifted toward spiraling drunk driving and injury statistics. Consumer groups have persuaded states to impose stricter control over alcoholic beverage sellers. Tougher dramshop and liability laws help people recover compensation for injuries caused both to patrons and third parties by intoxicated drivers.

Common Law Liability

The common law placed the entire responsibility for intoxication on the consumer, and not the seller of alcoholic beverages. The theory behind this was that it was the *voluntary consumption* of alcohol by an able-bodied person that caused intoxication, and not the sale or service of alcohol.[33] Therefore, since the person knew or should have known the consequences of drinking, he or she voluntarily assumed the risks of intoxication, including any injuries suffered as a result.[34] In practice, neither the intoxicated patron nor any innocent third person injured by that patron could recover from a licensed seller of alcohol.

A number of states have reversed this policy, either by court decision or by their legislatures passing statutes governing alcohol-related injuries.[35]

Liability to Third Parties: Dramshop Acts

Mostly, liability is covered through state *dramshop* acts or by court interpretation of state liquor laws. Disobeying state laws regarding the sale of alcohol leaves operators not only open to criminal prosecution, but to liability to injured parties as well, since persons bringing suit are often able to use the violation of the law as support for their claim. State laws especially pertinent to operator liability are those involving sales of alcohol to minors, intoxicated persons, and known or habitual alcoholics. However, the common law of most states still excuses operator responsibility to the so-called able-bodied patron.[36]

A minor is not presumed to be an able-bodied patron. Sales of alcohol to minors, forbidden by all states, may make operators liable to the minor or to the minor's survivors for injury or death claims. The law presumes that a minor does not have the maturity or drinking experience to know the risks of intoxication.[37]

The able-bodied definition does *not* usually include known alcoholics, and operators may be responsible for injuries to these patrons, as well as any third parties they may injure. Some states even allow relatives of an alcoholic to give written notice to licensed operations, ordering them not to serve liquor to the alcoholic relative.[38]

Dramshop acts typically require the injured third party to prove the following: 1) an illegal sale of alcohol by a licensed seller; 2) that the seller caused or contributed to the intoxication; and 3) that the sale resulted in injury to the victim. (See **Appendix E.**)[39]

Injured third parties may have a choice of legal remedies to support their claim: 1) common law negligence in failure to foresee the effects of lack of supervision of the premises; 2) dramshop act liability; or 3) common law negligence for failure to comply with statutes designed to curb illegal sales of alcohol.[40]

The common law ruling exempting licensed operators from liability to able-bodied patrons is not altered by dramshop acts unless the statute specifically creates such liability.[41] In a few states, the common law of negligence is applied whether or not the sale is illegal. These states hold that operators must *foresee* that any sale to an obviously intoxicated patron whom it is known intends to drive a motor vehicle creates a *reasonably foreseeable* risk of harm to other motorists or pedestrians.[42]

Liability can be expensive. Some states permit recovery of compensatory as well as punitive damages. Other states permit recovery of compensatory damages, and death claims to compensate the estate of the deceased third person.[43] A few states place a monetary ceiling on liability.[44]

Branigan v. Raybuck
Supreme Court of Arizona
667 P.2d 213 (1983)

Facts: Three underage boys were killed in a motor vehicle accident. The surviving parents sued for the wrongful death of their children alleging that the operator of the Good Times Inn had breached a duty of care by furnishing liquor to the boys and that this had been the cause of the accident. The boy who drove their pick-up truck had a post mortem alcohol blood level of 0.23, well above the legal limit. It was established that the Inn did not ask the boys for identification, and that teenagers frequented the establishment because the inn had a reputation for not checking IDs.

The State Supreme Court overruled its long-standing common law rule that consumption, not service of alcohol, is the proximate cause of injury sustained by the drinker and by third persons, thus insulating the alcohol server from liability. The case was sent back to the trial court with instructions to hear the claims on the basis of the legal principles set forth by the Supreme Court.

Reasoning: The court first voted that in a companion decision (*Ontiveros v. Borak,* *667 P.2d 200*), it decided to hold a tavern owner liable for injuries caused by the illegal sale of alcohol to a customer whose intoxication inflicted harm upon an innocent third party. The rationale for this change in the common law is the increasing number of injuries and deaths to patrons as well as innocent third parties caused by drunk driving resulting from sales to underage or obviously intoxicated patrons. The Arizona statutes prohibiting sales of alcohol to either underage or obviously intoxicated patrons serves to support this view, and permits the court to establish a legal duty upon licensed vendors to respond in damages if they fail to follow the statutory duty or a common law duty that the court creates independently of statute. Because the violation of a statutory dramshop duty is considered negligence *per se* (in and of itself) in Arizona, the Supreme Court ruled that a vendor could establish that his or her violation was excusable; i.e., that the minor appeared to be of age and his identification appeared to be genuine or that the person served appeared to be sober. The court pointed out that these were questions for the jury to resolve upon remand for trial.

Finally, the court ruled that under Arizona constitutional law, questions of patron contributory negligence and/or assumption of risk were solely defenses for the jury to resolve, even in the face of a court ruling to the contrary as a matter of law.

Conclusion: Arizona joined a growing number of states in imposing liability on commercial vendors who serve underage or intoxicated patrons whose intoxication harms innocent third parties.

Arizona also joined a minority of states that permit the patron to sue commercial vendors for injuries the patron sustains. (The Arizona legislature later reversed this rule of drinker recovery by statute as set forth in *Carrillo v. El Mirage Roadhouse, Inc.* 793 P.2d 121 [1990].)

The following case demonstrates the California legislature's response to sweeping changes in dramshop liability law holding both commercial vendors and social hosts liable for illegal sales and service to minor and intoxicated patrons resulting in harm to third parties as well as drinkers.

Calendrino v. Shakey's Pizza Parlor Company, Inc.
California Court of Appeals
198 Cal. Rptr. 697 (1984)

Facts: Calendrino, a minor, was served a number of mugs of beer by Shakey's, a licensed alcohol vendor. Afterward, he went to a private party where he drank even more alcohol until early the next morning. While intoxicated, he left and accepted a ride from another intoxicated drinker. The drinker was involved in a single car accident, injuring Calendrino.

Calendrino sued Shakey's on the ground that Shakey's served him alcohol while intoxicated, thereby making Calendrino unable to care for himself, causing him to accept a ride from an obviously intoxicated person.

The court ruled against Calendrino, saying that he failed to state a cause of action.

Reasoning: In 1978, the California legislature virtually immunized commercial vendors and social hosts from liability for serving alcohol to minors or intoxicated patrons resulting in harm to the patrons or to innocent third persons. This action was in response to the decisions of the Supreme Court of California, which had created a common law cause of action against both groups of servers on behalf of intoxicated minors as well as third parties. Only one exception to this blanket legislative immunity was created: any person who has suffered personal injury or death has a cause of action against a licensed vendor who served an intoxicated minor where the service of alcohol is the proximate cause of the personal injury or death sustained by such person.

The court held that the term "any person" provides a cause of action only for other persons injured by the intoxicated minor and not for the intoxicated minor himself.

Conclusion: This case and the legislation on which it is based are among the most favorable to licensed vendors and social hosts. The majority of states do allow third parties injured by intoxicated patrons the right to sue licensed vendors. Some states permit third persons to sue social hosts if the intoxicated patron causing the harm is a minor. A few states also permit the intoxicated minor to recover for his or her own injuries.

This overview suggests that dramshop liability is well established in most states, with California representing the exception. This fact of life requires food and beverage operators to exercise care in the way they conduct alcohol service policies.

Defenses against Liability for the Sale of Alcohol

There are few defenses to liquor liability to third parties.

As for liability to patrons, if the patron is an adult and not an alcoholic, there are the defenses of *contributory negligence* or *assumption of risk*. Contributory negligence is based on the premise that the adult patron should know when to stop drinking.

Assumption of risk is a similar defense; patrons should assume risks if they consume too much alcohol. However, what is too much for an already drunk patron? After a certain point the patron might not be able to determine that, and it might be up to the manager, the bartender, or other service employees to do so. (See **Exhibit 4.2.**)

Exhibit 4.2 Steps to Serving Alcohol Responsibly

1. Never serve alcohol to anyone under 21 years old or anyone who is visibly intoxicated.
2. All guests' identification should be checked before serving. Acceptable documentation is a driver's license with a photo or other photo identification.
3. When a patron shows visible signs of intoxication (slurred speech, impaired judgment, trouble walking or with other motor functions, or disorderly conduct), a manager or employee should inform the patron that alcohol service is no longer legal and offer a nonalcoholic drink.
4. If a visibly intoxicated person tries to leave, a manager or employee should offer alternative transportation. If the patron still attempts to drive, every effort should be made to get the patron's keys. If, despite all efforts, the intoxicated patron still leaves in his or her car, the employee or manager should note the car's license number and call the police.
5. Incident reports should be filled out and kept for all incidents involving alcohol service.
6. All alcoholic drinks should be measured using standard equipment and utensils.
7. No guest should ever be served more than one drink at a time.
8. No guest should be served several drinks within a very short time period.
9. All employees who serve alcohol should take part in responsible alcohol service training.
10. Any employee violations of policies concerning alcohol service should be disciplined and/or terminated.
11. No promotional activity should be used that promotes irresponsible or excessive drinking. Instead, promote food, entertainment, music, contests, games, etc.
12. Promote and suggest food and nonalcoholic beverages to all patrons.
13. Promote responsible alcohol service and responsible drinking to patrons by serving "designated drivers" free soft drinks, posting signs, etc.

Adapted from the *Responsible Alcohol Service Manager Coursebook*. Copyright © 1991 by The Educational Foundation of the National Restaurant Association.

In one landmark case, *Ewing v. Cloverleaf Bowl*, the California Supreme Court ruled that a bartender was liable for willful misconduct for serving a patron who had just turned 21, could not fully appreciate the danger of drinking large quantities of liquor, and died from acute alcohol poisoning.[45] In this case, the court pointed out that the bartender: 1) knew the patron had just turned 21 and was an inexperienced drinker; 2) despite that, served him 10 shots of 151-proof rum, as well as several other drinks; and 3) exercised willful misconduct in his actions.

There are virtually no defenses to knowingly serving alcohol to minors. At minimum, an operation will encounter a lawsuit or lose its liquor license; at worst, both managers and employees may encounter a criminal penalty, including jail. All efforts toward responsible alcohol service should be made.

The following cases illustrate liquor liability under certain conditions.

Grasser v. Fleming
Court of Appeals of Michigan
74 Mich. App. 338, 253 N.W.2d 757 (1977)

Facts: Fleming, a tavern owner, served alcohol to Grasser, an intoxicated, elderly, habitual alcoholic, contrary to an agreement with Grasser's family not to serve him. After leaving the tavern, the alcoholic died when he lost his balance and fell eight feet to the ground while walking on an unguarded, narrow concrete projection to a bridge. Grasser's family filed a wrongful death claim against Fleming, who moved to dismiss the claim on the grounds that Michigan's Dramshop Act was the exclusive remedy, and no action by an adult drinker was recognized at common law by the Michigan courts. The trial court denied the motion to dismiss, and this decision was appealed and upheld.

Reasoning: In this case, the court said that the state's dramshop act did not prohibit "common law cause of action for gross negligence or willful, wanton and intentional misconduct in the sale of alcoholic beverages under the circumstances of this case."

The court pointed to an exception to the general rule of a patron's liability for his or her own actions, when "the customer was in such a helpless state as to have lost his free will . . ." The court said there was no reason to dismiss the case under the circumstances and that Grasser's family had "a cause of action for gross negligence and willful, wanton, and intentional misconduct independent of the dramshop act . . ."

Conclusion: This case illustrates the rule that an operator may be liable for selling liquor to known alcoholics. Alcoholics are often given the same special consideration as minors, because of their inability to control their drinking.

1. Any willful (deliberate), wanton (reckless), grossly negligent (indifferent to the consequences), or intentional misconduct, which intoxicates the person, may give that person or his or her family the right to sue.
2. Involuntary intoxication, when established, will also impose liability, even though the drinker is normally not able to recover for his or her own injuries.
3. Involuntary intoxication will not allow operators to escape liability automatically as a matter of law by using the defenses of contributory negligence and assumption of risk. Contributory negligence and assumption of risk remain jury questions.

Chausse v. Southland Corp.
Court of Appeals of Louisiana
400 So.2d 1199 (1981)

Facts: Three teenage girls were riding in a car driven by an intoxicated, 16-year-old boy who was sold alcohol at the defendant's licensed tavern. A two-car collision resulted in the death of one of the girls, injuries to the two other girls, and injuries to two people in a second car. The passengers from the second car sued Southland Corp. and were awarded damages. The girls in the 16-year-old's also sued, but damages were denied them because the court found that the girls, by getting drunk and driving with someone they knew, or should have known was drunk, contributed to their own injuries and assumed the risk of riding with a drunk driver. The plaintiffs appealed that decision, and the appeals court reversed the lower court's decision.

Reasoning: The appeals court found that minors served illegally do not contribute negligently to their own injuries or deaths. In such cases, the law does not prohibit recovery by survivors or the minors themselves. The court pointed to Louisiana's prohibition of liquor sales to minors as evidence of the legislature's intent to prevent risks like the one associated with the accident, and said that even in the absence of a dramshop act, recovery could be obtained on behalf of a minor, if not an adult.

Conclusion: This case illustrates the rule of the common law: that licensed operators who sell liquor to minors are responsible for injuries or death inflicted on patrons and third parties, whether or not a dramshop act exists.

The key issues dealt with are the defenses by the tavern operator of contributory negligence and assumption of risk on the part of the girls. The violation of the Louisiana statute making it a crime to sell liquor to minors imposes strict or absolute liability, making those defenses unavailable to food and beverage operators. A Louisiana high court ruling that makes these defenses valid in cases involving adult drinkers does not apply to minors.

SUMMARY

Foodservice managers are confronted with a variety of laws that govern the day-to-day conduct of business. This chapter dealt with one of the most critical in terms of economic survival—legal liability to patrons and others with respect to the service of foods and beverages.

Training is the key to prevention of liability for the service of food and alcohol. Knowledge must be shared with employees directly responsible for serving food and alcohol.

Operators can be held liable for personal injuries to whomever eats or drinks in the buyer's household. This liability is the *minimum liability* provided by the Uniform Commercial Code. They also may be liable for personal injuries to persons or to corporations for economic losses, such as lost profits. However, the patron may now sue the grower or producer. Operators should adopt and maintain strict sanitation standards and train employees thoroughly.

Alcoholic beverage sales can result in liability for injuries caused by illegal transactions. Managers are duty-bound to avoid sales to minors, intoxicated persons, and known or habitual alcoholics. There also may be liability to third persons who are injured or killed by patrons as a result of such sales.

QUESTIONS

1. Grant Mooney buys two chocolate malteds at Diane's Lunchroom for himself and his friend, Karen. The malteds are contaminated with a chemical Diane used to clean the mixer, which is not designed for that purpose. Both Grant and Karen become ill. Grant sues on the basis of the negligence theory and wins damages. If Karen sues on the basis of this theory, she would be unlikely to get far. Why? What theory should she use instead?

2. What protection does the Uniform Commercial Code give to foodservice operators?

3. What are the best ways for foodservice operators to prevent liability for the service of food?

4. In what cases can servers be held liable for patron injuries due to the service of alcohol?

5. Sheila Longyear was walking down the street near her home in California when she was hit by a car driven by Richard Pettit, who had been served alcohol earlier at Earl's restaurant and bar. Richard was 16, intoxicated, and as he let it be known, driving to a party from Earl's. What grounds does Sheila have for suing Earl? What would you suggest to Earl to prevent this problem in the future?

NOTES

1. *Webster v. Blue Ship Tea Room, Inc.*, 347 Mass. 421, 198 N.E.2d 309 (1964).

2. For example, Kentucky in an early statute, prohibited the reading of English common law cases and reliance upon such cases as legal authority in any of its courts. Act of February 12, 1808, reprinted in 3 *The Statute Law of Kentucky* 457 (Littell 1811). The New Hampshire judges declined to listen to citations of cases from "musty old worm-eaten (English) books," and stated that "not (English) Common Law—but common sense" would control their decisions. Warren, *A History of the American Bar* 227 (1911). Also see *Wagner v. Bissell*, 3 Iowa 396 (1857), a representative decision where Judge Trumbull said: "however well adapted the rule of the common law may be to a densely populated country like England, it is surely but ill adapted to a new country like ours."

3. What these statements indicate is that the common law, unless adopted as the controlling law by state statute, was American common law based on the customs, condition, and the usage of the people of each state.

4. See N. Dowling, E. Patterson, R. Powell, *Materials for Legal Method 5-7* (2d. ed. 1952) (Judicial Decisions and Persuasive Authority of Precedents from other Jurisdictions).

5. The *law merchant* was a specialized court, originating in Roman civil law, that heard commercial cases only, for the benefit of persons engaged in commercial businesses in which the prevailing customs of merchants

were recognized and enforced. The body of laws developed by the law merchant was international in scope, administered and enforced in France, Italy, Spain, and other trading nations, as well as in England. As early as 1543, the English common law courts absorbed the law merchant into the common law. As a general rule, the law merchant concepts were applied by the English common law judges in all law cases, unless the concepts were found to be unreasonable, in which case the common law judges refused to accept or apply them. In those cases the applicable common law rule was substituted. W. Walsh, *A History of Anglo-American Law* 362–68 (2d. ed. 1932).

6. *Seixas v. Woods*, 2 Caines 48, 2 Am. Dec. 215 (1804) (sale of wood).

7. *Id.*

8. Intentional wrongdoing or negligence was the traditional common law tort theory of recovery. *Seixas v. Woods, supra*, note 6, stresses the rule that absent proof of fraud or other misconduct, "the purchaser purchases of foods and beverages imposes liability upon a seller to a consumer of such products.

9. *Merrill v. Hodson*, 88 Conn. 314, 91 A. 533 (1914) (sale of food).

10. See Dowling, *et al., supra*, note 4, at 167 nn. 14 & 15 (synthesis of cases on the liability of a seller or manufacturer of goods).

11. Under a legal theory called Conflicts of Law, the place of injury traditionally determines the applicable law of liability. However, the rule favored by a growing number of courts is to apply the law of the place having the greatest *contacts* involving the injured party and the occurrence for which that party seeks to recover. See M. Levine, *Business and the Law 8–11* (1976).

12. *Chysky v. Drake Bros. Co.*, 235 N.Y. 468, 139 N.E. 576 (1923).

13. *Greenburg v. Lorenz*, 9 N.Y.2d 195, 173 N.E.2d 773, 213 N.Y.S.2d 39 (1961).

14. R. Covington, E.B. Stason, J. Wade, E. Cheatham, T. Smedley, *Legal Methods* 159–61 (1969) (*noting Chysky v. Drake Bros. Co., supra*, note 12).

15. The traditional rule is contained *Chysky v. Drake Bros. Co., supra*, note 12.

16. See Covington *et al., supra*, note 14, at 260–63, quoting Schnader, *The New Uniform Commercial Code: Modernizing Our Uniform Commercial Acts*, 36 A.B.A.J. 179 (1950).

17. *Id.* at 264–67.

18. Louisiana has adopted only Articles 1, 3, 4, and 5.

19. This rationale is noted in *Matthews v. Campbell Soup Co.*, 380 F. Supp. 1061 (S.D. Tex. 1974), and by the New York Court of Appeals in *Randy Knitwear, Inc. v. American Cyanamid Co.*, 11 N.Y.2d 5, 181 N.E.2d 399, 226 N.Y.S.2d 363 (1962) (a nonfood case), citing and following *Greenberg v. Lorenz, supra*, note 13.

20. See J. Sherry, *The Laws of Innkeepers,* 3d. ed., sec. 15:2 (rev. ed. 1981). See also *Matthews v. Campbell Soup Co., supra*, note 19.

21. See D. Noel and J. Phillips, *Products Liability in a Nutshell*, 25–28 (1974).

22. *Spencer v. Good Earth Restaurant Corp.*, 164 Conn. 194, 319 A.2d 403 (1972) (glass particles in chow mein): *Cushing v. Rodman*, 82 F.2d 864 (D.C. Cir. 1936) (pebble in roll).

23. *Musso v. Picadilly Cafeterias, Inc.*, 178 So.2d 421 (La. App. 1965).

24. See Sherry, *supra*, note 20.

25. See *Matthews v. Campbell Soup Co., supra*, note 19.

26. See *Webster v. Blue Ship Tea Room, Inc., supra*, note 1 (liability based on negligence as well as breach of warranty).

27. *Matthews v. Campbell Soup Co., supra*, note 19.

28. Restatement (Second) of torts sec. 402, representing the rule adopted by the majority of state courts. A minority of state courts, including California and Texas, reject this negligence rule and apply the strict liability rule.

29. See Noel and Phillips, *supra*, note 21.

30. *Matthews v. Campbell Soup Co., supra*, note 19.

31. *Wintroub v. Abraham Catering Service*, 186 Neb. 450, 183 N.W.2d 741 (1971).

32. Sherry, *supra*, note 20, secs. 15:4–15:5.

33. See *Vesely v. Sager*, 5 Cal. 3d 153, 486 P.2d 151, 95 Cal. Rptr. 623 (1971) in which the Supreme Court of California overruled its prior common law decisions exonerating the commercial seller or dispenser of alcohol from all responsibility arising out of the patron's intoxication.

34. See *Recent Developments, Common-Law Negligence Action Held to Lie Against Tavern Owners for Injuries Resulting from Illegal Sales of Liquors*, 60 Column. L. Rev. 554 (1960).

35. *Vesely v. Sager, supra*, note 33, abrogated the common law rule by its authority to alter its prior decisions. Other states, such as New York, have adopted dramshop acts making a licensed owner or operator liable to third persons injured as a result of the owner or operator's illegal sales to a patron. See N.Y. Gen. Oblig. Law secs. 11–101 (McKinney 1978 & Supp. 1982).

36. *Mitchell v. Shoals, Inc.*, 19 N.Y.2d 338, 227 N.E.2d 21, 280 N.Y.S.2d 113 (1967).

37. *Ewing v. Cloverleaf Bowl*, 20 Cal. 3d 389, 572 P.2d 1155, 143 Cal. Rptr. 13 (1978) (patron had just attained 21, legal drinking age). The court ruled that the victim could have assumed the risks of voluntary intoxication, but not alcoholic poisoning that resulted in his death.

38. Sherry, *supra*, note 20, at sec. 15:8, noting that Illinois, Iowa, Minnesota, New Jersey, Ohio, Oregon, and Washington impose common law liability on commercial vendors of alcohol. Nebraska and Wisconsin reject common law liability.

39. *Id.* interpreting the New York Dramshop Act. Each state is free to establish its own legislation, but these factors are representative of what the courts would require to sustain liability.

40. See *Paul v. Hogan*, 56 A.D.2d 723, 392 N.Y.S.2d 766 (4th Dep't 1977) (discussing the three theories and applying New York Law).

41. *Robinson v. Bognanno*, 213 N.W.2d 530 (Iowa 1973).

42. *Grasser v. Fleming*, 74 Mich. App. 338, 253 N.W.2d 757 (1977); *Vesely v. Sager, supra*, note 33; *Rappaport v. Nichols*, 31 N.Y. 188, 202–3 156 A.2d 1,9 (1959); *Berkeley v. Park*, 47 Misc. 2d 381, 262 N.Y.S.2d 290 (Sup. Ct. 1965).

43. See N.Y. Gen. Oblig. Law, Secs. 11–101(1) (McKinney 1978 & Supp. 1982) (any injured party); *id.* 11–101(2) (specifying surviving husband, wife, or child); *id.* 11–101(4) (specifying surviving father or mother).

44. A few states place a monetary ceiling on liability. Only California insulates licensed operators from liability, except in the cases of illegal sales to minors who injure someone or kill third persons [Cal. Bus. & Prof. Code sec. 25692(a), (b), and (c) (West 1978)].

45. *Ewing v. Cloverleaf Bowl, supra*, note 37.

Foodservice Liability for the Safety of Patrons and their Property

OBJECTIVES

After reading this chapter, you should be able to:

1. Explain potential operator liability caused by unsafe conditions that lead to patron injury, or to loss or damage to property.
2. Outline the duty of operators to maintain safe premises.
3. Discuss the defenses against liability claims for injuries to patrons or their property.

Case in Point

Laura, a restaurant patron, walked up the stairs to the restroom located on the mezzanine level. On her way down the stairway, she slipped and fell down four stairs, cutting her leg on one of the lower stairs.

She sued the restaurant owner, claiming the management owed her a legal duty to maintain the stairway in a reasonably safe condition, which was violated by the presence of a raised metal strip on the stair which caught her shoe heel, causing her to fall and be injured. However, during court testimony, she admitted she could not remember what caused her fall.

Is this claim sufficient to result in a decision in Laura's favor? No. It is not enough for Laura to state facts, which, if proven, might give rise to liability. In every lawsuit for liability, the victim must prove that a cause-and-effect relationship existed between the incident causing injury and the defendant's responsibility. The mere fact that Laura tripped and fell does not make the restaurant liable. An owner or operator of a business is not an insurer against all accidents occurring to patrons on the premises. Since Laura admitted during her testimony that she could not remember what caused her fall, there is no liability.[1]

LIABILITY FOR SAFETY

Liability is the responsibility one person has to another that is enforceable in court. The foodservice operator's liability consists of the duties and obligations owed to patrons for safe premises, a safe environment, and safe food.

Some lawsuits arise out of statutory duties imposed by government regulatory agencies. However, many legal duties are created by the common law and enforced by the courts independently of any statute. We will deal with these common law duties.

TORT LAW

In common law, a *tort* is a legal wrong caused by one person that harms another person's reputation or property. The tort arises from the responsibility one person owes to another. For example, restaurant operators are responsible for patron safety while the guests are in the restaurant. If an operator violates that duty, either intentionally or through negligence, the operator has committed a legal wrong—a tort—and the patron may hold him or her liable. A tort may not be a crime, depending on state and local laws. A tort involves a lawsuit for compensation to the victim for damages (harm inflicted). A crime involves a suit brought by the state to punish a wrongdoer by fine or imprisonment. Intentionally threatening another can be both a tort and a crime.

Tort law rests on two foundations: a legal wrong has been committed, and *compensation* is owed to the person who is injured as a result of the wrong.

THEORIES OF RESPONSIBILITY AND LIABILITY: INTENTIONAL TORTS, NEGLIGENCE, STRICT LIABILITY

There are three types of torts, and these three types of torts also constitute the theories for the nonfood area of foodservice operator liability, including liability for personal acts, those of employees, and those of other patrons for injuries and property losses or damage suffered by customers.

The first type is an *intentional tort*, in which one person means to harm another. A tort caused by *negligence* is one in which the wrongdoer, through ignorance or neglect, failed to prevent harm to another. The third type of tort, *strict liability*, is imposed by law, regardless of whether the incident was caused by negligence or intentions. Strict liability applies only to extraordinary circumstances and is rarely used in nonfood cases.

Intentional Tort

The degree of damage or harm inflicted in an intentional tort is irrelevant. If an employee intentionally strikes a patron, the fact that he did not intend to break the patron's arm does not excuse compensation. The intentional striking is the tort, not the *amount* of harm (the broken arm) intended.

There are two kinds of intentional torts: 1) wrongs against a person, and 2) wrongs against property.

Personal Torts   An *assault* is any act or threat that creates a reasonable likelihood of harmful or offensive contact. The completion of a harmful act toward someone is a *battery*. A threat to strike a patron is an assault, whereas the act of striking the patron is a battery.

False imprisonment, another personal tort is interference with someone's right to move without restraint. Threats, as well as use of physical force, are included within the definition. An operator does have a right to detain a patron who tries to leave without paying, since this wrongdoing on the patron's part justifies the restraint. But if the patron is innocent of any wrongdoing, then he or she may sue.[2]

Mental distress is defined as a personal act, outrageous and extreme, that caused emotional harm.[3] Canceling a patron's reserved wedding reception without any notice would be an example. *Defamation* involves injury to a person's reputation. It is a false statement or one made in reckless disregard of the truth. The statement must be given to a third person, either in writing or by word of mouth. A written defamation is called *libel*. An oral defamation is called *slander*, and proof of damage to the victim is usually essential.[4] The common law definition of slander includes statements that another person has committed improprieties or has engaged in serious criminal misconduct while operating a business or profession.

A tort similar to defamation, called *slander of title* or *disparagement of goods*, consists of false statements uttered about one's product, business, or property. Statements by a competitor that you use tainted food items in your restaurant is an example. Communication to a third person and proof of damage caused by the false statements must be proved.

Misrepresentation (also called fraud or deceit) involves making deliberate false statements or concealments of the truth for personal gain.[5] *Defamation by computer* is a fairly new tort and involves false information disseminated either by a computer company and/or by the company that owns the information, such as a credit collection agency. The company at fault is responsible. Erroneous computer printouts about failure to pay bills would fit this category.

Invasion of a right of privacy includes the use of a person's name or picture for commercial purposes without permission, and public disclosure of private facts about another.[6] If the person is a private citizen and not a public figure, liability is more likely for this tort.

Property Torts   *Trespass to land* involves wrongfully interfering in or invading another's property rights. No damages need to be proven, since the tort is the invasion or interference itself. A property owner need only prove ownership or legal right to occupy the land to obtain relief.

Conversion is taking, use, or retention of another's property without legal justification. For example, an operator could not borrow a car for personal use and then, without permission, use it for business. The injured party may recover the fair market value of the property.[7] However, an *intentional trespass* may be justified. Many states, for example, permit a garage owner who is not paid for work done on a car to keep the vehicle until the bill is paid, and even to sell it if no payment is made. A similar right is provided innkeepers to ensure payment of guest charges. They may keep a guest's property until the room charge is paid.

To prove a claim of *interference with another's contract relations*, there must be proof that the defendant caused the violation of a contract for personal gain. For example, suppose you were to persuade a famous chef to breach an employment contract with your competitor. This would establish the tort of interference with another's contract, and any losses suffered by your competitor could be recovered from you. If you did not interfere with the contract, but merely benefited from a contract broken by either the chef or by your competitor, you would not be liable.[8]

Tort Requirements   For a tort to exist, there must be court determination that someone suffered an injury as a result of the failure of another person to meet a required standard of care. The following criteria must be established.

1. There must be a *legal duty* of care owed by the party causing the harm to the injured party. Tort law equates a duty of care with the *standard of reasonable care owed under the circumstances of each case.* In other words, what a reasonable person would expect another to do under the particular circumstances.
2. The defendant must be found to have done or failed to do something, thereby violating the duty of care owed to the injured party. For example, an operator may commit a tort either by striking a patron personally, or by failing to break up a fight between two patrons in which one patron is injured, who then sues for failure to act.[9] Liability under the circumstances may rest on a number of factors: the nature of the act (how serious or morally wrong); how the act is performed (intentionally or accidentally); the nature of the injury (serious or slight); and the ability of the defendant to pay.[10]
3. There must be *causation*. The wrongful act or failure to act must cause the harm for which the victim seeks compensation. If there is any other reasonable explanation for the injury that does not involve the operator in any violation of a legal duty, the operator cannot be held liable. Some-

times the cause of the harm is beyond a manager's control, and normally this fact will remove liability. But if an operator contributes in any substantial way to someone's injuries, he or she may still be held liable to that person. For example, if an arsonist were suddenly to set several restaurant booths on fire, and a patron were injured, only the arsonist would be liable to the victim. However, if the operator's failure to report the fire contributed to the victim's injury, then some courts would impose liability on the operator as well. The fact that the operator did not cause the fire does not excuse him or her from liability for the operator's failure to report it, if that failure to act aggravated or worsened the injury or harm.[11]

4. Legal *damages* must exist, or a loss or injury that would justify recovery for that loss. Without damages, there can be no compensation, *and no liability*. The scope of damages, that justify recovery is broad. Physical injury, loss of physical security, and loss of freedom of movement, as well as loss or injury to property, are *tangible* forms of damage. *Intangible* forms of damage include interference with the right to privacy and damage to one's good reputation and to personal dignity.[12] The victim is required to state and prove *all* of the four essential elements of a tort to recover under the tort theory of liability.[13]

Negligence Theory of Liability

Negligence is the most common liability theory. Negligence is actually an *unintentional tort*, and is usually a result of carelessness. Negligence liability applies to liability for either property damage or personal injury. Most legal claims for injuries are based on this theory, since most people do not intend to cause others harm.

To prove negligence for injuries, a patron must meet four conditions.

1. The patron must prove that the restaurant operator had a *legal duty to protect him or her from injury*. For example, someone who is trespassing after business hours or is there for illegal purposes is likely to be unable to prove the owners owed him or her a legal duty. However, a patron who slips on food in a restaurant might meet this element.

2. The patron must prove that *actions or failures to act on the part of the operator or the employees created an unreasonable risk*. If a server spills salad oil on the floor, he creates an unreasonable risk for any patron passing by, and the owner might be liable.

3. The patron must prove that the negligence was *proximate cause or contributed to by the operator*. In the Case in Point, Laura Parker could not prove proximate cause, because the mere presence of a raised metal strip and the fact that she fell were not enough to establish the owner's liability.

4. The patron must prove *legal damages*. A patron who is injured enough to require medical attention has suffered legal damages. A patron who is embarrassed but not physically injured by a fall will more than likely be unable to prove legal damages.

If a server knocks over a pot of hot coffee, which was too close to the edge of the table to begin with, and the coffee scalds a patron, the owner may be liable for negligence. Just because the server did not intend the coffee to spill does not excuse the operation's responsibility to prevent it. An example of *negligence for property damage* would be carelessly spilling grape juice on a patron's clothing and ruining it. Carelessness can result in liability, despite people's best intentions.

The negligence theory of liability covers every possible type of event, including injuries caused by improperly maintained carpeting; defective doors, windows, furniture, and fixtures; and violations of statutory safety standards governing fire doors, flammable materials, and exterior and interior lighting.[14]

Strict Liability

Strict liability theory makes operators responsible for certain acts *regardless of whether they were at fault*. The fact that they acted reasonably under the circumstances does not matter.[15] This theory is not usually used in liability claims in nonfood areas, as only a few state courts will apply it for that use.

Strict liability normally applies to extraordinary, hazardous activities, such as the use of fireworks at a Fourth of July celebration. In nonfood cases, the theory involves three findings: 1) the activity is potentially extremely harmful; 2) the activity involves a degree of risk that cannot be avoided or minimized by reasonable care; and 3) the activity is not commonly performed in the area.

Compensation for the Victim under the Three Theories

The common law negligence doctrine requires compensation only to the victim. No punitive damages are required because by definition, negligence does not involve the performance of an *intentional* legal wrong. Punitive damages are not awarded in contract or negligence actions, but only in intentional tort actions.

The purpose of the intentional tort doctrine is to compensate the injured party *and* punish violators, making them think twice before undertaking the same activity again.[16] This is similar to the purpose of the criminal law: to punish and to deter misconduct.[17] Punishment for liability takes the form of punitive damages, similar to a criminal fine, which are awarded in addition to compensatory damages. In other words, there are two payments. First, the wrongdoer pays the victim punitive damages because of the wrongful act. Second, the victim is paid the court-fixed cost of the injury suffered. Crimes are punished by society, whereas punitive damages are paid only to the injured victim.

The negligence doctrine and the strict liability theory allow only for compensatory, not punitive, damages.

Sometimes proof of negligence or of an intentional tort is not readily available, yet the law wishes to impose liability. The law does this by applying the doctrine of *res ipsa loquitor* ("the thing speaks for itself"), and uses circumstantial evidence.[18] For example, should a mirror fall from a wall of your foodservice premises and injure a patron, the patron, unable to prove you were at fault, might still win the suit if able to prove that: 1) the mirror was within your exclusive control and supervision; 2) the injury suffered would normally not have happened but for your fault; and 3) the patron did not voluntarily cause or contribute to his or her own injury.[19] Another example would be that circumstantial evidence exists that the mirror was not sufficiently attached to the wall. This rule shifts the burden of proof to you to prove that the injury was caused by circumstances beyond your control. You are given the opportunity to dispute the presumption of fault this rule creates. Proof that the mirror fell as the result of an earthquake might be ample. If you fail to provide such proof, you will be held liable.[20]

DUTY TO PROVIDE SAFE PREMISES

In general, the law requires operators to exercise reasonable care to protect patrons from personal injuries or harm to property caused by poorly maintained premises or lack of safety precautions. Patrons must be protected against injuries due to defective, improperly constructed, or improperly supervised premises.[21]

All public accommodations (including food services and lodgings) must adhere to local statutes regarding building codes, fire regulations, and safety. Not only does an operation avoid fines and criminal prosecution by keeping your premises up to standard, but it can usually avoid liability as well. Taking shortcuts with regard to building materials and safety precautions is never profitable.

Hospitality managers are not required to take extreme or expensive measures to guard against each and every accident. They are expected by the law only to exercise *reasonable care* in the supervision of employees and the management of premises to help ensure patron safety. For example, if a manager notices that a rug in the lobby needs to be tacked down but does nothing about it, he or she may be liable if someone trips over it. Likewise, if the manager does tack it down tightly and someone still falls, he or she is less likely to be held responsible by performing the minimum duty the law requires. The only other option would have been to rip out the rug and start over, but the expense would probably not have justified such an action. In a similar example, if an owner sees that an employee dropped a bottle of catsup and swept the area before going back to work but did not mop it all up, the owner may be liable if a patron slips on it.

The test courts often use to determine a standard of care is called *foreseeable risk*. If an operator foresees a risk but does nothing about it, he or she can be held liable for a patron's injury as a result of the risk. Cleaning spills, removing obstacles, fixing hazards, and posting warning signs can help defend against liability claims. However, even these reasonable standards for adult patrons might not apply to children, since they lack the experience to appreciate and avoid dangers.[22] For example, if an operation has unmarked sliding clear glass doors, an adult patron might not be able to collect damages for injuries caused by walking into the doors. This rests on the commonsense theory that the adult did not look where he or she was going, should have known better, and thereby contributed to the accident. But a minor might recover because of a lack of experience with such doors.[23] However, if the minor were old enough to understand the risk, or had been warned previously about it, then the operator might prevail.[24]

To establish negligence, an injured patron must prove the violation of a legal duty. Legal duties for safety include exercising ordinary care to protect patrons against unreasonable risks of harm.[25] The patron must prove both that the operator knew or should have known of an unreasonable risk, such as failing to clean up a spill; and that the operator failed either to warn the patron, or to remove the risk. The *reasonable person test* is applied to both requirements. The law requires only that the operator do what a reasonable person would do under the circumstances. In practice, this burden of proof excuses the operator from liability when he or she has too little time to anticipate the risk and do something about it.[26]

The following case illustrates the rule that where an operator has knowledge of known concealed perils or ought to have known of their existence in the exercise of reasonable care, the owner must warn patrons of such conditions. Failure to warn constitutes negligence if, as a result, the patron is injured.

Allgauer v. Le Bastille, Inc.
Court of Appeals of Illinois
428 N.E.2d 1146 (1981)

Facts: Allgauer and some friends dined at the defendant's restaurant. The entrance to the premises was located at the top of a steep stairway, with a drop-off between the last stair and the entrance door. After eating, the group left and Allgauer stepped out onto what she thought was a level platform instead of a drop to the first step, causing her to lose her balance and fall, injuring herself. She sued claiming that the entrance was dangerous. The operator responded that the entrance was constructed in conformity with the building code and that Allgauer contributed to her own injuries by not looking where she was going. The trial court granted the defendant's motion for summary judgment and the plaintiff appealed. The court of appeals reversed and remanded the case for trial.

Reasoning: The court first acknowledged the general rule that the patron carries the burden of proving all elements of negligence: 1) an existing legal duty; 2) breach of that duty; 3) that the breach proximately caused her injuries; and 4) damages. Normally the duty required is that of reasonable care under the circumstances, and that duty is breached when the restaurant owner, upon proper notice, creates an unreasonable risk of harm and fails to warn the patron of that risk. If that knowledge is equally available to a reasonable patron, the patron has a duty to protect himself or herself, and cannot recover if he or she fails to do so.

However, a corollary rule exists, which requires a reversal of this case. Where the operator is or ought to have been aware of a concealed peril, to a greater degree than a patron in similar circumstances, the operator is under an *affirmative legal duty* to warn the patron. On the facts presented, the court of appeals held that the placement of the entrance door to the restaurant stairway could have created such a peril, resulting in injury to the patron, and that the jury should have been afforded the opportunity to decide these questions. The court rejected the defendant's argument that by ascending the stairs without difficulty the plaintiff was precluded from arguing that the door and top stair were dangerous. The court stated that this issue was also a jury question.

Conclusion: This case represents an exception to the rule compelling the plaintiff to prove all the elements of a slip and fall negligence claim. Normally these issues are questions for the jury to resolve. Only where no reasonable jury could render a verdict for the victim will a court take the case from the jury and dismiss the claim as a matter of law. Here the reviewing court found facts in dispute that could allow a jury to rule for or against the victim. In such a case, the proper procedure is to let a jury decide these questions.

What is noteworthy in this case is that on the issue of whether the patron had sufficient knowledge of the risk to have avoided injury, the owner's responsibility to warn is greater where the risk is concealed and the owner should have been aware of the problem. Here the court held that the owner has an affirmative duty to warn.

The concealed peril rule, if applicable, increases the likelihood of patron recovery because the rule weakens the owner's argument that a patron equally aware of danger as the owner cannot recover or can recover only a lesser sum. This argument rests on the proposition that the patron should be careful and look where he or she is going. However, this issue is also for a jury to decide.

Reschamps v. Hertz Corp. [and the Ramada Inn]
Court of Appeals of Florida
429 So.2d 75 (1983)

Facts: Reschamps was injured in an automobile accident with another vehicle driven by a hotel guest leaving the hotel driveway. Reschamps sued the hotel on the negligence theory of liability, arguing that the hotel owners failed to maintain their property in a safe condition. The negligence consisted of allowing unnatural shrubbery to grow so as to obstruct the view of the guest as he attempted to exit the driveway. The hotel moved to dismiss the claim for failure to state any liability. The lower court granted the motion. The court of appeals reversed and sent the case back to trial, stating that the facts pleaded were enough to allow the plaintiff a jury trial.

Reasoning: The law requires a hotel keeper, like any other occupier of land, to maintain premises in a reasonably safe condition for use by guests. This duty applies not only to the interior of the building but to the grounds, which the guest is invited to use. A hotel driveway constitutes such grounds. The hotel keeper's duty extends to responsibility to a guest for the guests losses as well as those of a nonguest, since it is reasonably foreseeable that an accident might result in injuries to both parties. In this case, the hotel guest sued by the plaintiff *inpleaded* (brought into the case) the hotel owner. These facts are property for the jury to decide.

Conclusion: This case warns that a hotel or restaurant owner must take the same reasonable care of the exterior as the interior of the premises. The entire premises, meaning all areas under the control and supervision of the owner, fall within that scope of responsibility. This rule does not automatically result in a finding of liability, but merely permits a jury to resolve factual issues needed to render a verdict either for or against one of the parties. If parking facilities are provided for guests or patrons, access to those facilities must be adequately maintained and supervised. Guests and patrons should be told of any unusual problems in or about a driveway so that they can take precautions for their own safety. Although the facts of this case are unusual, the reviewing court was satisfied that a jury should be permitted to decide the necessary factual questions.

DUTY TO POLICE CONDUCT OF PATRONS, EMPLOYEES, AND THIRD PARTIES

Every foodservice operator, as an owner or occupier of business premises, is under a duty to exercise reasonable care for the safety of patrons. This general duty includes the duty to protect patrons from intentional injuries or other harm inflicted by other patrons, employees, or third parties.[27]

Third-Party Actions

The foodservice operator is not totally responsible for the safety of patrons with regard to acts of third parties or other patrons. In a liability case involving a third party, the common law requires the patron to prove that the operator knew or should have known that the misconduct of a third party would cause the patron injury. If a manager knows or should have known that a sequence of events can cause others harm, then the law requires the manager to make every reasonable effort, as soon as possible, either to prevent the misconduct, stop it, or remove the offending person. At the least, calling the police or summoning other help is required, especially in situations involving weapons.[28]

Service of Alcohol   Operations that serve liquor must furnish a certain number of guards, bouncers, or security personnel to control the actions of others on the premises. No minimum number of security personnel has been established by the courts; each case is examined on its own merits. One factor may be the existence of prior incidents calling for better security measures, which, if ignored, leave operators open to liability. The location of security personnel is also important. They should mingle with patrons to identify and defuse potential problems.

The risk of assaults, fights, and general misconduct are known to increase where alcoholic beverages are sold. This is due to the effects of intoxication on normally reasonable persons. The operator's duty to protect patrons is viewed by most courts as coinciding with such known risks.[29] The greater the risks created, or allowed to continue, the greater is the likelihood of operator liability.

Employee Misconduct

The intentional misconduct or negligent actions of employees who injure patrons is governed by the same rules that apply to third-party misconduct. An employee is always liable for his or her own torts. To make the employer liable, there must be proof that the employee was acting within the scope of the operator's authority.[30] This legal rule is called *respondeat superior*. This doctrine is critical to those bringing legal action, since without proof the employee was acting with the employer's authority and knowledge, the latter may escape liability.[31]

However, employers have an independent legal duty to hire and retain competent and responsible employees. If the manager knows or should have known that an employee might injure or harm a patron, and the employee does, the patron can sue on that basis.[32] In such cases, it makes no difference whether or not the employee was acting within the scope of the manager's authority.

Because foodservice owners and operators are assumed to be better able to pay for injuries inflicted by their employees than the employees are, some courts tend to stretch the rule to its outer limits.[33] This "deep pockets" tendency means that there is a greater potential for liability, as the following case illustrates.

Riviello v. Waldron
Court of Appeals of New York
391 N.E.2d 1278 (1979)

Facts: Riviello, a regular patron of the Pot Belly Pub, a tavern owned and operated by Raybele, lost an eye because of the alleged negligence of Waldron, a Raybele employee. The accident occurred when Waldron was flipping his Boy Scout knife during a lull in his work as a short-order cook. Waldron was making idle conversation with Riviello and another patron, as he had done many times in the past. Waldron was called away to prepare an order and when he returned, Riviello unexpectedly came in contact with the blade of the knife which Waldron still had in his hand.

The lower courts found in favor of Riviello against Waldron. The issue for decision by the Court of Appeals was whether Raybele was also liable under the theory of *respondeat superior*, or whether Waldron's negligence was outside the scope of his employment, which would excuse Raybele from any responsibility for Riviello's negligence. The Court of Appeals held that the acts of Waldron were within the scope of employment, and reversed the contrary ruling of the intermediate appeals court.

Reasoning: The legal issue presented was whether Waldron was acting within the scope of Raybele's employment so as to bring Raybele within the doctrine of *respondeat superior*. In noting that the original narrow construction of the scope of employment requirement had expanded, the court concluded that the test has come to be whether the act was done while the employee was doing work for the employer, no matter how irregularly, or with disregard of instructions. This determination is normally for the jury to make. The court pointed to the following factors to be weighed: " . . . the connection between the time, place, and occasion for the act; the history of the relationship between employer and employee as spelled out in actual practice; whether the act is one commonly done by such an employee; the extent of departure from normal methods of performance; and whether the specific act was one that the employer could reasonably have anticipated."

The court concluded that, in finding for Riviello on all of the above factors, in particular that Waldron's activities did not so depart or deviate from his normal activities, Raybele was responsible for Waldron's acts.

Conclusion: This is an example of the current judicial tendency to interpret literally the factors used to determine employer *respondeat superior* liability. Much of this change in attitude is due to the escalation of employee-produced injury; concern that the average innocent victim cannot recover his claim from the wrongdoing employee because of lack of resources; and the use of cost accounting and insurance coverage by employers to spread the impact of such costs.

Schwingler v. Doebel
Supreme Court of Minnesota
309 N.W.2d 760 (1981)

Facts: Schwingler, a patron at the Earl and Dorothy Tavern, sued Doebel (a patron), the owning corporation, and the sole shareholder and bartender, Earl McQuiston, for Doebel's assault on Schwingler. Schwingler said he saw Doebel serve beers to patrons and assumed he was employed there, although he was not. After throwing a glass of beer in Schwingler's face, Doebel followed him to his car and assaulted him. McQuiston was tending bar at the time.

Schwingler claimed that all defendants were responsible, either because Doebel was acting within the scope of his employment or, independently, because as owners and operators of the tavern they failed to maintain a reasonably safe place for their patrons. The trial court awarded a *default judgment* (a judgment awarded where the defendant fails to appear and defend the action) for Schwingler against Doebel, and at the next trial, found for Schwingler against the remaining defendant, McQuiston. He appealed.

The judgment of the trial court in favor of the plaintiff (Schwingler) against the bartender and the owner/operator of the tavern was reversed and remanded to the trial court for entry of judgment in favor of the defendant.

Reasoning: The appeals court said that the fact that Schwingler thought Doebel was an employee, did not justify imposing liability on McQuiston and the owner. The court pointed out that Doebel had acted out of a "deep personal animosity" toward the plaintiff, and not within the scope of his "volunteer employment." They further pointed out that Doebel was outside the tavern when he assaulted Schwingler. The court said that while Doebel had an assault record, he had, according to testimony, never been unruly at the tavern either previously or on the night of the incident, and the bartender had no way of foreseeing the incident. Finally, there was no evidence that the bartender could have stopped the incident, as there was no verbal threat inside the tavern.

Conclusion: The case is important because it illustrates a situation in which the appeals court refused to apply the doctrine of *respondeat superior*, sparing the owner substantial liability. The case upholds the rule that both the employee, as the wrongdoer, and employer, under *respondeat superior*, may be responsible for injuries inflicted by the employee on a patron, if, in fact, the employee was liable and *was acting within the scope of the employment.* It also establishes that neither the employee nor the employer was liable because the person making the assault was not a regular, full-time employee; the assault took place off the premises; and the bartender was motivated by personal animosity toward the patron.

The case also applies the rule of *foreseeable* unreasonable risks and concludes that the sudden, violent assault of the patron off the premises was not foreseeable by the owner/operator, requiring the conclusion that the premises were not unsafe or disorderly.

Removing Unruly Patrons   There are times, especially in operations with bars, when unruly patrons will have to be removed. When removing a bothersome patron, a manager or employee should use only such reasonable force as required. The use of unnecessary force will expose both employee and employer to liability. If a patron will not leave through vocal persuasion or by gently guiding him or her out, call the police and let them handle it. Unless the patron is harming or threatening to harm others, the operator could be liable for unnecessary force used in removing the person. Erring should be done on the side of caution.

Liability for Inaction

Operators are considered just as responsible for *failures to act* as for actions. If a manager does nothing after a patron is injured, or watches a patron being hurt by another person, he or she may be held liable for the inaction. At minimum, if unable to provide help directly, the operator should summon help quickly and make the injured patron as comfortable as possible.

All employees should know how to obtain help. Emergency numbers should be posted near the telephone and near service work stations. Managers must also be alert to other aspects of the operation which themselves could be hazardous during the emergency (theft of bar inventory due to outbreak of fire).

Taking appropriate action quickly can make a difference between no liability, low liability, and losing everything.

What the Law Expects

Business persons who serve the public directly are expected by law to anticipate and react to trouble much more rapidly than other businesspersons.[34] Hospitality managers deal with many different people in the normal operation of business. This fact requires them to stop any disturbance by removing the people responsible or summoning help quickly.[35] When alcohol is served, managers and employees must be constantly on the lookout for trouble.

Patrons are entitled to assume that public premises are and will remain reasonably safe and that supervision of the premises will foster congenial surroundings and good food and drink.[36] The law does not expect operators to anticipate every risk. It does require them to remove or warn of such risks once they are known.[37] No set formula exists for doing so. The best rule of thumb is: At the first sign of trouble, call for help.

DEFENSES AGAINST LIABILITY CLAIMS FOR INJURIES

The only defense against strict liability claims is to prove that the injury did not take place on the premises, but somewhere else.

There are three general defenses to liability claims based on torts or negligence. These defenses may be used simultaneously or alone. The critical thing to remember is that the burden of proving the defense is on the defendant.

The first defense is that the original negligence was not the cause of the harm because it was followed by an *unforeseeable cause* that was the direct cause of the harm. For example, if flammable materials are stored in an operation, it is foreseeable that a patron might be injured if the materials are ignited by a careless employee, and the owner or manager might be liable. However, if the materials are ignited by a natural disaster, such as lightning, the operator might escape liability. When an operator either creates a foreseeable and unreasonable risk of harm *or* contributes to an unforeseeable risk by careless conduct, such as failing to call the fire department or installing flammable furnishings in violation of the building code, then he or she may be held liable. The test of the liability is whether careless conduct made the harm more serious.[38] Since each case is judged on its own facts, no general rule is available that will cover every situation.

A second defense to liability claims for intentional torts or negligence is that of *contributory* or *comparative negligence*. Under common law, everyone was expected to use reasonable care in looking out for one's own well-being. If the victim contributed at all to his or her injuries or losses, the common law made the defendant's proof of that contribution a complete defense to the claim. This was true regardless of the degree of negligence.[39] To balance the law between the injured party and the defendant, most states have passed statutes that create a *comparative negligence rule*. Under this rule, a jury must assign a percentage to the victim's carelessness in comparison with the defendant's carelessness and then reduce the victim's total recovery by the dollar amount of that percentage.[40] For example, if a patron won a judgment of $1,000 against a restaurant owner, but the jury found that the patron was 10 percent negligent, the patron would recover only $900 ($1,000 less 10 percent) from the owner.

The third defense is called *assumption of risk*. When a patron knows that a risk of harm exists and voluntarily takes that risk, the patron assumes that risk and may not recover. For example, if an adult patron walks up a stairway that is clearly blocked with a rope and a sign, the owner may not be held liable because the patron ignored the sign and assumed the risk. If a child walks up the same stairs, however, the owner may be held liable, because the child could not be expected to recognize the risk. The owner must prove that the patron knew of a risk and took voluntary action based on that knowledge.[41] There are exceptions, however. If a risk is forced on a patron, such as unlawful blocking of a fire exit, and a fire forces the use of that exit, the operator is responsible. This is true even though under normal circumstances the exit was adequate.[42]

In emergencies, assumption of risk is not considered a defense.[43] Nor does it apply to situations involving a violation of statutes intending to protect patrons from risks, such as one requiring an adequate number of fire exits.[44] The knowledge of possible risks is evaluated by courts on the basis of the particular patron's age, maturity, and general intelligence.[45] If a restaurant caters to families with young children, the sign and the rope to warn people away from the stairs would be inadequate. Stronger steps would be needed to protect the operator against liability and small patrons against injury.

All of the defenses described above must be proved by the defendant in court. They are not automatically granted. If they fail, the plaintiff will win the lawsuit.[46]

DUTY TO SAFEGUARD PROPERTY OF PATRONS

Unlike innkeepers, foodservice operators are required to exercise reasonable care only in protecting patron property handed over to them for safekeeping.[47] (Innkeepers must ensure guest property against virtually all risks, subject only to ceilings imposed to protect them against excessive liability.[48] Here innkeepers are required to comply with the innkeeper's liability statutes to be protected in such cases. See Chapter 6.) Only the hotel guest is entitled to strict liability for the loss of or damage to property.[49]

To protect patron property, foodservice operators must do the following.

1. Supervise the premises to prevent thefts or other losses of property.[50]
2. Supervise the service of food and beverages by employees and try to halt careless actions that might result in damage to patron property.[51]
3. Guard carefully property deposited for safekeeping. There is an additional responsibility to return the property in the same condition as when deposited.[52]

The voluntary transfer of personal property to another for safekeeping is called *bailment*. Checking a coat in a checkroom is a typical bailment. An operation is under no legal duty to accept any property for safekeeping. Once it does so, however, the operation is liable for loss or damage unless it proves that the loss of or damage to the property was caused by circumstances beyond anyone's control, such as a flood or fire.[53]

Unlike innkeepers, foodservice operators do not have to accept patron property just because they admit patrons. Once they have admitted the patron, however, *and* accepted the property the restaurant is liable for the full established price of the item at the time it was lost.[54]

The extent of an operator's responsibility depends in part on the nature of the operation. A first-class French restaurant is expected to exercise more care in safeguarding patron property than a quick-service operation. Patrons of the former would expect to have a coat checkroom, whereas patrons of the fast-food operation would expect to keep their coats with them.

When patrons keep property under their own supervision, the operator is not liable for loss of or damage to the property unless the patron can prove that failure to exercise *reasonable care* in safeguarding it.[55] In every case where the disappearance of or damage to the property cannot be explained, the party having the burden of proof loses that case.[56] If a patron keeps the coat while eating, and the coat is lost, the patron loses. When the coat is left with an attendant, in a bailment case, the operator loses.

The key element in the creation of a bailment is *voluntary acceptance* of the property for the purpose of exercising *exclusive* supervision and control. Simply stated, a bailment cannot be imposed without the operator's consent, except when an employee finds patron property left behind and turns it over to the management for return to the patron.[57] This is an example of bailment.

Operators also have responsibility for the *contents* of a patron's bag, briefcase, or similar container *if* they are told the nature of the contents and/or their value, and they voluntarily accept the contents, as well as the bag, for safekeeping. When the operator is not told of the contents or their value, he or she is liable only for the value of those items which may reasonably be expected to be present—a scarf and gloves in an overcoat, but not a $1 million negotiable stock certificate. This commonsense rule varies in its application with the size, kind, and class of establishment. The larger the size, the higher the class or nature of the business, the greater is the potential liability for property losses. This means wealthy clientele may be expected to carry larger amounts of valuables than patrons of a more modest establishment.[58]

Is a bailment created when an operator permits patrons to hang their own coats and leave other articles on open racks? No, since the requirement of exclusive control and supervision is not present.[59] However, for protection, there should always be a prominent sign or notice stating "Not responsible for property unless checked with the Management." A printed notice on each menu is also desirable. This does not ensure against liability, but will go far toward strengthening a defense case should the patron sue.[60]

If an employee takes a patron's property and hangs it or deposits it out of sight of the patron, a bailment may be created,[61] since the patron is unable to exercise control and supervision over the property.

If an operator voluntarily accepts patron property for deposit in a checkroom or cloakroom, can the operator issue a claim check or stub conditioning acceptance of the article on the patron releasing the operator from a liability claim? No. Public policy in some states forbids those who make bailments in the normal course of their business to waive liability.[62]

Many first-class operations park their patrons' cars for them. This is considered a bailment, and the restaurant owner may be liable not only for the loss of the car itself, but for any property in it.

In the case to follow, a bailment by operation of law was found to exist between a hotel guest and the Shamrock Hilton, whereby the hotel was found liable for the loss of the guest's purse and contents.

Shamrock Hilton Hotel v. Caranas
Court of Appeals of Texas
488 S.W.2d 151 (1972)

Facts: Caranas and her husband, paying guests at the defendant's hotel, ate dinner at the Shamrock Hilton hotel dining room. After their meal, Caranas left her purse behind. The purse was found by a hotel busboy who, in compliance with hotel policy, delivered it to the restaurant cashier. The cashier delivered it to an imposter posing as Caranas. The purse was never recovered. Caranas sued the hotel claiming negligent delivery of the purse to an unknown person and sought to recover damages for the value of purse and contents. The court of civil appeals affirmed a judgment for the plaintiffs on liability and the extent of damages they could lawfully recover.

Reasoning: The court first dealt with the question of whether the Texas limitation of liability statute applied to limit the hotel's responsibility to $50. It did not, because the statute does not apply to a case of innkeeper negligence, and negligence was properly established in this case.

Second, the court ruled that a constructive bailment was created by the hotel employee's delivery of the purse to the restaurant cashier. The fact that Caranas did not deliver the purse, which would be necessary in a true bailment by agreement, was overcome by the fact she would have wanted the hotel to safeguard her misplaced purse until she could claim it later.

Under a bailment theory of responsibility, the hotel's failure to return the article upon demand created a presumption of negligence by the hotel. The hotel offered no evidence to show that the purse had been lost due to circumstances beyond its control. This meant that the hotel's liability was established as a matter of law. The bailment was found to be one for mutual benefit, regardless of the fact that no charge was made for returning guest property left in its dining rooms. This meant that the duty of care governing the case was ordinary care under the circumstances.

Third, the court ruled that it was reasonably foreseeable that Caranas would have brought expensive jewelry with her in her purse, because of the status of the Shamrock Hilton and the class of people that would frequent its dining room. Therefore the value of the jewelry in the purse as well as the value of the purse itself was properly part of the jury verdict on the issue of damages in this case.

Conclusion: This case cautions all hospitality managers that once they adopt a policy of requiring employees to return misplaced guest or patron property for safekeeping, they are presumed responsible if the property cannot be returned on demand. It is the manager's responsibility to explain that failure to do so was beyond his or her control. Moreover, managers may be responsible for contents as well.[63] This is usually a jury question. The Texas Court of Civil Appeals was merely affirming the jury verdict in this case. In other cases, the jury would be free to find the contrary.

One solution to the problem of accepting valuables of patrons for safekeeping is to refuse to do so when their value is obvious (a mink coat offered to be checked should be politely but firmly refused). The more uncertain problem is where the patron hands over a brief case or bag without disclosing its contents. Only New York State limits the liability of restaurant keepers in such cases. All other states follow the Texas rule.

Defenses against Liability Claims for Property

The most significant defense to liability claims for property is to establish whether or not a bailment was created. If a bailment was created, the operator must prove that he or she exercised reasonable care to protect the property.

Where no bailment is created, the patron has the burden of proving negligence. However, the operator must still prove that reasonable care was used and that he or she did nothing to place the property in a risky situation. If an employee sees someone taking a coat known to be another patron's and does nothing, the operator may be liable for the loss.

COMMON LAW EXCEPTIONS TO LIABILITY

Foodservice operators (and operators of most other types of business) are not liable for events over which they have absolutely no control. Such acts include natural catastrophes such as a flood, tornado, or hurricane. However, advance notice of such catastrophes requires the operator to act to protect patron property using reasonable care. Also included are *acts of a public enemy*, either through war, civil disorder, or martial law, over which the foodservice operator has no control.[64] Another general category would include airplane crashes or spilled dangerous gases or liquids—other humanly caused acts—for which others, not the foodservice operator, would be liable.

INSURANCE GOVERNING PUBLIC LIABILITY

Liability insurance protects businesses against the legal claims of patrons. There are four types of insurance important to foodservice operators.

1. *Fire insurance* is intended to protect against the risks of fire and other related perils, such as windstorms, lightning, or earthquakes, caused by natural elements, arson or other human activity.

2. *Casualty insurance* is intended to protect your personal property from acts other than fire and the elements. These acts are usually caused by humans, and include strikes and robberies.
3. *Dramshop liability* insurance protects operators against claims made by third parties in alcohol related accidents.
4. *Liability insurance* shifts the costs of defending a lawsuit and the value of any judgment against the operator to the insurance carrier (insurer).

Evidence of insurance is not admissible in court to establish financial responsibility. Managers are held to a standard of overall responsibility, not proof of financial ability to pay.

To qualify for liability coverage, an operator must have an *insurable interest* in the premises to be insured. Insurable interest is defined as whether the insured will suffer any loss in the event the risk insured against happens. This means that the property must be owned or leased (not rented) to qualify for coverage. This is a protective device to both the insured and the carrier, and helps prevent unethical or illegal "accidents." For example, a competitor cannot take out a policy on a restaurant in order to burn it down and collect on the claim.

Insurance: Know What You're Getting

It is important to allocate resources intelligently to avoid excessive liability in one area.

Insurance is a contract, which means that both sides—the carrier and the operator—must agree on all terms and conditions. The fact that most insurance policies are enforced as written, places an especially heavy responsibility on the insured to understand all the terms and conditions.[65]

In all types of insurance, the carrier assumes the risk of the covered situations in exchange for payment of a *premium*. The size of the premium is usually based on the nature of the operation, volume of business, frequency of legal claims made, and the size of such claims.

The *policy* is the contract between the insured and the carrier containing the terms and conditions by which the agreement is governed. The *face amount* of the policy is the maximum payable on a claim, regardless of economic loss. In the case of a claim, the insurer is obligated to pay only the amount of the economic loss, *not the total face amount of the policy*. The insurer is not obligated to pay any claim in excess of the face amount of the policy.

Because understanding of insurance terms by most insured parties is limited, the law usually applies the *reasonable person* standard to insurance language.[66] This means that any vague language in the policy may be interpreted most strongly against the carrier.[67]

In case of a dispute between the insured and the carrier over the coverage of a policy, the insured party must prove that: 1) there was a liability claim; 2) the claim occurred while the policy was in force; and 3) the claim was of a kind covered by the policy.[68]

Exclusions   The carrier always has the right to *exclude*, or refuse to cover, certain types of liability claims. The most obvious exclusion is the intentional injury to a person or property. Criminal misconduct against a patron or patron's property also falls into this area.[69]

However, negligent, unintentional acts by the insured or by employees, which might result in a liability claim, should be covered. In practice, when the carrier denies a claim, the insured has the burden of proving that it was covered in the language of the policy.[70]

Exceptions   An *exception* in the policy means that, although the claim is otherwise covered, an exception exists that excuses the carrier from paying under certain circumstances. For example, a carrier may not cover an alcohol-related incident if a liquor license is not valid at the time the incident took place. Without a valid license, the carrier *excepts* itself from coverage on that claim. However, the carrier must prove the existence of the exception and the insured's failure to comply with the policy terms. Usually, the courts view exceptions strictly against the carrier. This means that if the exception is unclear, it will not be enforced.[71]

Carrier Defenses to Claims

False, material representations made by the party covered are grounds for the carrier electing to cancel the policy without further liability. The seriousness of the misrepresentation is tested with the question: Would the policy have been issued *but for* the falsehoods? If the statements are false but not serious, then the carrier may not avoid some payment. All paid-up premiums must be returned to the insured when the carrier cancels on this basis.[72]

Concealment is another carrier defense in liability insurance disputes. Here an insured's failure to disclose relevant facts must be fraudulent as well as serious enough to justify denial of a claim. The test in such a case is as follows: 1) Is there reason to think that the fact concealed was material? 2) Would the carrier have issued the policy had the truth been known?[73]

Another kind of defense that might excuse performance of the carrier would be a policy requirement that a health permit be in effect when a patron sues to recover on the grounds of food unfitness.

Notice of a claim is often an express condition written into a policy. This means that the insured must give notice of a claim to the carrier within a reasonable time. The insured must notify the carrier of any occurrence that *might* give rise to a legal action and forward to them any legal summons and other papers served as soon as possible.[74]

How Many Policies?

Unless an insurance policy prohibits it, an operator may obtain more than one policy from different carriers covering the same liability claims. This does not permit the operator to recover the full or *face amount* of the claim on every policy. A policyholder can recover only once for the full amount of that claim, limited to the maximum amount stated in your policy.

An operator may carry a *basic public liability* policy and an *excess liability* policy for protection in the event the basic policy money limits were insufficient to cover the liability claim in full.

LIABILITY PREVENTION

All liability in a foodservice operation is traceable to risks, most of which can be prevented. Taking steps to *minimize* risk will help *prevent* costly liability, and *maximize*, or at least *maintain*, profitability.

A manager must be aware of *where* potential risks are. At the very least, regularly inspect premises for dangerous conditions and repair them. If local building or fire inspectors point out violations, take steps to correct them immediately. These inspections may prevent liability, as well as a fire.

In every type of business, it is important to keep costs down and profits up, and the two are related. Business costs should be kept to a minimum, but this does not mean failing to keep food products and premises safe.

SUMMARY

Foodservice operators owe patrons a safe environment, free from threats of harm to themselves or their property. Failure to provide a safe operation may result in a *tort*, or legal wrong, which, in turn, may result in costly liability.

There are three types of torts: *intentional* torts, torts caused by *negligence*, and torts created by the courts, which may impose *strict liability*.

The damages for torts due to negligence or strict liability are *compensatory*. Damages for intentional torts can be both compensatory and *punitive*.

The duty the law imposes on foodservice operators to provide safe premises is one of *reasonable care*. The law expects operators to do only what is reasonable under the circumstances and what customers can reasonably expect. It is also important to obey local laws regarding building and fire safety, since an injured patron may use violation of such a law as grounds for a lawsuit.

Foodservice operators should exercise reasonable care in the selection and supervision of employees and in supervision of patrons' activities while on the premises. The doctrine of *respondeat superior* makes operators liable for the misdeeds of employees, if the actions take place in the course of employment.

There are several defenses to negligence claims: *unforeseeable cause*, *contributory (comparable) negligence*, and *assumption of risk*.

Foodservice operators have more leeway when it comes to protecting patron property. Generally, they must supervise employees to prevent careless actions that might damage patron property, and supervise the premises to prevent theft. Greater care must be taken to safeguard property accepted as a *bailment*.

Foodservice operators must shop carefully for insurance and select only what is necessary to prevent substantial losses arising from liability claims.

QUESTIONS

1. What is an intentional tort? How does this type of tort differ from injury due to negligence? Provide an example of an intentional tort.

2. A patron strikes another patron in a restaurant, injuring him. The manager does not summon help or try to stop the unruly patron. Can the manager be held liable? Why or why not?

3. How does the law hold a manager responsible for employee actions that injure patrons? Give an example where a manager may be held liable for an employee's action against a patron. Name the principle that applies here.

4. What must a patron establish to prove negligence? Give a foodservice example in which all the elements of negligence are present.

5. A patron puts a fur coat on her chair and leaves it there to go to the rest room. It is gone when she returns. Another patron leaves her fur coat with the attendant in the coat checkroom. It disappears. In which instance would the operator probably be held liable? Why?

NOTES

1. *Campbell v. Bozeman Community Hotel*, 160 Mont. 327, 502 P.2d 1141 (1972) (stairway to hotel dining room).

2. *Keys v. Sambo's Restaurant, Inc.* 398 So.2d 1083 (La. 1981).

3. *Cacares v. Anthony's Villa*, File No. 65063 (County Court, York District of Toronto, Ontario, Feb. 23, 1978). See also *Counce v. M.B.M. Co.*, 266 Ark. 1064, 597 S.W.2d 92 (1980) (wrongful firing of restaurant employee).

4. *Mid-America Food Service, Inc. v. ARA Services, Inc.*, 578 F.2d 691 (8th Cir. 1978).

5. *Whipp v. Iverson*, 43 Wis.2d 166, 168 N.W.2d 201 (1969).

6. *Nader v. General Motors Corp.* 25 N.Y.2d 560, 255 N.E.2d 765, 307 N.Y.S.2d 647 (1970).

7. *Staub v. Staub*, 37 Md. App. 141, 376 A.2d 1129 (1977).

8. *Wear-Ever Aluminum, Inc. v. Townecraft Industries, Inc.* 75 N.Y. Sup. Ct. 135, 182 A.2d 387 (1962). See also *Imperial Ice Co. v. Rossier*, 18 Cal.2d 31, 112 P.2d 631 (1941).

9. *Eastep v. Jack-in-the-Box, Inc.* 546 S.W.2d 116 (Tex. Civ. App. 1977) (fast-food establishment); *Kimple v. Foster*, 205 Kan. 415, 469 P.2d 281 (1970) (tavern).

10. See J. Sherry, *The Laws of Innkeepers*, 3d. ed., sec. 9:4 (rev. ed. 1981).

11. *Northern Lights Motel, Inc. v. Sweaney*, 561 P.2d 1176, *reh'g* 563 P.2d 256 (Alaska 1977).

12. *Sunday v. Stratton Corporation*, 136 Bt. 293, 390 A.2d 398 (1978). (Vermont Supreme Court disallowed assumption of risk doctrine in downhill ski-slope injury to skier.)

13. Sherry, *supra*, note 10, sec. 9:5.

14. *Id.* secs 9:8–9:17.

15. H. Lusk, C. Hewitt, J. Donnell, A.J. Barnes, *Business Law and the Regulatory Environment* 76–77 (1982).

16. W. Prosser, *Law of Torts* 15–16 (4th. ed. 1972).

17. W. LaFave and A. Scott, *Criminal Law* 21–25 (1972).

18. Sherry, *supra*, note 10, sec. 9:7.

19. *Deming Hotel Co. v. Prox*, 142 Ind. App. 603, 236 N.E.2d 613 (1968).

20. Sherry, *supra*, note 10, sec. 9:7.

21. All owners and occupiers of land who invite the public to use their facilities are subject to this rule of law, since injury due to carelessness is most likely to occur in such facilities. Foodservice establishments are for the most part open to the public, but private clubs are also governed by this rule. The Restatement (Second) of Torts, sec. 343 (1977), which states this rule, has been repeatedly applied to hotels and restaurants. See *Winkler v. Seven Springs Farm, Inc.*, 240 Pa. Super. 641, 359 A.2d 440 (1976), *aff'd per curiam*, 477 Pa. 445, 384 A.2d 241 (1978) (resort); *Dillman v. Nobles*, 251 So.2d 210 (La. App. 1977) (bar lounge); *Withrow v. Woozencraft*, 90 N.M. 48, 559 P.2d 425 (1976) (motel); *Bauer v. Saginaw County Agricultural Society*, 349 Mich. 616, 622–23, 628, 84 N.W.2d 827, 833–34, 839 (1957) (private fairground); *Deming Hotel Co. v. Prox, supra*, note 19 (hotel restaurant); *Afienko v. Harvard Club of Boston*, 365 Mass. 320, 312 N.E.2d 196 (1974) (defective window washing hooks and bolts; club liable to employee of independent contractor).

22. *Gault v. Tablada*, 400 F. Supp. 136 (S.D. Miss. 1975), *aff'd*, 526 F.2d 1405 (5th Cir. 1976); *Haft v. Lone Palm Hotel*, 3 Cal.3d 756, 478 P.2d 465, 91 Cal. Rptr. 745 (1970).

23. *Peterson v. Haule*, 304 Minn. 160, 230 N.W.2d 51 (1975) (fast-food establishment).

24. *Williams v. Travelers Ins. Co.*, 235 So.2d 600 (la. 1970) *reh'g denied*, 256 La. 818, 239 So.2d 345 (1970) (private swimming club).

25. Sherry, *supra*, note 10, sec. 9:5.

26. *Winkler v. Seven Springs Farm, Inc., supra*, note 21; *Campbell v. Bozeman Community Hotel*, 160 Mont. 327, 502 P.2d 1141 (1972) (hotel dining room).

27. *Eastep v. Jack-in-the-Box, Inc. supra*, note 9. Authorities are noted in Sherry, *supra*, note 10, sec. 11:13.

28. Sherry, *supra*, note 10, sec. 11:13.

29. *Alonge v. Rodriquez*, 89 Wis.2d 544, 279 N.W.2d 207 (1979) (bar lounge).

30. *Riviello v. Waldron*, 47 N.Y.2d 297, 391 N.E.2d 1278, 418 N.Y.S.2d 300 (1980) (tavernkeeper).

31. Sherry, *supra*, note 10, sec. 11:8.

32. *Bradley v. Stevens*, 329 Mich. 556, 46 N.W.2d 382 (1951); *Tobin v. Slutsky*, 506 F.2d 1097 (2d Cir. 1974) (liability predicated upon violation of innkeeper's contractual duty to guest).

33. *Riviello v. Waldron, supra*, note 30.

34. See *Eastep v. Jack-in-the-Box, Inc., supra*, note 9; *Alonge v. Rodriquez supra*, note 29. See also Sherry, *supra*, note 10, sec. 11:13.

35. *Heathcoate v. Bisig*, 474 S.W.2d 102 (Ky. 1971) (barroom).

36. *Alonge v. Rodriquez, supra*, note 29.

37. *Eastep v. Jack-in-the-Box, Inc., supra*, note 9.

38. *Ford v. Jeffries*, 474 Pa. 588, 379 A.2d 111 (1977) (residential file).

39. *Karna v. Byron Reed Synidate #4*, 374 F. Supp. 687 (D. Neb 1974) (dangerous condition known to hotel guest).

40. *Peterson v. Haule, supra*, note 23.

41. *Hunn v. Windsor Hotel Co.*, 119 W.Va. 215, 193 S.E. 57 (1937); *Scott v. John H. Hampshire, Inc.*, 246 Md. 171, 227 A.2d 751 (1967).

42. *Montes v. Betcher*, 480 F.2d 1128 (8th Cir. 1973) (hotel lakefront dock).

43. *Darby v. Checker Co.*, 6 Ill. App. 3d 188, 285 N.E.2d 217 (1972) (hotel fire).

44. *Northern Lights Motel, Inc. v. Sweaney, supra*, note 11.

45. *Gault v. Tablada, supra*, note 22.

46. See *Bazydlo v. Placid Marcy Co.*, 422 F.2d 842 (2d. Cir. 1970) (ski toboggan).

47. *Shamrock Hilton Hotel v. Caranas*, 488 S.W.2d 151 (Tex. Ct. Div. App. 1972) (loss of purse from hotel restaurant caused by negligent misdelivery to unauthorized third person).

48. Sherry, *supra*, note 10, sec. 16:2; *Minneapolis Fire & Marine Ins. Co. v. Matson Navigation Co.*, 44 Hawaii 59, 61, 352 P.2d 335, 337 (1960); *Hulett v. Swift*, 33 N.Y. 570, 572–75 (1865).

49. *Summer v. Hyatt Corp.* 153 Ga. App. 684, 266 S.E.2d 333 (1980). (Hotel guest does not lose guest status by using hotel restaurant. As such, guest may not recover for loss of purse containing valuables because guest did not comply with statutory requirements for deposit of valuables.) *Diplomat Restaurant, Inc. v. Townsend*, 118 Ga. App. 694, 165 S.E.2d 317 (1968).

50. *Montgomery v. Ladjing*, 30 Misc. 92, 61 N.Y.S. 840 (1899), cited and followed in *Kuchinsky v. Empire Lounge, Inc.*, 27 Wis. 2d 446, 134 N.W.2d 436 (1965).

51. *Block v. Sherry*, 43 Misc. 342, 87 N.Y.S. 160 (1904).

52. *Wentworth v. Riggs*, 159 A.D. 899, 143 N.Y.S. 955 (1st Dep't 1913); *Johnston v. B. & N., Inc.*, 190 Pa. super. 586, 155 A.2d 232 (1959).

53. *Shamrock Hilton Hotel v. Caranas, supra*, note 47, and *Johnston v. B. & N., Inc., supra*, note 52.

54. Except New York State, which allows foodservice operators to limit the ceiling for liability for lost or damaged property.

55. *Wentworth v. Riggs, supra*, note 52.

56. *Black Beret Lounge & Restaurant v. Meisnere*, 336 A.2d 532 (D.C. App. 1975).

57. *Shamrock Hilton Hotel v. Caranas, supra*, note 47; *Forte v. Westchester Hills Golf Club, Inc.*, 103 Misc. 2d 621, 426 N.Y.S.2d 390 (1980) (unattended cloakroom).

58. *Shamrock Hilton Hotel v. Caranas, supra*, note 47.

59. *Black Beret Lounge & Restaurant v. Meisnere, supra*, note 56.

60. *Id.* See also *Forte v. Westchester Hills Golf Club, Inc., supra*, note 57.

61. Cf. *Apfel v. Whyte's, Inc.*, 110 Misc. 670, 180 N.Y.S. 712 (1920) (No responsibility where restaurant employee took coat and hung it in close proximity to patron's table.)

62. *Ellerman v. Atlanta American Motor Hotel Corp.* 126 Ga. App. 194, 191 S.E.2d 295 (1972) (bailment of guest vehicle) is authority for this principle of public policy preventing professional bailees from disclaiming responsibility for bailed property by contract.

63. However, New York has seen fit to give restaurant keepers a degree of protection by including them in its innkeepers's liability statute.

64. Sherry, *supra*, note 10, sec. 17:8.

65. *Stark v. Grange Mutual Ins. Co. of Custer County*, 203 Neb. 154, 277 N.W.2d 679 (1979) (fire policy on business premises).

66. *SFI, Inc. v. United States Fire Ins. Co.*, 453 F. Supp. 502 (M.D. La. 1978), *aff'd*, 634 F.2d 879 (5th Cir. 1981) (burglary policy on business premises).

67. *Steigler v. Insurance Co. of North America*, 384 A.2d 398 (Del. 1977) (fire policy on residence); *Morrison Assurance Co. v. Armstrong*, 152 Ga. App. 885, 264 S.E.2d 320 (1980) (workers' compensation policy).

68. *Keddie v. Beneficial Ins., Inc.* 94 Nev. 418, 580 P.2d 955 (1978) (fire policy on commercial fishing boat).

69. A theft of patron property by you as owner or occupier of the premises is a typical example. Your deliberate burning of your own building, the crime of arson, will prevent you from recovering under your policy of insurance. See *Steigler v. Insurance Co. of North America, supra*, note 67.

70. *Citizens Ins. Co. of America v. Tuttle*, 96 Mich. App. 763, 294 N.W.2d 224 (1980), *rev'd*, 411 ich. 536, 209 N.W.2d 174 (1981) (no-fault insurance).

71. *Youse v. Employers Fire Ins. Co.*, 172 Kan. 111, 238 P.2d 472 (1951) (no ambiguity in exception limiting coverage to hostile fires).

72. *Hawkeye-Security Ins. Co. v. Government Employees Ins. Co.*, 207 Va. 944, 154 S.E.2d 173 (1967).

73. *Id.*

74. *Pini v. Allstate Ins. Co.*, 499 F. Supp. 1003 (E.D. Pa. 1980), *aff'd*, 659 F.2d 1070 (3d. Cir. 1981) (fire policy on residence).

Hotel Liability for the Safety of Guests and their Property

Common Law Exceptions to Liability

Preventing Liability

Summary

Questions

OBJECTIVES

After reading this chapter, you should be able to:

1. Explain unique aspects of innkeeper liability caused by unsafe conditions that lead to guest injury, or loss or damage to guest property.

2. Explain the different rules governing liability caused by unsafe physical premises and inadequate security causing injuries to guests by other guests, hotel employees, and third parties.

3. Review the defenses to liability claims for damage to or loss of guest property.

4. Discuss the unique statutory limitations of liability that shield innkeepers from excessive guest property loss or damage claims.

Case in Point

Roger and Nora Smith and their 14-year-old daughter, Mary, stayed in a suite of rooms at the Carlton Manor Family Resort in the Catskill Mountains of New York state. The Carlton management was aware that the dead bolt lock on the access door to the Smith suite was inoperative, and that there had been a number of burglaries of other hotel rooms related to defective dead bolt mechanisms during the past year. The management did not inform the Smiths of this problem, but did order electronic locking devices for all guest room doors, which had not yet been installed.

One afternoon, Mary was resting alone in the suite while her parents were finishing lunch in the resort dining room. An intruder burst into the suite and, threatening her with a knife, sexually assaulted her. The intruder identified himself later to police as an off-duty resort front-office employee who, being aware of the dead bolt problem, had also committed some of the prior guest room robberies. In all other cases, the guest rooms were vacant when the employee entered them.

It came out later that the man's references had not been checked, a violation of the resort's written employment policies. Had these policies been followed, the references voluntarily given at hiring would have revealed a history of criminal assaults and robberies for which the employee had been convicted and imprisoned.

Is the resort legally responsible for Mary's physical and emotional injuries resulting from the assault? The courts are divided on this question. The majority of courts would apply the theory of *respondeat superior*, (let the employer respond in damages for the wrongful acts of employees) to determine the resort's responsibility.[1] Under that theory, the employee activity must be wrongful, and must arise out of and during the course of his employment. The fact that the employee was off-duty and acting strictly to satisfy his own motives might excuse the resort.[2] However, a minority of states reject this doctrine, applying the rule that the resort guarantees the fitness of its employees as part of its contractual obligation to its guests.[3]

Finally, all states recognize that a resort keeper that violates its own hiring polices may be held liable for employee misconduct resulting from such violations if they are related to the injuries suffered by the guest.[4]

Your rights and responsibilities as an innkeeper for the safety and security of your guests, highlighting the differences between the foodservice operators rights and responsibilities to patrons (see **Chapter 12**), will be explained below.

LIABILITY FOR GUEST SAFETY

In general, an innkeeper has the same degree of responsibility to exercise reasonable care for the safety and security of guests[5] as the foodservice operator.[6] However, the circumstances surrounding the duty to exercise reasonable care are generally broader and more varied in the case of innkeepers, because the amenities available to the hotel guest are likely to include services that the foodservice operators cannot provide. The law defines a hotel guest as occupying overnight accommodations on a transient or day-by-day basis. Foodservice operators provide a specific foodservice experience during particular time periods. No overnight accommodations are provided. Moreover, the hotel may offer its guests swimming, sauna, health club, and a variety of other services not available in a foodservice setting.

Historically, the innkeeper's primary responsibility at common law was to protect the safety and welfare of guests from the criminal elements that frequented the highways used by travelers, particularly at night.[7] This responsibility still exists even though the reasons for the law have long since ended.

Out of this early concern for those seeking sleeping accommodations, the law developed a corollary rule that the innkeeper was required to admit all who sought such accommodations as long as they were in a fit condition to be received and able to pay the reasonable charges of the inn.[8] However, the rule applies only to registered hotel guests, and excludes any others, such as tenants, lodgers, visitors, rooming house occupants, or diners. Until the advent of civil rights laws, other people could also be excluded, such as patrons of a food service, theater, or movie, since these establishments were viewed as places of entertainment and conviviality rather than places of shelter.[9] However, these people now are protected as patrons.

Because of the wider range of services provided the hotel guest, creating a potentially greater possibility of risks leading to legal liability, it is important to understand how the innkeeper-guest relationship is created. The three elements necessary to do so are an intention to become a guest, the communication of that intention to the innkeeper, and acceptance of the guest by the innkeeper.[10] In some cases an illegal purpose on the part of the person seeking admission will defeat a claim of guest status,[11] but if an illegal act occurred after the person has been accepted and is unrelated to the illegal or concealed purpose of the guest, the guest may sue the innkeeper for any breach of duty owing the guest.[12]

What, then, are the legal rights of patrons of a hotel-food service? Are only hotel guests entitled to seek legal recourse for legal wrongs committed by an innkeeper? The answer is no. A patron of a hotel is generally entitled to ordinary or reasonable care on the part of an innkeeper, as in the case of a foodservice patron.[13] However, the hotel guest has a greater degree of protection than that afforded a patron.

For example, the hotel guest has a legal right to receive courteous and considerate treatment from the innkeeper,[14] a legal right not available to a hotel or foodservice patron. Likewise, a hotel guest has a right to greater protection from wrongs committed by a hotel employee than is normally true of a hotel patron. Lastly, the hotel guest may be given greater protection from the risk of assaults or other wrongs committed by third-party intruders than is true in the case of a hotel patron. These increased legal protections will be discussed below.

NEGLIGENCE THEORY OF LIABILITY

Normally, hotel and foodservice operators do not intentionally wish to injure paying customers. Likewise, the law generally does not impose strict or insurer's liability upon such establishments,[15] because to do so would discourage the exercise of reasonable care by such patrons to the detriment of their guests and patrons and impose too high a cost on them for doing business. Exceptions to this general theory of responsibility exist in the case of food and beverages served to customers. Here the law imposes a warranty of fitness for human consumption. This means that the operator is responsible even where the operator commits no fault or legal wrong causing the injury.[16] Therefore, in the majority of cases, negligence or a fault theory of liability must exist and be established to impose legal responsibility for injuries suffered by the innkeeper's guests or patrons. To summarize the essential elements of the theory, there must be a duty imposed on the operator, a breach or violation of the duty, the violation must *proximately cause* or contribute to the harm, and legal damages must be found to exist.[17] The burden is on the person seeking to recover to set forth and prove these elements.[18]

The prevailing standard, or duty of care, is generally held to be ordinary or reasonable under the circumstances. A violation of that duty occurs when the innkeeper has notice of an unreasonable risk of harm, which requires warning or the removal of the risk. Warning or notice can be actual or constructive.[19] *Actual notice* is notice given the innkeeper directly by another guest, patron, employee, or a passerby. *Constructive notice* is notice the innkeeper should have obtained in the exercise of reasonable inspection of the premises.

Clearly one of the difficult areas of proof is that of proximate cause. This element is usually determined by whether or not the risk was reasonably foreseeable to the innkeeper. That question is viewed by the courts in terms of the invitation extended to the guest by the operator. For example, an innkeeper who advertises his establishment as a family resort represents that it is safe for minors who may not have the same ability to respond to risks as their parents or guardians.[20] A young child may not have the ability to read and respond to warnings of risks that an adult would under the same set of circumstances. In most cases these are questions of fact for a jury, unless the court concludes that no reasonable jury could have found for the injured parties under those circumstances. As to the negligence of the minor excusing liability by the innkeeper, all courts adopt the rule that an infant under the age of five cannot commit contributory negligence as a matter of law.

So far, we have examined the common law duty of care that exists as one element of proof under the negligence theory of liability noted above. However, the law also creates mandatory statutory or regulatory duties that must be obeyed. No issue exists as to whether these duties are reasonable under the circumstances. To illustrate, when fire safety regulations adopted for the protection of hotel guests and patrons are violated and proximately cause injuries to guests, these violations are said to be proof of negligence *per se* or standing alone, regardless of reasonableness.[21] This means that the innkeeper is conclusively found to be negligent in such cases, with the court or jury required only to hear and determine damages. About one-third of our states adopt this rule. Another third of the states rule that a statutory violation, if found to be causally related to the injuries suffered, creates a rebuttable presumption of negligence, forcing innkeepers to prove that the injury was caused by circumstances beyond their control. The remaining states view a statutory violation involving guests or patrons as merely evidence of negligence to be weighed by the jury, as in common law cases.

DUTY TO PROVIDE SAFE PREMISES

As a general rule, an innkeeper is required to use ordinary or reasonable care in constructing, furnishing, and maintaining the physical premises that the guest is invited to use. However, the violation of a statutory duty related to these responsibilities creates the additional burden of a finding of *per se* negligence or a rebuttable presumption of negligence.

In the case of a power outage or other emergency, operators and employees should follow the hotel's emergency plan of notice to guests and evacuation of premises, to the extent permitted by the circumstances.

In the following case, a key issue was whether compliance with existing statutory fire safety codes and statutes conclusively established the innkeeper's nonliability for the injuries suffered by its guests resulting from a hotel fire.

Northern Lights Motel, Inc. v. Sweaney
Supreme Court of Alaska
561 P.2d 1176, rehearing 563 P.2d 256 (1977)

Facts: A guest died of carbon monoxide poisoning from gases released into his room during a hotel fire on the defendant's premises. His *personal representative* sued to recover for his death, alleging negligence in not maintaining his hotel room ceiling with plasterboard but another fire board, which permitted the fire to enter and burn more rapidly than would have been the case if plasterboard had been installed. Under the fire code, the fire department concluded that the motel was an existing building under state fire and building codes, meaning that its current construction was considered adequate without the addition of safeguards required for new construction (sprinklers and smoke detectors). However, the fire chief testified that had he known of the inadequate ceiling construction in the wing in which the deceased's room was located, he would have ordered additional safety devices.

Reasoning: In affirming a jury verdict for the deceased guest, the Supreme Court of Alaska first determined that the trial court has properly applied the negligence *per se* standard governing the violation of statutory duties. That is, the statute in question, the Alaska Uniform Building Code, was passed to impose duties on innkeepers, protect hotel guests, and protect against the hazard of fire.

Moreover, the high court ruled that the *deceased* could introduce evidence of additional safeguards, the lack of which constituted a violation of an innkeeper's common law duty to provide reasonably safe premises for the housing of its guests. The lack of any community standard of safety did not stop introduction of proof of such additional standards for the jury's consideration where, as here, the risk of unreasonable harm was clear to a reasonable person.

Conclusion: This case illustrates the point that violations of statutory duties can and often are deemed proof of negligence *per se*, where the proper foundation for introduction of the statute has been provided. Even where the negligence *per se* rule is not adopted, other courts will adopt the rebuttable presumption of negligence rule, thus shifting the burden of proof of non-liability to the innkeeper. In either case, the jury is precluded from weighing the reasonableness of the statute under the circumstances, since that decision has been made for it by the state legislature or local government. This means that innkeepers must know and comply with all statutory duties or face a greater likelihood of liability in these cases.

Additionally, the lack of any community safety standard under the circumstances, or the fact that the statute has in fact been complied with, is no automatic assurance that the courts will rule in an operator's favor. Each case must be examined on its merits, and courts differ as to whether statutory compliance will excuse the hotel from liability. Remember that independently of structural fire safety requirements imposed by statute, innkeepers have a common law duty to warn guests of fire, notify the fire department promptly, and prevent the fire from spreading.

The next case deals with the issue of constructive notice in a ski resort.

Paul v. Kagan [and Ramada Hotel Operating Corp.]
New York State Appellate Court
3rd Dept. 461 N.Y.S.2d 489 (1983)

Facts: Paul, a female guest at Kagan's resort, rented a snowmobile from the defendant's resort. The snowmobile was operated over a course maintained by the resort. Paul was a passenger in the snowmobile, which overturned without warning while being operated by the defendant's son. The cause of action against Kagan was dismissed.

The sole issue on appeal was whether the proof of notice of unreasonable risk of harm to Paul was sufficient to sustain a jury verdict entered for Paul by the trial court.

The New York Appellate Court found the proof of constructive notice sufficient to allow the jury to find in her favor, and affirmed the judgment below.

Reasoning: The key to resolving this appeal was the question of whether the resort had constructive notice of the defect in the snowmobile course which caused the accident, sufficient to support a finding of negligence. No actual notice was proved. Constructive notice, the court stated, means notice of a concealed defective condition that has existed long enough for the operator to have known of its existence, such that the operator violated his duty of reasonable care to warn course users of its presence or to remove the risk.

Moreover, Paul sufficiently proved that the condition was dangerous and the resort could have reasonably foreseen that injury could result since corrective measures were not taken. In all other respects, Paul established the essential elements of her claim based on negligence existence of duty, violation of that duty, and injury as a result of that violation of duty. Whereas a resort operator is not an insurer of the safety of its guests, it is required to exercise reasonable care for their safety. On the evidence presented and the court's instructions to the jury on the law, the jury was authorized to find for the plaintiff.

Conclusion: The case illustrates the point that hotel operators cannot rely on lack of evidence of actual notice of a dangerous condition to excuse them from liability in the maintenance of recreational areas furnished to resort guests for their use. If constructive notice can be established, based on an operator's duty to inspect and maintain the area in a reasonably safe condition, the jury is free to impose liability on that ground. Moreover, reviewing courts are not likely to reverse such jury verdicts unless the evidence is clearly insufficient to support the verdict or because the court's instructions on the law are so wrong as to require reversal.

This case illustrates the *additional management principle* that where the operator has greater knowledge of a concealed risk than the guest, such circumstances will likely result in a finding of liability in favor of the guest.

DUTY TO POLICE THE CONDUCT OF GUESTS, EMPLOYEES, AND THIRD PARTIES

The duty to exercise reasonable care to protect hotel guests extends not only to providing safe facilities but also to secure guests from misconduct by others that result in their injury or death.[22] The duty to provide reasonable guest security is examined below.

Third-Party Criminal Misconduct

A major cause of innkeeper liability claims is inadequate security measures that result in assaults or other criminal misconduct inflicted upon guests by intruders. A prior history of similar misconduct on the premises is often the key to establishing notice to the innkeeper of an unreasonable risk of harm, an essential component of common law negligence liability. Some courts require very close similarity of prior incidents to satisfy the notice requirement, but a growing number permit some deviation. For example, prior robberies caused by defective locks on guest room patio doors were held sufficient to notify the innkeeper that greater precautions were required to protect a guest from assault facilitated by an inadequate patio door lock.[23]

Some courts consider criminal misconduct an independent unforeseeable risk that the innkeeper could not prevent regardless of the security precautions taken, and thereby find no liability. However, in the majority of cases the courts do not rest liability on the innkeeper's ability to prevent harm, but on whether or not the innkeeper took reasonable steps to protect the guest, regardless of whether the steps would have prevented harm.[24] In summary, if the risk of harm was reasonably foreseeable by the innkeeper, then if all other negligence requirements are met, the likelihood of guest recovery is increased. When there is no such proof of foreseeability, the likelihood of innkeeper liability is greatly reduced.

Employee Misconduct

Two rules have been developed to deal with guest liability caused by hotel employees' misconduct.

The majority rule imposes liability on the innkeeper for the misconduct of his employees, both carelessness and intentional infliction of physical harm, by applying the doctrine of *respondeat superior*.[25] That doctrine, defined as "let the employer respond for the wrongs of employees," requires proof that the employee committed a legal wrong, and that the wrong arose out of and in the course of employment. A basic assumption of the doctrine is a further finding that the act of the employee was done for the benefit of the employer, regardless of whether the employer approved or disapproved the act in question. Therefore, if the employee is not liable, the employer cannot be liable. However, if the employee acts to further the interests of the employer, the fact that the act was forbidden does not prevent the employer from being held liable. If the act is done purely to satisfy the personal motives of the employee, unrelated to the job or the interest of the employer, the doctrine would excuse employer liability. Only the employee would have to compensate the victim. This rule applies only in personal injury cases. In contract cases, the employer is liable only if the employee committing the wrong had actual or apparent authority to commit the act. If there was no such authority, the employer escapes liability.

A minority of states, Massachusetts, Nebraska and New York, reject the application of the doctrine of *respondeat superior* and apply a contract rule of strict innkeeper liability for employee misconduct inflicted upon guests arising out of the innkeeper-guest relationship.[26] This means that regardless of whether the employee was on duty or off, whether the act was merely careless and not intentional, or whether the act was within the scope of employment, the guest is entitled to recover. Possibly the only situation in which no employer liability would be found would be where the employee act occurred off the premises and the employee was off duty.

The following case illustrates the general rule of innkeeper responsibility for third-party misconduct.

*Orlando Executive Park, Inc. v. P.D.R. [d/b/a (doing business as) Howard
    Johnson's Motor Lodge]*
Supreme Court of Florida
433 So.2d 491 (1983)

Facts: A woman registered and occupied a room at the defendant's motor lodge at night. While walking to her room from the parking lot, she saw a man standing behind the registration desk. While continuing to her room located on the ground level, the man behind the desk followed and assaulted her, rendered her semi-conscious, and robbed her of her jewelry. He escaped and was never apprehended. Prior assaults on guests had occurred. The one guard on duty had been hired to patrol the parking area, and not the motor lodge building. Whether or not the three guards recommended by the security service could have prevented the assault was viewed by the court as immaterial, since it was established that additional guards would have provided a deterrent to this attack, and also that the likelihood of the accident recurring was slight. The court, having stated the general rule of reasonable care owed to guests for their safety by innkeepers, reasoned that the jury could properly conclude that the inadequate security measures proximately caused the assault.

The victim proved the lack of TV monitoring devices in the public areas to deter criminal activity, openness and availability of the premises to anyone to enter, the inadequate number of security staff, inadequate security standards to protect guests, and failure to warn the victim of prior criminal activity on the premises.

Florida's Supreme Court affirmed a judgment rendered for the victim.

Reasoning: The high court found that the outside security staff was inadequate, that the motor lodge had been advised by its security force to hire two more guards, which was not done, and that it was likely the additional guards would have greatly lessened the risk of attack and robbery upon the victim. The court specifically rejected the argument that the robbery and assault were an independent cause of harm for which the defendant was not responsible. Additionally, the court stated that the lack of industry security standards does not require a finding of nonliability. The victim's evidence of reasonable standards that could have been employed to deter the attack and robbery in this case was sufficient to allow the jury to find the motor lodge negligent. As long as reasonable measures would have made the chances of this attack slight, the jury could properly find that the lack of such measures proximately caused the victim's injuries.

Conclusion: This case warns that the security measures employed by an innkeeper, depending on the size and layout of the premises, must be adequate, and that adequacy does not depend on whether the assault could have been prevented, but rather whether the measures could have acted as a reasonable deterrent to such misconduct inflicted upon guests. This means that where notice of prior misconduct reasonably similar to this case is established, the innkeeper may not ignore these warnings by arguing that the misconduct was a "supervening independent" cause for which it is not responsible. That argument is being rejected by a growing number of courts. What is considered adequate security depends on your own situation. At the very least you must evaluate your security measures, based on your past experience, and try reasonably to respond to them. This includes evaluation of the neighborhood surrounding your property, the development of a plan, implementation and training, cooperation with local police and reassessment, as well as readjustment based on changing conditions.

The following case represents the minority rule regarding employee misconduct for which the employee may be liable. However, it lays down a basic test of overall responsibility that is being adopted by other courts.

> **Tobin v. Slutsky**
> **United States Court of Appeals, 2nd Circuit**
> **506 F.2d 1097 (1974)**
>
> **Facts**:   A minor, a guest at a New York resort, was assaulted on the roof of the premises by an off-duty employee, who entered the hotel elevator occupied by the victim, forced her at knifepoint to the roof, and warned her that if she cried out or told anyone of the incident he would severely injure her. The resort argued that the employee was acting outside the scope of employment and on his own, rendering the resort free of liability for the guest's injuries under the doctrine of *respondeat superior*.
>
> The Circuit Court of Appeals rejected this argument and affirmed a finding of liability against the resort.
>
> **Reasoning**:   The reviewing court first concluded that under New York law, the innkeeper-guest relationship imposes a contractual duty on the innkeeper to secure the employment of reasonable hotel employees, and the fact that they were acting outside the scope of the authority under the *respondeat superior* doctrine is immaterial. Here the resort advertised that it was suitable for families, and the jury could properly rely on the advertisement as proof of the resort's representation as to its duty to protect minor guests. Massachusetts and Nebraska also adopted this narrower rule of employer liability.
>
> **Conclusion**:   The Circuit Court held that, as a general rule, an innkeeper is required to provide reasonable security measures for all invited guests, appropriate to the size, class, and quality of the operation. Although the innkeeper is not an insurer of guest safety, the jury could properly decide whether it was reasonable to have no policing in the lobby or the elevators to prevent such incidents.

Higher Standards of Care

Although the reasonable care standard imposed on innkeepers for hotel guest security as well as safety is the majority rule, a number of states are adopting a higher standard of care in cases of assaults inflicted on guests by intruders. Although the higher standard of care does not reach the level of making innkeepers insurers of guest security, it points to a judicial trend toward increasing the hotel industry's responsibility to deter or minimize such incidents. This rule of higher care has not been extended to govern defects in the structure, furnishings, or supervision of the physical premises where the harm to the victim consists of negligence not related to assaults by third parties. Nor does the rule apply to the foodservice industry.

The case below applies the higher standard of care rule adopted in Louisiana and extends the area of responsibility to include a public walkway immediately adjacent to the hotel's front doors.

Banks v. Hyatt Corporation
United States Court of Appeals, 5th Circuit
722 F.2d 214, rehearing denied 731 F.2d 888 (1984)

Facts: A hotel guest was shot to death within four feet of the entrance doors to the hotel by an intruder. The hotel had employed a perimeter patrol to police the area, but did not increase the size of the patrol after incidents of third party assaults on guests and passersby had increased prior to the shooting of this guest. The court below had entered a judgment for the victim's representative, and the Circuit Court affirmed that judgment on appeal.

Reasoning: The reviewing court first examined the evidence of hotel security to determine whether the jury could properly find that the hotel's perimeter patrol was adequate to deter the shooting death of the guest. In doing so, the court applied the law of Louisiana (the location of the incident being New Orleans), and found that the Supreme Court of Louisiana had recently adopted the higher-than-ordinary-care standard of responsibility regarding incidents involving third-party premises liability of hotel keepers in that state.[27] The court then found the evidence, governed by this standard, sufficient to allow the jury to find for the victim. It used a cost prevention analysis to determine that the innkeeper, rather than the victim, was best able to determine the cost-justified level of accident prevention.

Finally the court rejected the hotel's argument that since the assault did not occur within its four walls, it could not be held responsible for the victim's death. Since the victim was attempting to enter the hotel at the time of the shooting, the hotel's duty of care extended to him. Four feet would not insulate the hotel from liability.

Conclusion: This case highlights the judicial trend toward increasing liability standards of care for hotel keepers in guest assault cases. This in turn means that we must review and assess our precautionary measures in combatting this problem. This duty is peculiar to the hotel industry and requires constant attention, even in the majority of states that rejected it. This is because it may indirectly influence courts in reviewing intruder misconduct claims under the ordinary and reasonable standard.

Liability for Inaction

The law includes within the definition of negligent conduct, not only acts but failures to act on the part of responsible parties.[28] This principle is illustrated by the innkeeper who, in defending a guest assault claim, is found not liable because the victim is unable to prove that the assault was caused by any inadequate security measures. However, if the victim proves its further claim that the hotel did not adequately provide rescue and medical aid, the failure to act may impose liability on that ground alone, even though the assault claim was dismissed. The hotel's failure to act in a reasonable manner, rather than its active negligence, is sufficient to create a question for the jury. This means that the innkeeper has a duty not only to take reasonable steps to secure guests from assaults by preventive measures, but also has a further responsibility to respond to calls for help to a guest in trouble in a reasonably prompt manner.

The case below makes this distinction clear.

Boles v. La Quinta Motor Inns, Inc.
United States Court of Appeals, 5th Circuit
680 F.2d 1077 (1982)

Facts: Boles, a guest at the defendant's inn, was assaulted and raped in her room by an intruder. She was bound and gagged by the intruder, but after his departure managed to reach the room phone and call the front desk. The operator on duty refused to send inn employees to help her and was slow in calling the police. Unknown to the victim, two relief managers stood outside her door listening to her screams but failed to comfort her while they waited for the police to arrive. She sued the inn to recover both for the assault and for the inn's failure to respond promptly to her telephone calls for help.

The reviewing court upheld the jury's verdict for the inn on the assault claim on the ground that the evidence of unreasonable security measures was insufficient to support a contrary verdict. However, the court ruled that the jury was justified in finding the inn liable for unreasonable delay in responding to Boles' cries for help.

Reasoning: An innkeeper, the court reasoned, has a duty not only to prevent assaults upon hotel guests by providing reasonable security measures, but also has an additional independent duty to render reasonable aid and comfort to a guest in distress. This duty to render aid and comfort does not depend on a finding of liability for the assault itself. It rests on the innkeeper's continuing duty to respond to a guest's distress, even if it is beyond the control of the innkeeper to prevent the assault. This duty arises in cases of calls for medical help or help in other emergency situations where the welfare of the guest is in jeopardy.

Conclusion: The duty to respond promptly to guest distress calls is a universal rule imposed as part of the innkeeper's responsibility to provide for the welfare of guests. The rule derives from the fact that guests place themselves in the care of innkeepers since they make the inn home, although for a limited period. Employees should be trained to respond to such calls in a courteous, as well as prompt manner; courteous and considerate treatment to guests is also imposed on innkeepers by the common law.

Duty to Warn of Concealed Risks

Normally the law requires persons to use ordinary care in avoiding known risks that are patent or obvious to a reasonable person under similar circumstances. This means that a hotel guest using hotel facilities must take notice of any warnings and avoid using facilities designated dangerous for use on that day. On the other hand, an innkeeper is under a duty to warn of dangerous conditions on its property that are *latent*, or not observable, to a guest exercising reasonable care in the situation. In such cases the knowledge of the innkeeper of admitted perils is weighed more heavily toward liability than the ignorance or lack of appreciation of danger to the guest. The defense to liability of the doctrine of assumption of risk on the part of a hotel or resort depends on proof that the victim has sufficient knowledge of the risk to avoid it and voluntarily assumes the risk in spite of the warning of concealed danger.

In the following case, the comparative knowledge of danger of the guest and the resort operator was examined to determine whether the resort was negligent.

Tarshis v. Lahaina Investment Corp. [d/b/a Royal Lahaina Hotel]
United States Court of Appeals, 9th Circuit
480 F.2d 1019 (1973)

Facts: A female hotel guest, having read in the hotel's brochure that the ocean on the west side of the island of Maui, adjoining the hotel, was safe for swimming, went into the water and was thrown on the beach by a huge wave, resulting in injuries. She sued, claiming that the hotel had failed to warn her that the ocean, apparently calm with only slight waves, was dangerous for swimming on that day. The hotel claimed that it had posted signs warning guests not to go swimming because of dangerous surf, and therefore the hotel had no further duty to the guest since she knew the risk and assumed it by swimming in the ocean in spite of the risk.

Reasoning: The trial court granted the hotel's motion for summary judgment without trial, holding that any reasonably intelligent person, such as the guest, should have known the danger and avoided the ocean. The reviewing court noted that the trial court had adopted the proper rule of law governing the case, which was that the hotel owed the guest a duty to warn her of dangerous surf along its beach front, but only if the risk was not known to her or could not have been known to an ordinarily intelligent person, and was known or ought to have been known to the hotel in the exercise of reasonable care.

The reviewing court concluded that the trial court erred in not allowing a jury to decide whether or not the ocean would have appeared dangerous to the guest. Therefore the reviewing court reversed and remanded the case for trial, giving the jury the right to decide either for the guest or the hotel, after hearing the evidence and receiving instructions from the court on the law.

Conclusion: This case reminds us that questions of negligence (did the hotel violate any duty and, if so, did it cause or contribute to the injury), as well as questions of whether the guest either caused her own injuries or contributed to the result, are normally questions of fact for the jury to decide. Only where no reasonable jury could have found for either the guest or the hotel is the court permitted to grant judgment to one side or the other as a matter of law, and these cases are rare.

This case also highlights the essential need to warn of dangerous conditions known to the innkeeper and not reasonably known to the guest. As more recreational activities, such as scuba diving, white water rafting, or bobsledding, are added to the amenities offered guests by resort operators, the need to warn or to have guests sign disclaimers of responsibility grows. This should be standard procedure in regard to activities that are, by their nature, dangerous to those who wish to engage in them. Innkeepers should review their insurance coverage to reduce exposure to claims.

DEFENSES AGAINST LIABILITY CLAIMS FOR INJURIES

Based on the rule that guests must exercise reasonable care for their own safety, the law provides two defenses to innkeepers as well as foodservice operators. These defenses must be set up by the innkeeper in the response (answer) and proved at trial. If the innkeeper successfully establishes these defenses, the common law rule completely relieves the innkeeper of liability, regardless of the fact that the innkeeper's own negligence is greater than that of the guest.

Contributory Negligence and Assumption of Risk[29]

In order to balance the rights and responsibilities of innkeeper and guest, the common law requires guests to be responsible for their own negligence or carelessness, so that the innkeeper is not held to strict or insurer's responsibility. The negligence of the guest must contribute to the injuries suffered, since a careless act or failure to act, unrelated to the harm inflicted, could not be said to affect the outcome and is of no value in deciding the case. This is embodied in the doctrine of *contributory negligence*. For example, a guest who carelessly sets fire to his guest room bed or carelessly jumps into a hotel swimming pool at night, when the pool is closed and has been drained, is responsible for any injuries sustained.

Assumption of risk, a doctrine similar in its result, differs from contributory negligence in that it requires knowledge of the specific risk of harm and voluntary action on the guest's part in spite of the risk. Where either the risk cannot be appreciated by reason of the age or level of intelligence of the guest, or the risk could not be understood by a reasonable person similarly situated, the rule may not apply. As in the case of issues of the innkeeper's own negligence, these are normally questions for a jury to decide. Only where no reasonable jury could decide otherwise (that is, where the proof is overwhelming in favor of the innkeeper raising the defense) will a court rule that the guest has assumed the risk as a matter of law.

A number of states have enacted laws that create a *comparative negligence theory*, permitting a jury to compare the amount of guest negligence as a percentage and deduct the value of the guest's own negligence from the total amount awarded the guest. Some states set a limit of 50 percent on the amount deductible; if the guest's own negligence exceeds that amount, the guest may not recover at all. Both of these defenses apply to claims based on innkeeper statutory violations of duty as well as common law negligence violations, as indicated in the following case.

Haft v. Lone Palm Hotel
Supreme Court of California
478 P.2d 465 (1970)

Facts: Haft and his 5-year-old son, both guests of the hotel, drowned in the hotel swimming pool. No one saw them drown. When last seen by another guest, they were laughing and playing in the shallow end of the pool. When next seen, 30 minutes later, both were submerged in the deep end. While there was no proof of direct or observed negligence, the reviewing court concluded that there was sufficient evidence that the hotel had failed to comply with the state health and safety code requiring the presence of a lifeguard or a sign advising guests of this fact; no depth markings along the edge of the pool; no sign warning that children were not to use the pool in the absence of an adult; no telephone numbers of the nearest rescue and medical facilities were posted near the pool; no diagrams were posted on the use of artificial respiration, or instructions as to the use of mouth-to-mouth resuscitation; and no life poles were provided.

The trial court held that the victims had failed to establish the necessary causation as a matter of law and permitted the jury to find that both victims were contributorily negligent, and their representatives were barred from any recovery.

The reviewing court first held that the victims had met their initial burden of proving the most serious statutory violation (failure to provide a lifeguard or post a warning sign), and that the hotel was required to prove that the violation was not the proximate cause of their deaths. If the hotel failed to prove the absence of primate cause, the element of causation of such deaths is established as a matter of law. The court further ruled that it was wrong not to instruct the jury that the child was not contributorily negligent as a matter of law.

Because of these errors committed by the trial court, the judgement for the hotel was reversed and the case was given a new trial.

Reasoning: As to the issue of proximate causation, the court reasoned that providing no lifeguard or a sign that none was provided constituted negligence *per se*, because it was possible to suggest that the presence of a lifeguard exercising reasonable care would have prevented the deaths of both victims. The hotel cannot use its own violation of the statute to show lack of proximate cause. Moreover, the hotel cannot claim that its failure to erect a warning sign that no lifeguard was provided limits its liability to harm caused by the failure to do so. The importance of the requirement for lifeguard service applies to this pool, and failing to erect a sign cannot be used to avoid this mandatory statutory duty.

On the issue of proximate cause, the alleged negligence of the father is not an intervening or independent cause either of his own death or that of his son. Likewise, the father's negligence, even if established by the hotel, cannot be used to deny any recovery to the son. The son's claim is independent of his father's claim, so that the father's negligence cannot affect the outcome of the son's own cause of action. This is settled law in California, but other states are free to adopt a contrary rule.

Finally, the supervision of the father defeats the argument that the son was sufficiently capable of recognizing the risk of swimming to allow the jury to decide whether or not he was guilty of contributory negligence in entering and using the pool. The court found no case where a young child was found negligent following the instructions of a parent.

Conclusion: A growing number of states and localities have passed statutes governing pool operators, including innkeepers. Violation of these statutes can have serious consequences, especially where death or serious injury occurs. It is important to review your swimming pool rules to be sure that you are in full and complete compliance. The courts use statutory violations as a means of imposing liability where none would exist if the case were tried on the common law reasonable care standard.

Bazydlo v. Placid Marcy Co. [d/b/a Hotel Marcy]
United States Court of Appeals, 2nd Circuit
422 F.2d 482 (1970)

Facts: A female hotel guest was thrown from a winter toboggan run when the toboggan failed to negotiate a turn and hit a hand-built obstruction. The jury found for the guest on the issue of liability and her contributory negligence. The hotel appealed, arguing that the court's charge to the jury on contributory negligence erroneously led the jury to believe that the guest's negligence must be substantial in order to prevent her from recovering damages. The reviewing court rejected that argument and held that the word *substantial* used in the charge referred to primate cause and not to the amount of negligence that the hotel had to establish, which was very slight. The judgment was affirmed.

Reasoning: This case reiterates the common law rule that any amount of contributory negligence, no matter how small, is sufficient to defeat a guest's cause of action, but that the connection between the guest's negligence and the harm done must be substantial in order for the resort operator to succeed.

Conclusion: Like the doctrine of assumption of risk, contributory negligence requires proof not only that a guest's negligence was slight but that it was a substantial factor in causing the harm. Proof of negligence without substantial proof of proximate cause is insufficient to make this defense work.

DUTY TO SAFEGUARD GUESTS' PROPERTY

The innkeeper has a unique common law responsibility to safeguard guests' property. Unlike the duty owed to patrons, tenants, and condominium and cooperative owners, which is based on negligence, or a *bailment relationship*, the innkeeper insures guests' property *infra hospitium* (within the walls of the inn) without the guest being required to prove fault or misconduct to recover. This rule is based on the early historical need to safeguard the goods of guests from innkeepers who, working with highwaymen, took advantage of the guest's vulnerability.[30] In its more current application, the innkeeper is either absolutely liable for loss or damage to guest property or, in a few states, is presumed to be liable in such cases with the burden imposed on the innkeeper to prove otherwise.[31] Subject to statutory laws limiting liability and common law exceptions to liability, this rule of absolute or *prima facie* liability (presumed but subject to rebuttal by the innkeeper) remains the law in all states.

In the following case, the Arizona Supreme Court interpreted its statute limiting innkeeper liability for losses of guest property, including jewelry from the guest room. The statute, adopting a customary pattern followed by all states, excused liability where statutory notice requirements were met, a hotel safe was provided, and the guest failed to deposit the jewelry in the hotel safe. However, the Arizona statute reimposed liability for such losses caused by the innkeeper's own *act*. How this word is interpreted is the subject of the decision below.

> ### Terry v. Linscott Hotel Corp.
> ### Supreme Court of Arizona
> ### 617 P.2d 56 (1980)
>
> **Facts**: While guests at the Arizona Hilton Hotel, property, including jewelry, was stolen from the plaintiff's guest rooms. The Arizona statute limiting innkeeper liability excused such liability where the guest fails to deposit valuables, in this case jewelry, in the hotel safe if statutory notice procedures are followed and a hotel safe is provided. However, the statute reimposes full liability if the loss is caused by the innkeeper's act. The trial court held the innkeeper not liable, and the Supreme Court affirmed that ruling.
>
> **Reasoning**: The Supreme Court first acknowledged that some states permit a full recovery by guests if they can prove that the loss was caused by the negligence of the innkeeper. Other states limit recovery even where innkeeper negligence is shown. The Arizona court made a distinction between *active negligence (misfeasance)*, and *passive negligence (nonfeasance)*, imposing full liability in the case of active negligence and no liability in the case of passive negligence. Here the word *act* in the statute was interpreted to mean active negligence. Since the guests failed to plead any negligent act, but only *negligent failures to act* (maintaining an inadequate security system), the statute was read to excuse the hotel from all responsibility, and the guests were denied any recovery.
>
> **Conclusion**: The majority of states excuse innkeeper responsibility in full where the guest fails to deposit the valuables in the hotel safe if the innkeeper complies with the statutory notice requirements. This is true regardless of the theory of liability on which the guest sues.
>
> Since states differ on this point, it is important to review your innkeeper limitation statute with a lawyer. This commonsense step will enable you to consider insurance coverage to cover any gaps in the statutory protection your law creates.

Statutory Limitations of Liability

Because of the risk of excess liability claims created by the common law, all states have passed statutes limiting the liability of innkeepers for loss of guest property.[32] In order to take advantage of these limitations, the innkeeper must maintain a hotel safe and must comply with certain notice requirements. These requirements are spelled out in each statute, and usually require *posting* a *conspicuous notice* in each guest room. Some statutes limit posting to the public rooms of the hotel. The word *conspicuous* is interpreted to mean language large enough for the average guest to read.[33] Note that guests are bound by a properly posted notice whether they have read it or not. This prevents a large loophole in the limitation law from being created. If the innkeeper were required to prove that the guest actually read the notice, the statute would be nullified. Thus the law presumes the guest has read the notice if it is conspicuous and properly posted.

Each statute also governs the type of hotel safe the innkeeper must install. Some states require a fireproof safe or one approved by fire underwriters. These requirements also vary from state to state, so you must check them to ensure full compliance. Any deviation may cost you your statutory protection.

In the case below, the term *infra hospitium* was extended to cover a car parked by a guest in an adjacent parking area, and the statute limiting liability was interpreted to make the innkeeper fully responsible for its loss.

Villela v. Sabine, Inc.
Supreme Court of Oklahoma
652 P.2d 759 (1982)

Facts: A guest's car trailer and contents were stolen from an innkeeper's open parking lot adjacent to the hotel, which was under the care of a hotel security guard. The guest parked the car at the direction of the hotel. The trial court asked the high court to answer two questions to enable it to decide the case: 1) whether or not the *infra hospitium* rule was a necessary condition to imposing absolute liability for losses of guest property at common law, applied to the loss of this vehicle; and 2) whether the vehicle was covered under the state's statute limiting liability for its loss. The high court answered yes to the first question and no to the second question.

Reasoning: The high court interpreted the statutory language "under the care of" broadly to apply to a vehicle that was not under the supervision and control of the innkeeper or its employees. No bailment needed to be created for the inclusion of the vehicle within the four walls of the inn. In this regard, the Oklahoma law differs from the majority of states, where the courts exclude automobiles from the *infra hospitium* rule when they are parked adjacent to the inn and in the absence of any bailment. This means that the majority of states require proof of negligence or misconduct to hold the innkeeper responsible for guest vehicles and contents. Where a bailment is created, the burden of proof of no liability is shifted to the innkeeper, but no absolute liability rule exists.

As to the second question, the court reasoned that no specific mention of cars was included in the innkeeper liability statute. No other objects were similar in kind to permit the court to interpret the word chattel or thing to include cars in this case.

Conclusion: Therefore, the innkeeper was held absolutely liable for the lost car and contents.

Just how adequate a hotel safe must be to satisfy the requirements of a state innkeeper limitation of liability statute was the subject of the next case.

Gonclaves v. Regent International Hotels, Ltd.
New York Court of Appeals
447 N.E.2d 693 (1983)

Facts: Two hotel guests each deposited jewelry worth $1 million in the hotel safety deposit boxes provided by the hotel, which conform with the New York statute. The boxes were housed in a room built of plasterboard with access controlled only by two hollow-core wood doors, only one of which had an ordinary residential tumbler lock. The room was not locked and was left unattended. A gang of professional thieves broke into the boxes, including those of the two guests. The guests sued for the amount each deposited. The trial court and the intermediate reviewing court denied recovery for more than the statutory ceiling of $500. The Court of Appeals reversed the decision and remanded the case for trial on the issue of whether the safety deposit boxes were adequate to satisfy the New York statute.

Reasoning: The court interpreted the New York statute strictly, and held that the hotel was required to provide a safe in keeping with its size, class and status as a luxury hotel catering to a wealthy clientele. Therefore, the lower court was in error to deny the guests an opportunity to prove that the safety deposit boxes in this case were inadequate.
   Finally, the court ruled that the safety deposit box agreements signed by the guests exonerating the innkeeper from liability were held unenforceable as being against public policy.

Conclusion: This case illustrates the importance of complying strictly with your state statute's requirements concerning the type of safe to provide for guest deposits of valuables. Failure to do so will cause you to forfeit the statutory limitations meant to protect you.

COMMON LAW EXCEPTIONS TO LIABILITY

Under common law, three public policy exceptions were created to excuse innkeeper liability for guest property. These exceptions apply even where innkeepers fail to comply with the limitations of liability statutes provided for their benefit. The first of these is *contributory negligence of the guest*, which the law regards as a loss chargeable in any way to guest carelessness or misconduct.[34] Examples of such negligence include failure to lock one's room door or window, or exhibiting money or valuables in public. In each case, the courts generally permit the jury to decide whether these acts are sufficient to constitute negligence. Only where the negligence is so great that no reasonable jury could find otherwise will the court rule that the conduct is negligent as a matter of law.

The second exception is a *loss caused by an act of God*,[35] which is defined as any extraordinary natural or physical calamity or happening without human intervention. If the innkeeper carelessly stored guest property, and that increased the likelihood of loss or damage, the fact that the initial cause was an act of God would not excuse innkeeper liability.

The third exception is a loss caused by an act of the public enemy.[36] Public enemy does not mean any loss caused by criminal elements; it involves acts of war or situations in which martial law is declared by the government to prevent looting or other unlawful activity in the wake of a natural catastrophe.

PREVENTING LIABILITY

In all of the above situations, whether the risk to the guest is physical injury or property loss, a proactive risk management program is essential. This, in turn, depends on whether you acted reasonably under the circumstances to remove or warn of unreasonable risks, not risks that are assumed as a daily part of guest activity. If there are no statutory duties of care that are mandated by the legislature or municipal authorities, you must be aware that your expansion of recreational and other activities may increase the risk of operation. These risks must be recognized and their costs weighed against their benefits, with the understanding that the greater the standard of amenities and service, the greater the likelihood that the courts will attach more responsibility to their operation. While innkeepers are not insurers of guest safety and security, some courts are raising the standard of reason with regard to liability for assaults and other misconduct.

SUMMARY

Innkeeper responsibility for guest safety rests generally on the same standard of reasonable care that applies to the safety of foodservice patrons.

Negligence is the usual theory of liability, requiring proof by the guest of a legal duty, violation of that duty, proximate cause, and damages.

A duty applicable to innkeepers and foodservice operators is that of maintaining safe premises. That duty may be statutory, such as compliance with building and fire safety codes, or it may be one of reasonable care under the circumstance.

Likewise, innkeepers have a broader responsibility to police the conduct of their guests, employees, and third parties. Here the duty of care includes the duty to provide guests with adequate security from assaults by intruders. A few states impose a higher standard of care in such cases. This higher standard of care requirement is unique to innkeepers; it has no application to foodservice operators. Employee misconduct toward guests also differs from that for foodservice operators. State courts that impose a contract theory of responsibility do so based on the innkeeper-guest relationship.

Otherwise the majority of courts apply the doctrine of *respondeat superior*, both to innkeepers and foodservice operators. Because of the greater likelihood of assaults and other misconduct occurring within a hotel than within a foodservice operation, the innkeeper is required to exercise due care in responding to distress calls from guests, not only for illness but also as a result of criminal misconduct by intruders or employees. Here the courts require prompt action, the absence of which may impose liability, regardless of innkeeper responsibility for the assault itself.

Both innkeepers and foodservice operators are required to warn guests and patrons of *known concealed risks*, not otherwise known to their victims. Once again, the exposure is greater to the innkeeper who operates a large, diversified property, including recreational facilities, than to the more modest operator.

Defenses against liability claims for injuries, contributory negligence, and assumption of risk are the same for both innkeepers and foodservice operators. Guests or patrons guilty of either contributing to or voluntarily assuming the risk of harm are precluded from recovering at common law. However, comparative negligence statutes apportion the cost of harm between the innkeeper or foodservice operator and the guest or patron so that the value of harm caused by the victim reduces the total recovery. Statutory violations proximately causing injury to both hotel guests and foodservice patrons are more likely to impose liability than the common law standards of reasonable care.

Losses of hotel guest property within the premises impose liability at common law on innkeepers. This is not the case with foodservice operators, where negligence or some other fault theory of liability must be established. Likewise, innkeepers benefit from *limitations of liability statutes* that substantially lessen their liability. Such statutes do not exist to protect foodservice operators (with the exception of restaurant keepers in New York State). Otherwise, innkeepers have absolute liability subject to statutory limits as to losses of guest property. Foodservice operators are generally held liable for their *negligent acts or omissions*, or on a *bailment theory* without benefit of any limitations of liability.

Innkeepers are required to comply strictly with the statutes limiting their liability. Maintaining a safe or safety deposit box and posting conspicuous notices in the proper places, all specified by the statute, are mandatory. Innkeepers who fail to do so will forfeit the statutory limitation, and they may be found absolutely liable for the value of the article lost or damaged.

Three common law exceptions exist to excuse an innkeeper's responsibility. The guest's *contributory negligence*, an *act of God*, or *act of the public enemy* will protect innkeepers where the statute limiting liability does not apply because the innkeeper failed to meet their requirements. These innkeepers' defenses must be set up and proven in order to successfully avoid liability. Liability prevention is essential to avoid or minimize legal liability in all cases. Risk management practiced company wide must be the rule, regardless of the size of the operation. This involves planning, training, supervision, and evaluation. The innkeeper is more vulnerable due to the size and diversity of operations, compared to the foodservice operator, but neither can ignore their duties under the law.

QUESTIONS

1. Under what circumstances are innkeepers held to a higher standard of care for the protection of his guests than the standard of care imposed upon foodservice operators?

2. Is innkeeper compliance with statutory standards the same as exercise of reasonable care under the circumstances? Why or why not?

3. What theories of liability to guests for the negligence or intentional misconduct of employees are imposed on innkeepers? Explain the elements needed to prove each theory.

4. State and explain the defenses available under common law against a guest seeking to recover for innkeeper negligence and what effect these defenses have, if accepted by the jury, on the guest's claim.

5. Are guest claims for property losses against an innkeeper treated the same as foodservice patron claims? Explain.

NOTES

1. J. Sherry, *The Laws of Innkeepers*, sec. 11:18 (rev. ed. 1981).
2. Sherry, *supra*, note 1, sec. 11:4.
3. *Id*, at 290.
4. Sherry, *supra*, note 1, sec. 10:9.
5. Sherry, *supra*, note 1, sec. 9:1.
6. *Roberts v. United States*, 514 F.Supp. 712 (D.D.C. 1981), citing *Smith v. Arbaugh's Restaurant*, 469 F.2d 97 (D.C. Cir 1972), cert. denied, 412 U.S. 939, 93 S.Ct 2774, 37 L.Ed. 2d 399 (1973).
7. Sherry, *supra*, note 1, sec 1:6.
8. *Doe v. Bridgton Hospital Association, Inc.*, 71 N.J. 478 483, 366 A.2d 646 (1976), cert. denied 433 U.S. 914 (1973).
9. Sherry, *supra*, note 1, sec. 4:2.
10. Sherry, *supra*, note 1, Chapter 5.
11. *Jones v. Bland*, 182 N.C. 70, 108 S.E. 344 (1921).
12. *Cramer v. Jarr*, 165 F.Supp 130 (D.Me 1958).
13. *Campbell v. Bozeman Community Hotel*, 160 Mont. 327, 502 P.2d 1141 (1972) (luncheon patron using rest room).
14. *Arky v. Leitch*, 131 Miss. 14, 94 So. 855 (1922). But see *Pollick v. Holsa Corp.*, 114 Misc. 2d 1076, 454 N.Y.S.2d 582, (App Term 1st Dept. 1982), *aff'd* as modified 98 A.D.2d 265, 470 NYS 2d 151 (1st Dept. 1984).
15. Sherry, *supra*, note 1, sec. 9:1.
16. See Chapter 11.
17. Sherry, *supra*, note 1, sec. 9:4, also see Chapter 12.
18. Sherry, *supra*, note 1, sec. 9:5.
19. *Pierce v. Motel 6, Inc.*, 28 Wash App 474, 624 P.2d 215 (1981).
20. *Tobin v. Slutsky*, 506 F.2d 1097 (2d. Cir. 1974).
21. *Northern Lights Motel, Inc. v. Sweany*, 561 P.2d 1176, rehearing 563 P.2d 256 (Alaska 1977).
22. Sherry, *supra*, note 1, sec. 11:1.
23. *Garzilli v. Howard Johnson's Motor Lodge, Inc.*, 419 F.Supp. 1210 (E.D.N.Y. 1976), the famous Connie Francis Case.
24. *Virginia D. v. Madesco Investment Corp.*, 648 S.W.2d 881 (Mo. 1983).
25. See citations in note 2, *supra*.
26. See citations in note 3, *supra*.
27. *Krazz v. La Quinta Motor Inns, Inc.*, 410 So.2d 1048 (La.1982).
28. Sherry, *supra*, note 1, sec. 11:10.
29. Sherry, *supra*, note 1, secs. 10:10 and 10:11.
30. Sherry, *supra*, note 1, secs. 16:1 and 16:2.
31. Sherry, *supra*, note 1, sec. 16:3.
32. Sherry, *supra*, note 1, sec. 17:10.
33. *Depoemelaere v. Davis*, 77 Misc. 2d 1, 351 N.Y.S.2d 808 (Civ. Ct. N.Y. Co. 1973), *aff'd* 79 Misc. 2d 800, 363 N.Y.S.2d 323. (1974).
34. Sherry, *supra*, note 1, sec. 17:28.
35. Sherry, *supra*, note 1, sec. 17:8.
36. Sherry, *supra*, note 1, sec. 17:9.

Maintaining Security

OUTLINE

Legal Responsibility

Summary

Questions

OBJECTIVES

After reading this chapter, you should be able to:

1. Describe various crimes against foodservice operators committed by invitees (guests and patrons), trespassers, and employees.

2. Outline procedures to prevent crimes and measures to deal with wrongdoers.

Case in Point

Tom Turner broke into and entered the hotel room of Cannon Astor, a wealthy real estate tycoon. Upon going through Astor's personal belongings, Turner found a credit card. He took the card to the Silver Room, a plush dining room situated atop the hotel, independently owned and managed by Jean Foster. Turner ran up a three-figure bill for food, drinks, and entertainment with an accomplice. At the end of their dinner, Turner presented the credit card, expertly forged Cannon Astor's name on the bill, and charged it to Astor's account. The maître d'hôtel verified the account by phone, was told that it was valid, and thanked Turner and his companion for their patronage and their generous tip, also charged to the Astor credit card.

What are Jean Foster's rights in this situation? Is she simply out the money if Astor reports the missing card and refuses to pay? Can she bring criminal charges against Turner? The answer depends on which state the restaurant is in.

Under common law, Turner committed a theft of property when he stole Astor's card. He could be prosecuted and/or fined only for that crime against Astor. His crime against Foster—theft of services—was not a crime under the common law, and Jean would be out the money. Her only recourse would be to try to persuade Turner's other victim, Astor, to pay the bill charged to his stolen credit card—an unlikely prospect.

Criminal prosecution for the theft of property at common law is insufficient to protect those who render services on the strength of the validity of a credit card.[1] In most states, this legal loophole has been filled by the creation of statutes protecting persons or parties who accept credit cards for payment in the normal course of their business operations.[2] In these states, the foodservice owner could prosecute Turner for theft of services.

CRIME AND PROFITS

Keeping costs down is important to the profitability of a hospitality operation. Operators who practice poor security leave themselves open to a neverending cycle of costly security problems, and, as a result, poor profitability.

If a restaurant or hotel has a reputation as a place where rowdies congregate and cause trouble, good customers will be driven away. If an owner gets a reputation as a soft touch or for running a "loose" operation, meaning a place that can be robbed easily, both amateur and professional criminals will mark him or her as an easy target for any type of fraud or outright robbery. In addition, if employees and patrons cannot resist stealing food and materials because it is so easy, the owner will find the cash flow outgoing as he or she scrambles to replace the inventory. In any event, the bottom line will reflect their carelessness.

Crime hurts; mostly it hurts the pocketbook. However, patrons and employees alike have been victims of violence during robberies and other crimes. Tighter security may not prevent every type of crime against operators, but it will, in most cases, keep an operation from becoming a revolving door for a variety of criminals.

The negative effect of criminal misconduct is two-fold. First, there is the economic loss. Whether the value of meals, lost property, and services is eventually recovered is of small comfort. Even if the owner will be reimbursed for the loss, he or she is immediately out the cash, and it takes time and effort to restore it.

Second, there is harm to one's personal business reputation. Unless operators take immediate steps to prevent wrongdoing and prosecute wrongdoers, they may become easy marks for professional rings of thieves who prey on the gullible without mercy, once given an opening wedge. No one can afford being known in criminal circles as a *soft touch*, since these artists can pick their targets clean and leave them empty-handed. Others must not be tempted to follow in their footsteps by, in effect, "giving away the store." Making it easy for the "pros" to commit crimes invites the amateurs to do likewise. This possibility multiplies the problem and can make operation unprofitable at best, or close an operator's restaurant doors at worst.

Security measures include the way you let your patrons pay for your service, how you let them conduct themselves when they are in your operation, and what you may do legally if they become unruly. Security measures also include the supervision of your employees to protect your patrons and your property. Finally, they involve steps needed to protect the business from criminals who have no business there.

Maintaining security is a legal matter and a personal business decision. This chapter presents positive steps operators may consider to protect their business, patrons, and employees from crime. The decision as to whether or not to use them is one each operator must make on the basis of local laws, the seriousness of the problem, and the economic feasibility of the solution.

GUEST AND PATRON CRIMES

Patron crimes vary but most of them are related to either theft of services or patron conduct while in the operation. Liability as well as criminal charges may result from the latter.

Theft of Services

Most patron crimes against foodservice operators occur when it is time to pay the bill; these are classified as thefts of services.

The law defines *theft of services* to mean either: 1) knowingly using a stolen, revoked, or canceled credit card to obtain services or inducing the supplier of services to accept payment on a credit basis; or 2) intentionally avoiding or attempting to avoid payment by failing to pay or by refusing to pay through misrepresentation of fact.[3] As a result of the increasing availability of credit as a method of payment, and its use in place of cash, the hospitality is being victimized by people who wish to obtain service and products without paying.

Credit Thefts   Thefts of services are mainly credit thefts; that is, the use of a stolen, revoked, or canceled card for the purpose of obtaining food, beverages and hotel accommodations on credit.[4]

Extension of Credit—Credit is defined as any method of payment—other than cash or in kind—at the time goods are transferred or services are rendered by the seller. Credit means payment or performance made over a period of time agreed upon by the parties to the transaction.

As a hospitality operator, you are entitled to receive cash in exchange for services and products provided to guests and patrons in your establishment. This is a legal right, not a matter of discretion which those served may exercise as they see fit.[5] In the same vein, you have the right to reject or accept a credit card or check in place of cash. The guest or patron cannot require you to extend credit as a condition of his or her business with you.[6]

Credit is a valuable, but not required, method of operating a business. Credit has the advantage of allowing managers to market services to more customers by allowing them to borrow the cost of food, beverages and lodging from you for a specified period of time established in advance. Credit benefits customers by enabling them to do business without the need to carry large amounts of cash, and, equally important to business customers, affords the ability to produce a written receipt for business and tax purposes.

The extending of credit also carries risks. Credit is granted on the assumption that the cardholder who uses it will pay the account on time and in the required amounts. Unfortunately, even in the best of circumstances, the customer may fail or refuse to pay. In that event, there are two courses of action available: 1) to terminate further credit, or 2) to take steps to collect the amount owed, either with or without court action. The first remedy is the least expensive and least time-consuming. The second is more expensive and time-consuming, but also more likely to result in payment of the bill.

The choice of whether or not to extend credit remains the operators decision, not the cardholder's or the state's. Refusal to do so at all is not a violation of the law.[7] Title II, the public accommodations section of the federal Civil Rights Act, does not make operators liable for denying credit on these grounds.[8] The federal Equal Credit Opportunity Act, which does prohibit discrimination in credit transactions, does not, however, apply to hospitality operators. Only financial lenders, such as banks and savings and loan associations, may not discriminate in extending credit on the traditional civil rights grounds of race, creed, color, sex, national origin, and marital status.[9] The reason for the difference in treatment is based on congressional recognition that a commercial hospitality operator's primary business is not that of extending credit, but of providing food and beverages to patrons in return for compensation.

The nature of the operation will determine whether management should establish a credit policy. In a quick-service establishment, with high customer turnover, a credit policy may not make sense because it is impossible to verify credit in advance under those conditions. However, in a hotel operation with heavy repeat business or established clientele, a credit policy will be appreciated. In any case, it is preferable to honor a national credit card rather than to set up a credit system. This is because: 1) a system using a national credit card is easier to operate; and 2) the burden of collection is placed on the issuing company, not the operation, as long as it complies with the issuer's requirements.

No credit card company can compel an operator to do business with it. Operators in turn cannot require a credit card company to do business with them. In the same vein, operators cannot compel guests or patrons to use credit in every case. The customer must consent to do so. (Only private clubs can compel members to sign for all charges. These clubs usually send a monthly bill to members.)

The following are guidelines for setting up credit as a method of payment.

1. Distinguish between individual credit and company or corporate transactions. Extending individual credit requires verification of that individual's credit standing *before* extending credit. This should be done independently of statements or claims made by the applicant.[10] Check credit references.

2. When a company or corporation applies for credit, it must do so through an authorized officer or agent, since an institution has no human existence of its own.[11] Officers and other agents may not legally establish authority to act on behalf of a corporation or company through their own statements or claims.[12] This means the manager must obtain a letter or corporate resolution establishing: a) that the person representing himself or herself as an officer or agent is in fact authorized to act; b) that the corporation has authorized that person to establish credit for the corporation; and c) that the corporation has authorized the specific purposes for which the credit is to be used. Otherwise the company can legally refuse payment and, in some cases, recover payments already made.[13]

The officer or agent who signs as that corporation's officer or agent is not individually liable to pay. This means that both the corporate officer and the corporation are off the hook, unless the corporation has given written authority to extend credit.[14]

When dealing with a corporate or company officer or agent, always have the individual guarantee payment of the account in writing. If a guarantee of payment cannot be obtained, try in every case to obtain a written guarantee of eventual payment; this is called a *guarantee of collectibility*.

3. Minors may disaffirm their credit transactions, which means that businesses extend credit to minors at their peril.[15] When dealing with a minor, or a person legally declared incompetent, always have a responsible adult guarantee payment.

   The difference between the two forms of guarantee is important. A *guarantee of payment* makes the guarantor primarily responsible for payment, as if he or she alone had opened the account. A *guarantee of collectibility* makes the guarantor only secondarily responsible for payment, meaning that you must sue the party in whose name the account was opened first, and then sue the guarantor in the event the first party cannot pay.

4. When in doubt as to the credit-worthiness of an individual or institution, do not extend credit unless the applicant can provide a financially responsible guarantor of payment.

5. Always keep careful records of each credit account opened, including the application itself, credit reports from reporting agencies, any unusual credit activity, and the card or charge account holder's payment record.

6. Always review and terminate further credit to any holder who exceeds established credit limits.

7. Vigorously pursue any seriously delinquent accounts and sue *if necessary*. A lawsuit should be brought only when all other reasonable collection efforts have been used without achieving results, and when the amount justifies the legal expenses.

8. Treat an *inability* to pay differently from a *refusal* to pay. Inability to pay can be resolved by a mutually satisfactory repayment schedule. Refusal to pay cannot, and requires sterner measures, especially when the refusal is unjustified. A refusal may be legally justified if services were not rendered on the date and time agreed upon.

Out-of-court methods of collection are preferable to litigation in court. Every overdue account that can be resolved by mutual agreement means continued patronage and repeat business. Litigation and criminal prosecution should be reserved for those who wrongfully refuse to pay or never had any intention of paying. Their intentions are to benefit at an operator's expense.

Handling Credit Cards—Normally, when an operator contracts with a credit card company to honor its card at the establishment, either the company or the cardholder will cover any losses suffered from the theft or misuse of the card if, but only if, the required card verification procedures are followed before accepting the card for charge purposes.[16]

Federal law limits the loss to a credit card holder resulting from the unauthorized use of the card to $50, if the cardholder complies with the conditions set forth in the Truth in Lending Act.[17] This means that, in most cases, operators may recover no more than $50 from the card holder, even if proper procedures are followed at the time a card is submitted for payment.[18] The act defines "unauthorized use" to mean "actual, implied, or apparent authority for such use" by the person obtaining credit with the card, and from which use "the cardholder receives no benefit."[19] Fraud, duress, or other wrongdoing caused by a third person constitutes unauthorized use of a credit card.

What procedures should be followed so that an operator can recover from the card company or issuer for losses suffered? The following steps should be followed by all employees who handle credit payments.

1. *Check the card* against a list of canceled cards furnished by the card issuer. This is essential, since the operator will be held responsible for the entire loss if the card is not verified to be current.
2. *Examine the card.* It should not be expired or not yet valid. The numbers should be legible and uniform.
3. *Check the signature* on the credit card receipt and on the card itself. If it looks suspicious, or they do not seem to match, do not accept the card, and instead request payment in cash or by other acceptable credit, such as a traveler's check drawn on a credible bank.
4. *Call the number* provided by the card issuer to determine whether the charge is within credit limits or will be treated as an overcharge. If the charge exceeds the credit limits established for that card, accept the card for charges up to that limit and require the card user to make up any difference in cash or by other authorized credit.
5. *Initial* the card receipt.

Never under any circumstances allow cashiers or other personnel to short-circuit these procedures when they become preoccupied with impatient customers or heavy customer traffic. Such excuses are not legal justifications for failing to check with the card issuer. The result is inevitable: the operation loses. The card issuer is not responsible to guarantee payment because the operation failed to fulfill the contract requirements. Failure to perform a material requirement of a contract excuses performance by the other party.[20]

What about patron charge accounts? Whether to extend charge accounts to patrons is a business judgment solely within an individual operator's discretion. A large operation may issue its own credit card for identification and billing purposes. The following basic rules should be followed.

1. Before establishing any credit, be sure the guest or patron has an acceptable credit rating. Always verify the applicant's name, address, employer, and bank account.

2. Place a dollar ceiling on the amount of credit authorized at any given time.
3. Make cards as tamper- and counterfeit-proof as possible.
4. Have applicants sign an agreement that they will be fully responsible for payment regardless of who is authorized to use the card *and* regardless of whether the card is lost, stolen, or otherwise wrongly used. As a private creditor, an operation is not required to use credit practices enforced by the federal government, since it is not a credit card company or lending institution.[21] To remove any doubt on this question, put this and any other terms and conditions in writing and have the agreement signed by the cardholder.
5. Police all accounts. Be especially watchful for sudden increases in any individual's spending habits.
6. Always require payment in full within 30 days of billing, unless other special credit arrangements are approved in writing.
7. Always post up-to-date lists of delinquent or closed accounts for cashiers. Adapt procedures similar to those set by national credit card companies. The best rule of thumb for employees is: When in doubt as to identity, signature, or credit limit, always check with management.

Checking the validity of credit cards protects both the establishment and the cardholder, and careful policies can bring the operator goodwill. Checking protects customers from unauthorized use of their cards, saving them much anxiety and frustration involved with having to inform you or the company of misuse. Honest, diligent customers will appreciate protection of their interests.

An operation should extend long-term credit only when it is properly financed to do so. Otherwise, there is risk of draining cash flow. Accepting national credit cards is less risky than instituting a credit system. An operation that opts to use its own credit cards must review them and cut off credit on overdue accounts. Many credit card users are businesspersons who know the value of prompt payment of their own invoices. A tactful but firm reminder should be enough to restore the account.

Passing Bad Checks Hospitality operators are especially prone to receiving bad checks and counterfeit money. The hospitality business usually depends on high turnover for profit, and it is easier for criminals to run a check or counterfeit money scam during peak business periods. Unlike credit card scams, bad checks and money are totally worthless, and recovery from either the wrongdoer or third party is either difficult or impossible.[22]

A bad check may be a totally fabricated check, a forged or raised genuine check, or a genuine check that cannot be cashed.[23]

A *totally fabricated check* is an imaginary check, meaning that the name of the account, person or party to whom the check is payable, and name of the drawer (person who wrote the check), are nonexistent. Even the name of the bank on which the check is drawn may be imaginary, especially if the bad check artist has a printer for making checks.

Exhibit 7.1   Guidelines for Examining Personal Checks

1. Accept no postdated; out-of-state checks; foreign bank checks; or checks marked "Payable in Canadian Dollars," "For Deposit Only," or "Payment in Full."
2. Accept no checks made payable to other than the person who will cash them.
3. Accept no checks over the stated credit limit.
4. Never cash any check that looks at all suspicious.
5. Never cash a prewritten check. Always require the check writer to make out the check in the presence of an employee to avoid any possibility of an altered or raised check.

6. Always require proper identification from the person using a check. Compare the signature on the check with the signature on the identification item. An excellent form of identification is a driver's license, especially in states where the license contains the photograph of the license holder.
7. Avoid cashing checks of minors, even those carrying a valid driver's license, unless the minor is known and creditworthy, or an adult agrees to guarantee payment. A minor may disaffirm his or her contract to pay for certain services, except where necessary or where a statute prohibits the minor to do so.
8. Have a check-cashing policy and train employees to adhere to it.

A *forged check* is genuine except for the signature of the drawer (check writer). Here, the name of a real person who has a real checking account at a real bank is imitated with a signature that the forger attempts to deceive a merchant into believing is that of the real drawer.

The *uncollectible* or *uncashable genuine check* is one drawn on a real account opened by the check passer, but deliberately written even though there are insufficient funds to cover it or the account has been closed.

Exhibit 7.1 lists proper procedures for examining personal checks. **Exhibit 7.2** shows a properly filled-out check.

Exhibit 7.2   Properly Filled-out Check

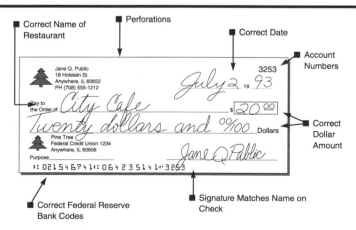

Reprinted with permission from the *Foodservice Security Manager Handbook*. Copyright © 1993 by The Educational Foundation of the National Restaurant Association.

Prosecution for the crime of passing bad checks requires proof beyond a reasonable doubt that the writer intended to obtain credit knowing the check was uncollectible. Failure of a bank to cash a check, standing alone, is not proof of intent knowingly to issue an uncollectible check.[24] The fact that the account on which the check was drawn has insufficient funds may create a presumption of knowledge, but that presumption is not conclusive, and may be denied by the accused.[25] In some states, no crime is committed if the check is redeposited and clears within 10 days after the bank's first refusal to honor the check.[26]

If a guest or patron offers to pay part of the amount owed on a check returned for insufficient funds, demand the entire amount. Acceptance of a partial payment converts the bad check to a promissory note, or a charge account, and the check writer is not obligated to pay back the money according to the established schedule.

Counterfeit Money and Foreign Currency   Counterfeit bills and coins will continue to be passed by professional counterfeiters, by amateurs as a joke, or by patrons who substitute genuine foreign money for U.S. money.

The counterfeit article may be a high-quality item, depending on the resources of the counterfeiter. However, even the best counterfeit bill can be detected if carefully examined. The special paper, steel engraving plates, and other techniques used in producing real money are virtually impossible to duplicate. Both the paper and the printing ultimately reveal less sharpness of detail in portraits and background, as well as broken lines not found in genuine bills.

A quick check to determine if a bill is counterfeit is to rub the back of the bill on a piece of paper. The green ink from the counterfeit bill will transfer to the paper.

Bogus coins, as compared with the real thing, are examined for the preciseness of the indentations along their outer edges. Counterfeit coins are not as precisely or evenly "reeded" as genuine coins—their indentations are broken or uneven—a sure warning to a cashier, waiter, or waitress that something is wrong. Counterfeit coins often can be cut or will bend with little effort.

Foreign currency, genuine where issued but not readily convertible in the United States, is generally distinguishable in most characteristics from U.S. money, and so should present no serious problem to anyone except the most indifferent employee.

In some countries, currency controls forbid currency exports outside the country of origin, in which case the currency cannot be converted. The practical problem of fluctuating conversion rates and time consumed in conversion are good reasons to be on the lookout for foreign currency, and to instruct employees not to accept it.

Counterfeit Coupons and Gift Certificates   Counterfeiting of a hospitality operation's coupons and gift certificates is not uncommon. All coupons and certificates should be checked as carefully as money and checks to be sure they are authentic. The following measures can help prevent duplication.

- Use bar codes that can be read only by the operation's sales registers.
- Number coupons and gift certificates and log those numbers in the sales system.
- Use perforated edges.
- Use pastel colors, which generally do not photocopy well.
- Use a very small repeating design in the background and a border that is difficult to duplicate.

Thefts of Cash and Shortchange Artists   When money is exchanging hands between patrons and employees all day long, there is plenty of opportunity for theft. **Exhibit 7.3** lists eleven procedures that can help prevent loss of cash from cash register systems.

Exhibit 7.3   Cash Register Security Procedures

1. Set up the register where it can be seen by several employees—servers, buspersons, and managers, for example. It should be visible.
2. Limit the number of keys that unlock the register cash drawer or its electronic functions. Each duplicate key should be assigned to an individual operator, such as a cashier, server, or bartender. A log of who has each key should also be maintained.
3. Provide the operator for each shift with a fresh cash drawer containing a standard amount of cash. Require the operator to close out the register at the end of his or her shift, and assign a manager to count the receipts. Receipts for the shift should be deposited in a drop safe.
4. Require only one person to work out of each cash drawer.
5. Require operators to ring up each payment immediately, to announce each item and the total to the customer, and to give the customer a register tape receipt.
6. Keep the register locked when not in use, and the cash drawer shut and locked between transactions. Operators should not use the register release lever or function key to open the drawer for a non-cash or "no sale" transaction. Only the manager should be able to do this.
7. Allow no personal items, such as coats, bags, or packages, to be left near the register.
8. Restrict operator access to the register tape. You may not want your operator to know the total receipts from the shift or the day.
9. Require management approval of all error corrections and all questionable transactions on the register tape, such as blank spots where transactions were not recorded, missing transaction numbers, over- and underrings, and "no sale," and "void" transactions.
10. Train all operators to recognize forged or altered checks, counterfeit money, legitimate money orders, traveler's checks, and drivers' licenses. Operators should also be trained to prevent short-change practices and schemes.
11. Perform periodic and random cash drawer audits.

Reprinted with permission from *Foodservice Security Manager Handbook*. Copyright © 1993 by The Educational Foundation of the National Restaurant Association.

Shortchanging can take several forms. Shortchange artists are most likely to strike during rush periods when cashiers are in a hurry and easily distracted. The best preventive measure is concentration. On the part of employees, only one transaction should be handled at a time. The cashier should always say out loud how much money has been handed over. This will be hard for the patron to contradict. Cashiers should also set the patron's bill on top of the cash register drawer while making change rather than putting it in with other money.

Walkouts   An owner or manager has the right to detain a customer for a reasonable time, and a reasonable investigation, when he or she has grounds to believe that the customer has not paid for what was received, or is attempting to take merchandise without payment.[27]

A number of states have enacted shoplifting statutes, which make it easier for operators to defend themselves against a patron claim of false imprisonment.[28] The shoplifting statutes basically say that when you have reasonable grounds to suspect a patron of shoplifting, you can escape liability for false imprisonment.

The following case, decided in a state that does not have a shoplifting statute, illustrates the available defenses.

Keys v. Sambo's Restaurant, Inc.
Supreme Court of Louisiana
398 So.2d 1083 (1981)

Facts: Keys, a patron, had become heavily intoxicated while celebrating the birthday of a friend, Wells, before stopping at Sambo's for a late night supper. Wells agreed to pay for the meal. After the two had ordered chicken dinners and eaten, Wells left and fell asleep in his car parked in Sambo's parking lot. A waitress informed Keys of Wells' departure. Keys told the manager that he would pay for Wells' meal and his own if Wells did not return. Later Keys attempted to leave Sambo's without paying for either meal. The manager caught up with Keys, before he had gone through the door, who refused to accept partial payment, and when Keys became belligerent, locked Keys in the men's rest room and called the police. When Keys became unruly after the police arrived, an officer struck Keys and took him to the police station. Keys was charged with disturbing the peace, failure to pay for a meal, and resisting arrest.

Criminal charges were dropped. Keys sued Sambo's for injuries suffered because of his arrest. The trial court found in favor of Sambo's and dismissed Keys' lawsuit. The intermediate reviewing court upheld the judgment of the trial court. Keys then petitioned the Supreme Court of Louisiana for final review, which was granted. That court upheld the decisions of both lower courts in Sambo's favor.

Reasoning: The Supreme Court found that: 1) the plaintiff committed a felony by refusing to pay; 2) the plaintiff attempted to defraud Sambo's according to state law; 3) the plaintiff did not dispute the attempts to defraud or refuse to pay; and 4) the plaintiff's injuries were a result of his attempt to resist arrest. These factors combined, the court said, justified affirming the lower courts' decision in favor of Sambo's.

Conclusion: This case is important for a number of reasons. First, the case notes the existence of criminal statutes in many states that make it a felony to receive food from a restaurant without intending to pay for it. This classification is important: where a misdemeanor is committed, a claim of false imprisonment may result. Second, the case establishes a foodservice operator's legal right to make a citizen's arrest when a patron attempts to leave under such circumstances. Third, the case establishes the operator's legal right to detain a person until the police arrive.

A word of caution is in order. A citizen's arrest is usually authorized only if: 1) the person arrested is a felon; 2) a felony has been committed, and there are reasonable grounds to believe the person arrested committed it; or 3) a misdemeanor amounting to breach of the peace is committed in the presence of the person making the arrest. A public disturbance, such as being drunk and yelling on a public street so as to disturb the good order and tranquillity of the neighborhood, would be a breach of the peace. Some states have extended the right to make a citizen's arrest to a criminal trespass of private property.

Police should be called when a crime has been committed or a threat or use of deadly force or bodily harm exists.

Once a customer walks out without paying, there is little an operator can do except call the police and let them take care of it. Do not go beyond the restaurant's premises to pursue a walkout: it is too dangerous. Note the patron's description, make of the car, and license plate number for the police. The best advice is to prevent crime by training employees to be alert at all times.

If an incident does take place, it is wise to try and settle with the customer peacefully on the premises. Most customers do not want to risk embarrassment, so unless they have, or feel they have, a legitimate reason for not paying (for example, poor food), they will usually pay, once given a friendly reminder.

Malicious Mischief

Most statutes define malicious mischief as intentional injury or destruction of the real or personal property of another.[29] Malice is not limited to situations where the person committing the crime intended to injure the owner of the property. Rather, malice has been extended to include situations in which the injury or destruction is committed wantonly and without justification against any property other than the property of the accused person.[30]

In the hospitality industry, any wanton, unjustified act resulting in damage or destruction of property, caused by anyone, including customers, intruders, employees, and the general public, is malicious mischief. A *wanton act* means one that is not provoked or caused by your own conduct and in response to that conduct, or not otherwise justified by law or public policy.

For example, if a patron of a restaurant were to become angry at a companion and, without warning or any provocation on your part, destroy a table full of dishware and glasses, the patron would be guilty of malicious mischief, since all of the elements necessary to establish the crime are present. The fact that the patron did not threaten or injure any person on the premises is immaterial, since the crime is against property, not a person. It must be differentiated initially from a criminal trespass because the patron was legally admitted. However, once the patron committed the unjustified destruction of property, the patron's status changed. Now the operator has the right to evict the patron as a trespasser, since the unjustified destruction of property was itself a crime. The operator would normally be required first to request the patron to leave, and upon a refusal have him or her removed with all reasonable force. But because of the seriousness of the patron's conduct, a citizen's arrest or request for a police arrest, would be justified, with either or both followed by a criminal complaint covering both the trespass and malicious mischief. The same result would hold true of a hotel guest who damaged hotel property.

More typically, malicious mischief results when a group of loud or rowdy patrons enters an establishment and, without warning, starts to "trash" tables, chairs, and furnishings. This wanton misconduct frequently occurs where a bar or cocktail lounge is adjoining. All operations should seek police assistance at the first sign of trouble.[31]

THIRD-PARTY CRIMES

According to the law, all businesses have the legal right to protect their premises and business from any unauthorized entry or unauthorized stay of any person resulting from misconduct or other wrongdoing.[32] The fact that it is a public place of business does not mean that any member of the public has the right of access or the right to remain under any and all circumstances.[33] Third-party crimes are considered those committed by people who have no business on the premises.

Trespass

The legal term for an unauthorized entry is a *trespass to land*. Such a trespass can take two forms: a civil trespass and a criminal trespass. A *civil trespass*, such as entering a private club without permission, is a tort, and the owner can recover compensatory damages for any loss suffered. If the trespass was malicious or made with intent to harm the owner or the premises, the operator may recover punitive damages.[34]

A *criminal trespass* such as a burglary differs only in that, to commit the crime, the trespass must be committed knowingly and consist of an unlawful entry or unlawful stay on the premises.[35] The trespass need not result in injury or damage to the premises in order for the owner to receive a legal remedy.

A foodservice operator, like an innkeeper, has the legal right to make a citizen's arrest without liability if criminal trespass can be established. This holds true even in those states that have no statutes authorizing an eviction for failure to obey a valid house rule, or for disorderly conduct or intoxication. Almost all states have criminal trespass statutes. However, the better procedure to follow is not to make a citizen's arrest, but to call the police and have them arrest the trespasser.

Burglary

Burglary refers to the breaking into and entering of a building, at night, with the intent to commit a felony or serious crime.[36] However, most states have eliminated both the "breaking" and "night" requirements.[37] Every foodservice operator should maintain a safe or strongbox on the premises for safekeeping cash until it can be deposited in the bank. By contrast, all innkeepers are required to maintain safes for the storage of guest valuables. Statutes limiting liability protect innkeepers against excessive loss or damage claims and apply to thefts by intruders and employees.

The following steps will help prevent burglary.[38]

1. A burglar-resistant safe will help prevent burglars from stealing the contents. Burglar-resistant features include the following.
   - Heavy steel walls and a door several inches thick
   - One or more tamper-proof deposit slots
   - Locks that require two keys to open
   - Programmable locks that can be set to open only at certain times, to open only after 10 to 15 minutes after the combination has been entered, to link with sensor devices, and to disburse cash and coins only after additional codes are entered

2. Install an additional protective device, such as a silent alarm system, to alert police to the presence of burglars before they get away. This will help keep insurance costs down and increase likelihood of apprehension. (See **Exhibit 7.4**.)

3. Train and supervise employees to never leave the safe open or unguarded, or fail to set the alarm system. Police the safe deposit area, and maintain a log of all persons who withdraw or deposit funds from the safe, including the date, time, name of person, what was removed or deposited, and the signature or initials of a responsible witness.

4. Every suspicious incident, including finding the safe unattended or open at unauthorized times, should be described in writing and kept on file for

Exhibit 7.4   Central Station Alarm System

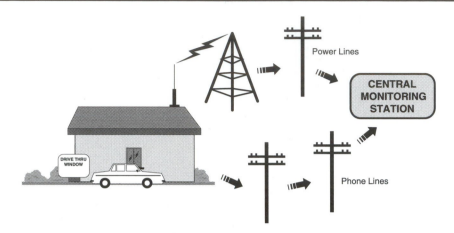

Reprinted with permission from *Foodservice Security Manager Handbook*. Copyright © 1993 by The Educational Foundation of the National Restaurant Association.

review and, if a loss occurs, reported promptly to the insurance carrier. Carelessness that is not corrected is an open invitation to crime, and may cause problems with insurance companies.

5. An armored-car or bonded-messenger service can be hired to deliver large sums of cash to the bank. If employees make the deliveries, more than one should be sent at a time. The number of trips, times of trips, routes taken, and people making deliveries should be varied so that would-be robbers cannot predict easy times to attack. The container containing cash should not be marked in any way. People making deliveries should make no stops on their way to the bank, park near the bank and only in well-lit areas, and take note of anyone in the vicinity.

Robbery

Robbery takes place when property is taken from a victim or in the victim's presence, and is accomplished by force or violence, or by putting the victim in fear of force or violence. Robbery is a felony. The typical scenario involves entry by armed intruders who force a manager or employee to open the safe or the cash register, and then lock them somewhere on the premises.[39]

The most important thing to remember during an armed robbery is that when given a choice between handing over money or your life, there's no question about which is more valuable. Procedures for employees to follow in case of a robbery should be posted on bulletin boards and included in an employee manual.

First, all employees should be instructed to cooperate with an armed robber and to make no sudden moves. Any person committing an armed robbery is bound to be under great tension. Employees should not do or say anything that will increase this tension. Second, they should give up any cash or supplies that the criminal demands. There should be no resistance, and no attempt to deceive the criminal concerning the amount or location of valuables. At the same time, no information should be volunteered.

All employees should be trained to carefully observe all physical characteristics of the criminal, only if it presents no danger to the employee. If an armed robber demands that employees not look at him or her, they should follow this instruction. Important physical characteristics include the following.

- *Height*—Notice the burglar's eyes. If they are above yours, the burglar is taller; if the eyes are below yours, the burglar is shorter. Estimate the difference in inches.
- *Eye color*
- *Skin color*—Even a masked criminal will have exposed skin around the eyes, hands, or neck.
- *Weight or body-type*—It is difficult to estimate what a person weighs, but, the burglar can be compared to another employee or a police officer.
- *Voice*—A voice is male or female, high-pitched or low, speech is slow or rapid, with or without an accent.
- *Right-handed or left-handed*—A burglar will probably carry a weapon in the favored hand. Remember which hand it was, and whether the burglar switched hands to perform some other act.
- *Dress*—Look at each item of clothing from top to bottom, including jewelry and eyeglasses.
- *How the burglar left*—Do not follow the criminal outside; look through the window. Try to get the make, model, and color of the car, and get the license number if possible. Notice which way the car turned.

After the criminal has departed, employees should not chase or drive after the robber. Each witness should write down all the facts as remembered. Each person should write without consulting others who were present. When questioned by the police, full and complete cooperation should be given.

There is no real protection against the threat of armed robbery by professional criminal rings. You are not expected to risk your life to protect the lives and property of your employees and patrons. By the same token, your employees and patrons are not expected to risk their lives to protect your life and property. Use of good security measures to protect your premises should protect your customers and employees, as well as your property, against reasonably foreseeable risks.

Civil Unrest

The threat of extraordinary, organized criminal activities confronts all public service businesses, including foodservice operations. Although, by their very nature, such civil disturbances cannot always be anticipated, operators must exercise reasonable care, and be prepared to implement and supervise procedures to cope with unusual problems. The absence of emergency procedures, as well as the failure to carry them out in a reasonable manner, may be evidence of negligence to a judge or jury should injury to patrons result.[40]

Your vulnerability to civil disturbances will affect your need to anticipate and deal with such situations. If you operate in a large city, and are close to areas used to hold demonstrations, the need for adequate disaster planning is obviously greater than would be the case if you were located in a remote, rural village.

Riots   A demonstration, harmless when contained, may turn into a riot when the activity gets out of control and tempers run high. Do not wait for an official declaration of a state of emergency, follow the directions of the local law enforcement agencies. If time permits, all flammable material outside an establishment's premises should be removed; employees should be assigned fire extinguishers to cover fire-prone points; and gas jets and other exposed fuel outlets should be rendered inoperative. Ideally, everyone should be evacuated, but if violence is already occurring, it may be safer to keep everyone inside. Lock the building and keep everyone away from doors and windows. Employees not involved in fire fighting should be assigned to give first aid and comfort to injured persons.

Communication is essential. The telephone should be used primarily to maintain contact with police and fire departments until an "all clear" is given or evacuation is ordered.

Terrorism   Terrorist activities may assume the form of bomb threats, hostage taking, and the holding of hostages on your premises. Although more likely to occur in hotels and convention centers, terrorism is not confined to any type of public property and can take place anywhere without warning. All bomb threats require attention, since the question will arise as to whether the threat is serious enough to disrupt business and evacuate patrons. In every case, a plan of action is required.

The following are suggestions for plans of action.

1. Train employees to get as much information from a terrorist caller as possible. The identity of the caller may be extracted from such clues as accent, background noises, and tone of voice. This information should be recorded and turned over to the authorities. The caller may even identify the size, type, and location of the bomb.
2. Any threat should be taken seriously. If a bomb or suspicious package is found, employees should not touch it. A supervisor should be told immediately and other employees warned. Nearby machinery should be turned

off and portable phones and beepers should be kept away because they may trigger a bomb.

3. Report all threatening calls to the police as soon as possible, first by phone and, when time permits, in writing. Let them advise you as to whether or not to evacuate the premises.

4. Cooperate fully with the authorities. Try to keep a floor plan available for examination, and have a responsible manager or employee work with the search team.

5. Anticipate potential threats when political leaders, celebrities, and other public figures are expected or present. Be ready to work with the FBI, Secret Service, and state and local authorities who may send an advance security party to canvass the operation and instruct staff. Even a relatively short visit warrants all appropriate safeguards.

If federal or other security authorities place armed personnel inside an establishment, the manager should request a written agreement stating that they will be liable for any patron, employee, or other party injured as a result of a shoot-out.

Hostage Situations   Hostage situations are rare but possible, particularly if an establishment has easy access to escape routes. If management suspects anything, police should be notified immediately. A manager with adequate warning of a possible incident who fails to warn of it or to evacuate the intended victim might be held responsible for any harm or injury. All that the law would require for a lawsuit is that the establishment's security measures be tested against the reasonable-care standard.

EMPLOYEE CRIMES

Employee theft of cash, food, beverages, and any other property, as well as theft of patron property, either alone or in the company of professional thieves, will occur unless proper precautions are taken.

The temptation to steal is always present. However, proper screening of job applicants and thorough supervision of their activities will drastically reduce incidents of stealing.[41] In large measure, employees will be honest and law-abiding, and should be treated as such. However, there are always a few employees who would take advantage of their employer. Following these measures will help employers prevent, detect, and eliminate employee misconduct.

Thefts of Cash

Cashiers and other cash-handling employees are in the best position to steal cash. The traditional method is to avoid ringing up or recording all cash sales on the cash register and then pocketing the difference between actual and recorded sales.

Inventory and Property Thefts

Inventory control involves making an employee responsible for items from the time they arrive until they are issued. At each stage, records should be kept and tallied against control forms. Paper controls must be checked by the manager or another employee.

Inventory control also includes strict security: locking all entrances and doors when rooms are not in use; allowing only authorized persons access to storage areas; allowing removal of inventory or stock only by written authorization; and daily record keeping.

Control of Employees to Prevent Thefts

Unless there are collective bargaining restrictions or local laws to the contrary, managers should require their employees to use one entrance and exit that is guarded during all hours of operation, possibly with a metal-detecting device or scale to check packages before entry or exit is permitted. Each employee should be required to sign a release allowing management to search his or her locker and to use visual monitors in designated work areas. Passes, which identify shifts and hours off and include a photograph, should be issued and checked.

Employee theft and other illegal conduct occurring during work should be made grounds for dismissal and for criminal prosecution. An employee contract may also contain the right to fire any employee disobeying security rules, such as drinking or taking nonprescription drugs on the job. Management should insist on this right at the bargaining table in operations that are unionized.

Failure to take effective and prompt steps to rid a workforce of dishonest employees will not only make the operation less profitable, but may also undermine reliable, honest employees, injure the overall reputation, and, finally, cause the business to suffer permanent harm. If allowed to continue unchecked, ultimately it may even lead to going out of business.

Crimes by Employees against Guests and Patrons

Crimes against customers by employees can be doubly serious. First, there is the potential harm to the business if word gets out that employees steal from or even verbally or physically abuse guests and patrons on the premises. However, management might also be held accountable for the actions of employees that result in injuries to customers.[42] (Chapters 5 and 6 on liability explain how this might occur.)

Thefts of Property   Theft of patron property by employees is a problem both in foodservice operations and hotels. This is especially true when the establishment checks patrons' coats and other belongings. Keeping a reliable employee on duty in the checkroom is the simplest way to avoid thefts. The operation also might set a limit on the estimated value of the items checked. In a fine-dining restaurant or luxury hotel, this might not be feasible.

Thefts of Cash   Dishonest employees usually will not slip into a customer's purse or coat pocket while the person is eating. There are easier ways to accomplish the same thing. They might shortchange customers, especially during rush periods. This is difficult to detect, but periodic observation of cashiers is an effective deterrent.

Verbal and Physical Abuse of Patrons   Servers especially can be victims of verbal abuse by rude patrons, but sometimes the tables are turned. Verbal abuse is not illegal unless a threat is involved, and then the patron may bring an assault charge against the employee. Of course, rudeness on the part of employees should not be tolerated, legality notwithstanding. It is simply bad business. Employees should be disciplined whenever it occurs.

Physical abuse is much more serious. If an employee injures a patron, management could be held liable along with the employee. (See **Chapters 5** and **6**.)

LEGAL RESPONSIBILITY

Managers are ultimately responsible for security and prevention of crime. The success or failure of an operation may depend in large measure upon the skill and diligence with which the manager plans, trains, and supervises employees in preventing antisocial or criminal behavior, and policing and controlling the persons involved.

Managers are not legally responsible for anticipating every conceivable occurrence that might result in injury to property, customers, or employees. They are required to exercise reasonable care to protect themselves, employees, and customers against unreasonable risks. If a manager fails or refuses to take reasonable precautions to prevent, deter, and halt criminal activity or to warn customers and employees of its presence, lawsuits could arise for any resulting injuries. (Insurance coverage for property losses or liability claims due to criminal misconduct is often conditioned on compliance with security requirements fixed by the insurance carrier or independently by law.)

The types of security measures used to protect a business and customers must depend largely on the seriousness of the problem. If a business is in a high-crime area, an alarm system, visual surveillance, or other security devices might be appropriate. Remember that costly security measures not justified by security problems, the area, or other factors will cost more than they help. Each operation's management must do what is necessary to make that business secure, and as with any other decision, it must take into account many factors.

SUMMARY

Crimes in the foodservice industry often take the form of credit thefts of services and the passing of bad checks. Hospitality operators are susceptible to such crimes if it is their policy to accept credit in any form. Operators are not required by law to extend *any* credit. If credit is accepted, an established credit card is usually more convenient than a personal system. All employees should be trained to follow the card verification procedures established by the credit card company. Failure to do so makes the operation liable for any loss, since reimbursement of credit extended on canceled, stolen, or expired accounts is possible only if the card's existence, validity, and credit limit were checked before extending credit.

Customers may also try to use counterfeit money. Shortchange artists may turn up during peak periods and attempt to confuse the cashier. Simply walking out without paying is another common method patrons may use.

Malicious mischief is serious in that it can include violence as well as damage to property. While operators could make a citizen's arrest, it is advisable to call the police once a crime has been committed.

Third-party crimes are committed by people who have no business on the premises in the first place and are there solely to steal property, hurt people, or both. There are a number of preventive measures operators can take, but they should be justified by the cost and size of the problem.

Theft leads the list of employee crimes and may be hard to detect. Employees may steal cash or inventory. Preventive measures require continuous supervision by employers.

QUESTIONS

1. Massimo Ciotto does a lot of credit business at his fine-dining restaurant, the Italian Aroma. He rarely has any trouble, since he uses only national credit cards and trains the cashiers. One Saturday night, however, patrons were backed up in the cashier line, blocking those trying to enter the restaurant. The cashier took a credit card from a patron for a dinner bill totaling $150, checked the signature, and ran it through the machine. What did the cashier do wrong, and what is the possible result?

2. What steps should a manager take to remove a trespasser from an operations premises?

3. A manager is training cashiers. List some tips to help them avoid being shortchanged.

4. Describe the difference between *robbery* and *burglary*.

5. A manager has found many discrepancies on a cash register tape. She knows she had a heavy turnover tonight because of the baseball game, but the cash just does not coincide with the volume. She has also noticed a number of "no sale" entries on the tapes. What most likely happened? How could this have been prevented? What should the manager do now?

NOTES

1. See J. Sherry, *The Laws of Innkeepers*, 3d. ed., sec. 22:1 (rev. ed 1981).

2. Cal. Penal Code, sec. 537 (West. Supp. 1982); N.Y. Penal Law, secs. 165:15–17 (McKinney 1975 and Supp. 1982) are representative statutes.

3. See N.Y. Penal Law, secs. 165:15–17 (McKinney 1975 and Supp. 1982).

4. See N.Y. Penal Law, secs. 155.00 (McKinney Supp. 1978).

5. Unless you demand cash payment for goods and services rendered, the buyer may pay for them by personal check or by any other reasonable means, such as a traveler's check drawn on a U.S. bank. If you demand cash when the buyer offers a check, you must give the buyer a reasonable time to obtain cash. This opportunity to obtain cash need not be given if you previously advised the buyer that the transaction would be all cash. See U.C.C. sec. 2–511.

6. *Id.*

7. Only when you choose to extend credit does section 296(a)(2) of the New York Executive Law require you to do so on a non-discriminatory basis.

8. The federal Civil Rights Act does not cover credit transactions.

9. H. Lusk, C. Hewitt, J. Donnell, A.J. Barnes, *Business Law and the Regulatory Environment* 1075–1076 (1982).

10. *Dudley v. Dumont*, 526 S.W.2d 839 (Mo. App. 1975). (The existence and scope of agency cannot be established by the declarations of the person or party claiming to act as agent.)

11. Lusk *et al., supra*, note 10, at 391–92. *Wild v. Brewer*, 329 F.2d 924 (9th Cir.), *cert. denied*, 379 U.S. 914 (1964). (A corporation does not enjoy the Fifth Amendment privilege against self-incrimination, even when it is claimed for the benefit of the sole stockholder. In most other respects, a corporation is considered to be a person apart from its stockholders. Corporate management is defined as business conduct by its officers and directors.) See also *First Nat. Bank of Boston v. Belloti*, 435 U.S. 765 (1978). (First Amendment rights of freedom of speech cannot be restricted by the legislature to corporate activities that materially affect its business, property, or assets.) However, the separate corporate identity of a fast-food franchise holder as a parent corporation to its 10 subsidiary corporations will be disregarded for inclusion of all subsidiary income as taxable to the parent company. *Wisconsin Big Boy Corp. v. Commissioner*, 452 F.2d 137 (7th Cir. 1971).

12. *Goldenberg v. Bartell Broadcasting Corp.*, 47 Misc. 2d 105, 262 N.Y.S.2d 274 (1965).

13. *Bentall v. Koenig Bros., Inc.*, 140 Mont. 339, 372 P.2d 91 (1962). (In the absence of special authority, a corporate president has no power, merely by reason of his or her office, to execute negotiable paper in the name of the corporation.)

14. *Id.*

15. Lusk *et al., supra*, note 10, at 157.

16. The credit card contract you execute mandates compliance with stated verification procedures, and is standard operating procedure for all national credit card companies.

17. Truth in Lending Act, 15 U.S.C. sec. 1643(a) (Supp. IV 1980).

18. *Martin v. American Express, Inc.*, 361 A.2d 597 (Ala. Div. App. 1979).

19. 15 U.S.C. sec. 1606 (1978 and Supp. IV 1980 and Supp. V 1981).

20. *Id.*

21. 15 U.S.C. sec. 1602(f) (1978 and Supp. V 1981).

22. Sherry, *supra*, note 1, sec. 22:9.

23. State laws must be consulted for the statutory definition of "bad check."

24. See W. LaFave and A. Scott, *Criminal Law* 678–81 (1972).

25. *Id.*

26. N.Y. Penal Law, secs. 190:15 (McKinney 1975) is a representative sample.

27. See 32 Am. Jur. 2d, *False Imprisonment* sec. 74 (1982).

28. The South Carolina shoplifting statute is set forth in S.C. Code Ann., secs. 16–13–110, –111, –120, –140 (Law. Co-op. 1977 and Supp. 1982). See *Faulkenberry v. Springs Mills, Inc.*, 271 S.C. 377, 247 S.E.2d 445 (1978).

29. R. Perkins, *Criminal Law* 333 (2d. ed. 1969).

30. *Id.*

31. Your negligent failure to do so may inflict liability upon you to patrons injured by such misconduct. See Sherry, *supra*, note 1, sec. 11:13.

32. See Chapter 5.

33. See Chapter 3, *supra*.

34. Lusk, *et al., supra*, note 10, at 60. See generally *Dial v. City of O'Fallon*, 81 Ill. 2d 548, 411 N.E.2d 217 (1980).

35. See, for example, N.Y. Penal Law secs. 140:05, :10, :15 (McKinney 1975 and Supp. 1982).

36. LaFave and Scott, *supra*, note 25, at 708–17.

37. *Id.*

38. *Id* at 692–704.

39. *Commonwealth v. Homer*, 235 Mass. 526, 533, 127 N.E.517, 520 (1920).

40. *Edwards v. Great American Ins. Co.*, 146 So.2d 260 (La. App. 1962) (innkeeper).

41. See generally Sherry, *supra*, note 1, secs. 11:4–11:8. Also see Chapter 6 for pre-employment screening procedures.

42. See Chapter 5, *supra*.

Staff Selection and Supervision

OBJECTIVES

After reading this chapter, you should be able to:

1. Explain the legal rights and duties of employers and employees with regard to civil rights laws, wage and hour laws, employee screening and surveillance, safety requirements, and union-management relations.

2. Discuss the basic legal requirements for dealing with employees.

Case in Point

Nancy DeCarlo, owner and operator of Pizzas Plus, a full-service Italian restaurant in Chicago, left her restaurant in the capable hands of her assistant manager, Don Frederick, two nights a week.

One night when Don was in charge, a female server told him that a patron had "fondled" her several times as she served his table and even when she passed by. She reminded Don that this was the third time she had complained about this customer and would appreciate it if he did something about it. Don said he did not know what he could do; the patron was neither drunk nor disorderly. Don told the server the patron was just having harmless fun and to ignore it.

Should Don have done anything to remedy this problem? Could Nancy be liable to the server? The answer to both questions is yes. Don should have taken steps to stop the patron's inappropriate behavior. What he viewed as harmless fun, she obviously saw as sexual harassment, and Don had the authority and the obligation to stop it. Employers are liable in some cases for the actions of patrons and co-workers, including sexual taunts, sexual touching, and provocative comments and gestures by co-workers and patrons.

LEGAL FRAMEWORK

The law provides a *framework* for relations with employees. Beyond that, it is up to the employer to clarify management policies and enforce them.

Fairness and consistency when managing hospitality employees is smart, both legally and economically.

EMPLOYMENT PRACTICES: FEDERAL CIVIL RIGHTS LAWS

Since 1963, a variety of federal and state equal employment opportunity laws have been passed.

The Equal Pay Act

Under the Equal Pay Act of 1963, employers are required to provide employees of both sexes equal pay for equal work, as long as the work performed is equal in skills used, effort, and responsibility, and is performed under parallel working conditions.[1] The courts have interpreted the word "equal" to mean *substantially equal*.[2] An employer may not evade the law by simply making minor changes in the work done. For example, if an operation employs two dishwashers, it would be very difficult to evade the law by saying that one worker does the breakfast dishes and the other does the lunch dishes. If, however, one dishwasher buses tables in addition to washing dishes, that may justify a difference in pay.

This law is administered by the Equal Employment Opportunity Commission (EEOC). The Equal Pay Act applies only when both men and women are employed in substantially similar jobs. It does not apply where only one sex is employed in a job.

Remedies for Violations of the Equal Pay Act   The Equal Pay Act requires employers to pay employees who have been discriminated against the same wage as other employees in similar jobs. The employer may not reduce the wage of any employee of the opposite sex to eliminate the inequality and to comply with the Act.[3] Neither is it permitted for the employees receiving unequal wages to obtain higher paying jobs as vacancies occur. Nor may the lower, unequal wages be frozen by a collective bargaining agreement adopted later.[4]

The Civil Rights Act

The Civil Rights Act of 1964, Title VII, prohibits all employer practices that discriminate on the basis of race, color, religion, sex, pregnancy, or national origin.[5] This law is far broader than the Equal Pay Act, since it regulates selection, referral, promotion, transfer, demotion, discipline, dismissal, separation, and pregnancy-related job benefits.

Hiring and management policies should be designed to prevent illegal or even subtle forms of discrimination. In this way, there is less of a chance of being held liable by an employee or group of employees who feel themselves treated unfairly in comparison to other employees. All policies should be fair, treat employees equally, and reduce chances for misunderstanding by employees, supervisors, and managers.

The law does not ask employers to hire or promote people who do not have the proper skills or characteristics to perform their jobs. It does require that all selection and supervision decisions are nondiscriminatory policies.

All foodservice establishments are covered by the Civil Rights Act if they employ 15 or more employees and are engaged in or affect *interstate commerce*. Service to out-of-state patrons, as well as purchases of goods and services from out-of-state suppliers, is sufficient to affect interstate commerce. This act is also enforced by the EEOC.

The EEOC is authorized to intervene in private lawsuits where it has reasonable cause to believe that an employer is engaged in a pattern or practice of discrimination that is intended to deny civil rights protection under the Civil Rights Act. These intentional violations trigger the remedies noted below.

Remedies for Violations of the Civil Rights Act   The Civil Rights Act of 1964 provides the following remedies for violations.

Back pay—The employee discriminated against is entitled to receive the difference between the actual pay earned and what should have been earned were it not for the discrimination. Back pay includes fringe benefits, regular and overtime pay, holiday pay, reasonable value of tips, and uniform allowances.[6]

Reinstatement—This means not only getting the job back but also any seniority. Reinstatement is not required in every instance, but is decided on a case-by-case basis.[7]

Court order stopping violation (injunctive relief)—The Civil Rights Act authorizes local courts to issue orders prohibiting employment discrimination. These orders are issued on a case-by-case basis.[8]

Damages—Employers are not normally required to pay compensatory or punitive damages under the Civil Rights Act, other than back pay. The employee must be able to prove financial loss to receive this.[9] This means that if the employee is able to obtain another job at the same or better wage level, the employer is not required to pay back wages.

When an employee can prove that he or she suffered humiliation, embarrassment, or discomfort arising out of an employer's unlawful discriminatory conduct, damages may be awarded under the Civil Rights Act.[10]

Attorney's fees—The legal fees of an employee who wins a judgment against an employer for discrimination can be recovered by his or her attorney under the Civil Rights Act.[11] Because of the length of time and the costs of preparing legal papers, not to mention the trial and appeal time involved, such costs can be staggering. Sometimes legal fees can exceed the cost of back pay, restoration of seniority, and other benefits the court may award the employee.

Retaliation against an employee for filing charges with the EEOC or for opposing unlawful employment discrimination is prohibited.[12] If retaliation is proved, the employee may obtain a court order removing any adverse or negative comments from his or her employment records. For example, an employer who demotes an employee from waiter to dishwasher for no other reason than that he or she filed a discrimination charge may find himself or herself on the receiving end of a court order.

Defenses under Federal Law   Civil rights laws establish two major defenses to any claim based on discrimination outlawed by the Act.

Bona fide occupational qualification (BFOQ)—A bona fide occupational qualification is any job-related requirement established in good faith that justifies a difference in treatment.[13] Suppose that an operator wishes to establish an authentic Chinese restaurant, serving genuine Chinese specialties in an authentic atmosphere. The operator refuses to employ non-ethnic Chinese servers over ethnic Chinese servers to make the atmosphere seem more authentic. Is the refusal to hire non-Chinese servers a violation of the Civil Rights Act? No. The operator has established a BFOQ defense. The choice of Chinese servers is both job-related and bona fide, meaning done in good faith for a legitimate reason.

The bona fide occupational qualification defense is very narrow, and difficult to prove.[14] It is not often used successfully by employers in employee civil rights cases.

Business necessity—This defense is a broader one than the BFOQ defense, since it involves the claim that the business itself requires the discrimination. For example, the fact that customer preference dictates hiring waitresses rather than waiters, however true, will not, standing alone, justify a refusal to hire men.[15] However, requiring waiters to cut their hair short but allowing waitresses to wear their hair long may be justified by customer preference. Grooming standards, which reasonably differentiate between the sexes, are generally allowed, as long as they are not motivated by an intention to discriminate on the basis of sex.[16] Why the difference in the two hypothetical cases? In the first case, sex is *immutable*, or unalterable. You are either male or female, and you are presumed not easily able to alter your sex. In the second case, sex is not an issue. What is at issue is hair length, a factor readily changeable by both sexes. Thus the courts have uniformly ruled that reasonable appearance and dress codes for employees that result in sex differentiation may be dictated by the public and are legally justified.[17] However, *they must be related to the job*. (A requirement that women bank executives wear uniforms and men executives wear customary business suits was struck down when there was no business justification, and the difference in dress was found to be demeaning to the women.)[18] In a foodservice operation, requiring a skimpy outfit for female cocktail servers, but not male servers might also be rejected as unjustified. The customer preference argument might not hold up, considering the demeaning nature of the women's outfits.

1991 Amendments   These amendments make it easier for individuals claiming discrimination to sue and win lawsuits. In 1989 the Supreme Court had increased the burden of proof that employment practices are discriminatory in *disparate impact* cases (where a challenged practice has an unintended discriminatory effect) by ruling that a person or class seeking relief must: 1) identify the specific practice or policy causing discrimination, and 2) prove that the employer had no justification for using it.

The amendments that eliminate the requirement to specify what particular practice was discriminatory, and shift the burden to the employer "to demonstrate that the challenged practice is job-related for the position in question and consistent with business necessity." The terms "job-related" and "business necessity" are not defined in the amendments, and it will be up to the courts to establish what facts prove their existence. Even these employer justifications, properly established, will not protect the employer if the alleged victim of discrimination can prove that there was an alternative practice available that would have a less severe impact upon the person or persons suing, and that the employer refused to adopt it.

Remedies for Violations   The amendments create both compensatory and punitive damage remedies in addition to the "make whole" remedies noted above. Victims of discrimination can now be compensated for other losses, as well as to punish their employer where the victim proves that the employer intentionally discriminated against that person or group. The amendments define these damages to include "future pecuniary losses, emotional pain, suffering, inconvenience, mental anguish, loss of enjoyment of life and other non-pecuniary losses." The amendments cap the amounts of damages based on the number of employees. Employers with 15 to 100 employees are held at $50,000; employers with 101 to 500 employees at $100,000; employers with 501 or more employees at $300,000. These remedies also apply to violations of EEOC sexual harassment guidelines. The Americans with Disabilities Act is now amended to provide similar damage remedies. However, an exception exists where the employer discriminates but provides a reasonable accommodation to a disabled person. In that case, no compensatory or punitive damages may be available if the accommodation is good faith.

These amendments clearly indicate a shift in employment discrimination policy in favor of accusers. Our response must be proactive and preventive. We must ensure that our policies are: 1) lawful and effective; 2) communicated to all employees; and 3) are carried out fairly and uniformly.

Age Discrimination Act

The Age Discrimination in Employment Act of 1967 prohibits discrimination against job applicants and employees 40 years old and above.[19] This means that employers may not create or increase burdens on persons in this group that are not imposed on younger employees. However, employers may bar minors from some jobs. Once again, the EEOC is the enforcement agency. Employers of 25 or more people are covered by this law.

Vocational Rehabilitation Act

The federal Vocational Rehabilitation Act of 1973 prohibits discrimination against job applicants or employees with mental or physical disabilities.[20] Federal government contractors and subcontractors are covered by this law. Also covered are private employers who receive any grant, loan, contract, or any other arrangement from the federal government.

Employers with government contracts are required to engage in efforts to provide employment for all persons protected by this federal law. This Act requires all government contractors with contracts worth $2,500 or more to take affirmative action to employ disabled workers.

The Americans with Disabilities Act (ADA)

The Americans with Disabilities Act (ADA) of 1991 provides the same protections to disabled individuals as those already provided under the Civil Rights Acts of 1964 and 1991. It covers all organizations employing 25 or more individuals. In July of 1994, the Act will govern all organizations employing 15 or more individuals. Private clubs are excluded.

Having a *disability* is defined as: 1) having a physical or mental impairment that substantially limits a major life activity (such as walking, seeing, hearing, or reading); 2) having a record of such impairment (such as a person who has recovered from a brain tumor); and 3) being regarded as having an impairment even when no limitations exist (such as a person with visible scars).

A *qualified individual with a disability* means a disabled person who, with or without reasonable accommodation, can perform the essential functions of the job being held or sought.

Recovering or rehabilitated alcohol and drug abusers are protected under this law. Persons with infectious diseases as well as other medical problems (cancer or epilepsy) are protected. Persons who are HIV-positive are covered, as are those who associate with AIDS patients.

The Act requires *reasonable accommodation* for applicants and workers with disabilities when such accommodations are needed and would not impose an undue hardship on the employer. This term might include: 1) making existing facilities readily accessible to disabled persons; 2) offering a flexible work schedule (without loss of working hours); 3) obtaining or modifying equipment; and/or 4) providing qualified readers or interpreters.

Undue hardship means that the accommodation cannot impose harsh difficulty or expense upon the employer. Factors to be considered when making an undue hardship decision are: 1) the size of the business, number of employees, number and type of facilities, and budget size; and 2) the nature and cost of the accommodation. These decisions must be made on a case-by-case basis so as to reflect the abilities of the disabled individual as well as the needs of the owner/operator.

It is important to note that if an accommodation imposes an undue hardship on the employer, the disabled employee must be given the choice of either providing the accommodation or paying for that part of the accommodation that causes the undue hardship.

One significant protection to employers is that the Act permits rejection of applicants and termination of employees who pose a "direct threat" to their own health or safety or to those of others in the workplace. However, this may be done only if there seems to be no reasonable accommodation to alleviate the problem. The assessment of risk must be based on facts, not on mere speculation, stereotyping, or remote possibilities.

Federal Contractors

A presidential order, Executive Order 11246 (1965), prohibits *all forms* of employment discrimination by federal contractors and subcontractors that hold contracts or subcontracts worth $10,000 or more. This order is enforced by the Office of Federal Contract Compliance Programs.

These federal subcontract rules would affect a foodservice company contracting with a federally funded public university to operate its student dining facilities. The food service would be barred from discriminating against job applicants and employees on any grounds covered by federal laws. It also would be required to take affirmative action to open up job opportunities for those persons protected by federal laws.

STATE EQUAL EMPLOYMENT LAWS

Only Mississippi has not passed a statute prohibiting employment discrimination. State laws follow the federal pattern and outlaw employment discrimination on the basis of race, creed, color, sex, national origin, age, or disability. These laws are important for two reasons. First, under federal law, the EEOC must first permit the state agency, if any, to attempt to resolve an employment discrimination suit before the EEOC steps in. Second, some state laws provide other means by which the employee can sue and recover from employers. These laws may also provide harsher penalties for employer violations.

Some states prohibit employment discrimination on the basis of marital status and arrest record, in addition to the areas covered by federal law.[21] San Francisco includes sexual orientation as a prohibited category of employment discrimination. This category is not included under federal law. Minneapolis includes status as a recipient of public assistance as well as sexual orientation in its discrimination laws.

Federal, state, and local laws may act together or independently in defining, regulating, and providing remedies for employment discrimination. Employers may be subject to three separate levels of law regarding employee rights.

Often the penalties for civil rights violations can multiply, such as when a woman who is also a member of a minority group is discriminated against.

The best course of action for any manager is not to violate the law in the first place. Even if an employer ultimately wins a discrimination case, the loss of time and money, and the unwanted publicity, can be devastating.

FEDERAL POLICY
ON HIRING ALIENS

In 1986 Congress enacted the Immigration Reform and Control Act (IRCA), which for the first time prohibited employers from hiring illegal aliens, rather than simply prohibiting employers from harboring them under prior immigration law.

IRCA imposes both criminal and civil penalties for hiring unauthorized aliens. There is a $3,000 maximum fine for each unauthorized alien when employers intentionally violate the provisions against unauthorized hiring and failure to verify eligibility and identity. A "pattern or practice" violation permits imprisonment for up to six months, or both a fine and imprisonment.

For unlawful employment, a first offense brings a civil penalty between $250 and $2,000 for each unauthorized alien. A second offense carries a penalty of between $2,000 and $5,000 for each unauthorized alien, while each subsequent offense brings between $3,000 and $10,000 for each unauthorized alien.

For *failure to verify employment eligibility* there is a civil penalty of between $100 and $1,000 for each unauthorized alien.

IRCA was used to prosecute the employer in the following case.

Steiben v. Immigration and Naturalization Service
United States Court of Appeals, Eighth Circuit
Missouri 932 F.2d 1225 (1991)

Facts: Steiben incorporated Wrangler's Country Cafe, Inc., under Missouri law. During its existence, Steiben, as proprietor and chief executive officer, exercised exclusive control of his business, including the hiring and firing of employees. Steiben personally hired the three unauthorized aliens named in the complaint brought by the Immigration and Naturalization Service.

In its complaint, the INS charged both Steiben and Wrangler's with violations of the employer sanction provisions of IRCA by knowingly hiring three unauthorized aliens and for failing to complete eligibility verification forms for these and 12 others. Steiben claimed that the INS lacked authority to issue its regulation that defined "employer" to include persons who act directly or indirectly as an agent of the employer, and that only the corporate defendant (Wrangler's) could be held liable. Steiben did not contest the violations contained in the notice.

Reasoning: The court rejected Steiben's argument, holding that the exercise of the regulatory authority was valid, and Steiben was properly charged and found guilty of violating the employer sanctions provisions of the law.

Conclusion: The Immigration Control and Reform Act removes any doubt over whether employers may hire illegal aliens: They may not.

Discrimination under IRCA

Employers of three or more employees may not discriminate in hiring on the basis of *citizenship* status against *authorized* aliens.

Employers of between four and 14 employees may not discriminate on the basis of *national origin* against authorized aliens.

Penalties for Discrimination

A first discrimination offense carries between $250 and $2,000 in penalties for each individual discriminated against. A second offense, between $2,000 and $5,000. Each subsequent offense carries a penalty of between $3,000 and $10,000.

Despite IRCA's prohibitions on hiring unauthorized aliens, Title VII of the Civil Rights Act continues to prohibit discrimination against *anyone* in hiring or employment, except citizenship discrimination. Citizenship discrimination is prohibited under the IRCA. IRCA has no effect on the Fair Labor Standards Act.

Discriminatory acts expressly forbidden include the following.

1. Asking job applicants for more or different documents than IRCA requires to establish identity and work authorization
2. Refusing to accept documents that reasonably appear to be genuine
3. Intimidating applicants or employees who say they intend to file discrimination complaints

The Office of Special Counsel of the Justice Department is given access to the I-9 eligibility verification forms as well as the INS and the Department of Labor.

CHILD LABOR

Federal law prohibits hiring minors under 14 years of age. However, some states allow minors between 12 and 14 years old to work during the summers or school vacations. Minors can be employed only in nonhazardous foodservice activities, such as cashiering, selling, assembling orders, light cleanup work, light kitchen work, and washing fruits and vegetables. Forbidden hazardous activities include the operation of heavy power-driven machinery, such as food slicers, grinders, choppers, cutters, and bakery mixers.

Child labor laws are enforced by the Department of Labor. Penalties for violations of federal child labor laws include maximum civil penalties of $1,000 per violation, and additional criminal penalties of $10,000 for first violations and $10,000 plus a six-month jail term for second offenses. The civil penalty can be imposed on unintentional violators by the Secretary of Labor without any court action. The criminal penalty is reserved for willful violators and requires court action.

In 1989, the Fair Labor Standards Act was amended to authorize the Secretary of Labor to raise the civil penalty to $10,000 per violation of its federal child labor and other requirements from the previous $1,000 ceiling. The Act restricts the hours employees age 14 and 15 may work, and prohibits all employees under age 18 from performing certain hazardous duties. Once again, these penalties do not require proof of an evil purpose or willfulness on the part of the violator. Willful violations fall within the criminal sanctions noted previously.

State labor laws normally prohibit minors from working where alcoholic beverages are sold. State laws also may be stricter in the work a minor may do. Some states have more restrictions on hiring minors than the federal laws, and age requirements also vary. *When a state child labor law differs from the federal law, the law offering the greater protection to the minor will be applied.*

EMPLOYEE SCREENING

An employer may be limited in employee screening practices by two major factors: 1) the Equal Employment Practices Act at the federal, state, and local levels; and 2) the terms and conditions of a collective bargaining agreement if there is a union contract.

The test adopted by the EEOC and by various other state and local agencies in evaluating an inquiry is whether it is *necessary to evaluate an applicant's ability to perform the job.* For example, questions about a criminal record, if not barred by union agreement or by law, will be measured by how critical or sensitive the job is. For some jobs, such as dishwashing, it could be argued that this question is unnecessary. However, for a cashiering position, the question may be justified. The existence of a criminal conviction record, standing alone, should not always disqualify an otherwise capable applicant. The conviction record would have to be related in a substantial way to the job. For example, a conviction for speeding would not be grounds for a refusal to hire a cashier. **Exhibit 8.1** lists guidelines for legal employment inquiries.

Involuntary disclosures of personnel records are allowed under federal and state laws for use by applicants for employment. This should be considered heavily by managers when they are making decisions as to what information is included in employee files.

Employee Testing

Hospitality operators may want to give job applicants various tests to determine whether they are fully qualified for a job. The EEOC has issued guidelines stating that tests must be *job-related, valid*, and *reliable*. The U.S. Supreme Court has upheld the EEOC guidelines, pointing out that tests should be used only to evaluate job seekers for a specific job and not to discriminate on any other basis, such as race or religion.[22]

You may want to test potential employees, especially those seeking positions as supervisors, buyers, cashiers, and chefs. These tests can be written or in the form of a demonstration of skills. All job applicants may be tested as long as the tests can be proven to be job-related.

Employer Security

There is a trend in some states to hold owners and occupiers of land open to the public to higher standards for the protection of customers against assaults by employees.[23] This trend requires careful screening and supervision of employees, since lapses in hiring and management policies can subject employers to legal action, if someone is injured by an employee. However, employers also have a corresponding responsibility to obey all equal employment opportunity laws when screening applicants. This apparent contradiction in responsibilities can be resolved by making use of whatever employment screening policies are *legally available*.

LEGAL EMPLOYEE SUPERVISION

Union contracts and employees' right to privacy may limit employers' right to monitor employee job conduct to prevent criminal activity.[24] However, employers may use all legitimate monitoring devices to prevent criminal misconduct.

The law, while it attempts to balance the rights of employers with those of employees with regard to hiring, is more uncertain when it comes to ongoing supervision and employers' rights to protect their operations from employee theft and other employee-related security problems.

Exhibit 8.1   Guidelines for Legal Employee Screening

GUIDELINES FOR LEGAL EMPLOYEE SCREENING

DISCRIMINATORY INTERVIEW QUESTION	LEGAL, JOB-RELATED INTERVIEW QUESTION/REQUIREMENT
Do you have children? Have you made arrangements for child-care while you work? Do you need to be with your children at night? Are you married?	We require a two-year commitment to our manager training program. Are you willing to travel? Are you willing to relocate if necessary? Are you willing to work at night? Weekends?
How old are you?	Are you over 21? (for positions that require serving alcohol) Are you over 18? (to comply with any state or federal labor laws)
Do you own a car?	Do you have reliable transportation to work? Can we count on you to be punctual?
Gender inquiry or reference.	*NO MENTION*
Do you have any disabilities? Are you healthy?	Do you have any physical conditions that would prevent you from performing vital functions required of this position?
Inquiry or reference to appearance, weight, etc.	*NO MENTION*
What groups or organizations do you belong to?	Are you a member of any professional organizations? Have you received any professional or academic honors?
Have you ever been arrested?	Have you ever been convicted of a felony? (This type of inquiry must be directly related to the position or type of employment sought.)

Reprinted with permission from *Screening, Interviewing, and Selecting Employees* skillbook (Management Skills Program). Copyright © 1992 by The Educational Foundation of the National Restaurant Association.

The hospitality industry is prone to employee pilfering of food and supplies. Pilfering can be hard to detect since it often involves small amounts over a period of time.

Stealing by employees is not a phenomenon unique to the foodservice industry, but it is often felt quickly in operations that must keep inventory costs down to maintain profitability. Chapter 7 on security contains management tips for preventing many security problems. When security measures involve questioning or scrutinizing employees, operators must use their best management judgment to determine what, if any, steps are really warranted by the size of the problem.

The constant challenge to managers is to increase security without violating employee rights of privacy or union collective bargaining agreements. Operators have the right to protect their property and business from criminal misconduct, based on a history of crime or location in a high-crime area.

Polygraph Testing

The federal Employee Polygraph Protection Act of 1988 prohibits private businesses from using lie detector tests to measure the honesty of potential, current, or former employees. A limited exception allows employers to test employees during an ongoing investigation of business-related economic loss or damage. Regulations under the Act forbid employers from administering polygraph tests solely because of a sudden inventory shortage. Testing is permitted only if an investigation focuses on specific missing items and finds evidence of wrongdoing. Testing is also prohibited in situations where the employee had access to stolen property, unless one employee had sole access to that property. The Act sets the minimum standard for use of polygraphs in states having no legislation, or legislation in conflict with its requirements. A $10,000 maximum fine may be imposed on violators for each violation.

Visual Surveillance of Employee Areas

When an operation has a history of burglaries, armed robberies, assaults, or vandalism, courts and union negotiators are more receptive to the installation of visual surveillance devices, such as closed-circuit TV monitors, to protect patrons, employees, and property. These devices may be used in work areas to prevent employee thefts or other illegal activities.

Generally, when the safety and security of the premises are threatened, the employee's right to privacy may yield to the right of the employer to protect both the business and its patrons.

When there is a justified need to take steps to prevent employee crime, operators should do so, particularly if the plan is cost effective as well as legal. Operators must insist on the right to take such steps at the bargaining table during labor negotiations. Visual monitoring systems are more appropriate in work areas, including receiving rooms, locker rooms, kitchen areas, and employee parking areas, than in employee washrooms, toilets, or lounges. However, employee assaults or sales of contraband in traditionally private areas might permit the latter. Union contracts and state laws are the sources for limitations here.

If an operation has a history of crime, or is located in a high-crime area, the manager may make a strong case for installing a complete surveillance system for the entire operation, including the outdoor parking area. Although this may be somewhat distasteful to management, patrons, and employees—because the system is expensive, intrudes on privacy, and is intimidating—the evidence to date supports its legality. Audio as well as visual surveillance is governed by state and federal laws protecting eavesdropping, but these devices do not protect against otherwise unlawful criminal activities.

Metal Detectors

Manufacturers and distributors of metal products often use metal detectors in their plants or warehouses. Some large hospitality operations may use them to prevent the loss of silverware. There are presently no state laws prohibiting their use, but union contracts may prohibit them. Legality aside, they are often difficult to use and may prove more costly than replacing lost utensils.

Inspection of Employee Work Areas

Constitutional protections prohibiting unreasonable searches of private areas, such as homes, without valid search warrants do not apply to private employers' inspections of work areas.[25] The work area is not considered a private area for employees.

Periodic inspections of work areas may be the best form of visual surveillance. However, it is important to remember that the objective of any type of inspection is to protect the business from theft and other crimes. Conducting inspections for any other reason, or using unwarranted tactics in the process, is unnecessary and will most likely result in low employee morale.

Sexual Harassment

Sexual harassment in the workplace has become a major concern in every industry.[26] The hospitality industry is no exception, and foodservice and hotel operators must develop effective policies to deal with a serious problem that could result in liability.

In 1980 the EEOC issued guidelines on this subject that broadly define both the forms of harassment and the scope of employer liability.[27] The guidelines make employers liable for harassment activities by supervisors, co-workers, and patrons.

The guidelines are based on court decisions that found violations of Title VII of the Civil Rights Act existed whenever an employee agreed to a supervisor's demand for sexual favors as a condition of employment.[28] Formerly, only if the employee was fired as a result of a refusal could he or she seek court relief. However, *direct employment consequences are no longer required to establish a violation of the guidelines.* The demand itself is enough. Moreover, the guidelines also hold employers responsible for tolerating a hostile environment (sexual taunts, lewd or provocative comments and gestures, and sexually offensive touching by co-workers and patrons).[29] Employers may be liable if they were given notice of the harassment, and the harassment was related to the employment activities of the victim.[30]

Protection against harassment by patrons or non-employees is also contained in the guidelines. Employers will be liable for harassment of employees by patrons, if the employer knew *or should have known* of the conduct and failed to stop it. Top managers are responsible for the failure of supervisors to stop harassment that they knew about.[31]

The guidelines specify the steps that employers should take to prevent sexual harassment.

1. Raise the subject of harassment with all employees.
2. Express strong disapproval and develop appropriate penalties.
3. Inform employees of their right to complain and how to do so.
4. Develop methods to make everyone fully aware of the problem and how to prevent it.[32]

The following case illustrates the consequences of management's failure to deal with the problem of one harassed employee.

Norma L. Rogers v. Loew's L'Enfant Plaza Hotel
United States District Court, District of Columbia
526 F. Supp. 523 (D.C. 1981)

Facts: In September 1979, Norma Rogers was hired by Loew's L'Enfant Plaza Hotel as assistant manager of the Greenhouse Restaurant. James Deavers, manager of the restaurant, was Rogers' immediate supervisor, with whom she was required to work closely in order to assure the smooth operation of the restaurant. Rogers alleged that after being employed a few weeks, Deavers began to make sexually oriented advances toward her, verbally and in writing, which extended over a period of two months. The defendant would write her notes and letters, pressing them into her hand when she was busy attending to her duties in the restaurant, or placing them inside menus that Rogers distributed to patrons of the restaurant, or even slipping them into her purse without her knowledge.

Rogers further claimed that Deavers would telephone her at home or while she was on duty at the restaurant, and made sarcastic, leering comments about her personal and sexual life. Rogers contended that she continually rejected Deavers' suggestions and rebuffed his advances.

During this period, Rogers received what she considered to be an abusive and violent telephone call from Deavers' wife, who had apparently discovered a letter written by her husband to Rogers. Ms. Deavers warned Rogers not to become involved with her husband. Extremely disturbed by this call, Rogers urged Deavers to tell his wife that there was no relationship, other than a working one.

Rogers alleged that on occasion Deavers would pull her hair, touch her, and try to convince her to spend a night or take a trip with him. She claimed that he offered her gifts and favors and at times used abusive, crude language.

Rogers added that Deavers sometimes excluded her from meetings of the Greenhouse staff, belittled her in the presence of the staff, and refused to cooperate with her or share work-related information, generally making it difficult for her to perform her job.

During this period, the hotel managers appeared unwilling to meet with Rogers to discuss the problem, and they did not offer to remove Deavers from his position even though they seemed to know of other, similar incidents.

Rogers brought suit against Loew's and its parent corporation for both compensatory and punitive damages. The defendants moved to dismiss on the grounds that the cause of action did not relate to a civil rights violation. They maintained there was no invasion of privacy, no proof of assault and battery, no damages were provided for emotional distress, and the parent corporation was not involved in the alleged activities.

The defendants' motion was granted in part and denied in part. The court held that Rogers had sufficient cause of action to bring suit for invasion of privacy, assault, battery, and infliction of emotional distress. But it held that neither civil rights statutes nor the Thirteenth Amendment related to the cause of action. The parent corporations, which did not control the hotel management, were dismissed as defendants, and Rogers could be entitled to punitive damages if the allegation were proved.

Reasoning: Pointing to the persistent phone calls, the court found that the plaintiff's invasion-of-privacy allegation could be upheld and was covered by the law.

As for the assault and battery claim, the court said that a defendant could be liable for assault and battery if that defendant acted with intent to cause harmful or offensive contact. The hair-pulling incidents were noted by the court.

The court found that there was an absence of consent on the plaintiff's part, since she specifically told Deavers that his advances were unwanted. This was sufficient to throw out the defendants' motion to dismiss the battery claim.

As for emotional distress, the court said:

In her complaint, the plaintiff has clearly alleged conditions and circumstances which are beyond mere insults, indignities, and petty oppressions and which, if proved, could be construed as outrageous. Emotional distress and physical harm could reasonably result from the conduct of Deavers, as stated, as well as from the conduct of the hotel management in response to plaintiff's plight. A cause of action for intentional infliction of emotional distress does, therefore, lie.

Conclusion: This case is an extreme example of sexual harassment resulting in litigation. However, it illustrates the need to keep the lines of communication open between employees and management, especially regarding sexual harassment. Rather than encouraging discussion of the subject, the management here went to great lengths to avoid the issue. If management discourages such harassment and an employee still sues, the employer will be on better legal ground than if the behavior were ignored. Win or lose, such court cases are time-consuming, costly, and usually avoidable. Remember that punitive damages are now recoverable for intentional or reckless harassment.

Minimizing Liability   Three guidelines will help minimize employers' potential liability for sexual harassment. First, adopt a clear, written employment policy against all forms of harassment. Second, appoint one person to whom complaints of harassment can be reported. Finally, establish an action plan for dealing with problems when they arise.

All employers should remember that what might seem like harmless fun to one person might be interpreted as harassment to someone else, and the employer is likely to be held liable.

Terminating Employees

At some point, a supervisor may have to face the fact that an employee is just not cut out for a certain job. Transferring the employee is one solution that could be considered. In some cases, however, the only solution is termination or firing. A number of states restrict employers' right to fire employees under certain circumstances.[33]

Firing an employee is distasteful to many managers, and need only be considered when all feasible attempts to rehabilitate the employee have failed. The following guidelines not only help keep employers on fairly solid legal ground, but provide the most humane methods for carrying out an unpleasant task.

1. *Give ongoing performance feedback.* Employees must be told whether they are not meeting specific standards. Their supervisors should help them come up with a plan for improvement.
2. *Keep detailed written records.* All job-related information should be documented, shared with the employee, and filed confidentially.
3. *Give a reason for the termination.* If an employee is terminated, he or she deserves to know why. Give several reasons, if possible.

4. *Give the employee a chance to ask questions.* Termination is a hard fact to face, and the employee must sort it out in his or her mind. Give him or her time to do that. Fired persons who are encouraged to talk about it will be less likely to take legal action than one who is whisked out the door. Tell them honestly about how you will handle their future job references, and about unemployment compensation.

5. *Stay calm.* Raising your voice will antagonize the employee and make the situation more unpleasant. Even if the employee raises his or her voice, the employer should remain unemotional.

6. *Inform the employee of his or her good points.* Most people are not completely bad employees but have one or two points that make them unsuitable for a job. By describing their good and bad points, an employer can salvage that person's self-esteem.

7. *Provide some sort of severance plan.* Unless an employee has stolen from the company or committed some similar activity that prompted the firing, there should be some kind of severance pay, final paycheck, or benefits payment.

8. *Keep a written record of the termination.* A written record of the termination should be added to the employee's file.

9. *Let go.* Once the employee has left, the employer's task is over. Carrying on a personal vendetta or bad-mouthing an employee to a prospective employer invites legal action. When it comes to employee references, you need only give the dates of employment, the salary, and, in some cases, the attendance record. In addition, consider two commonsense tips: 1) do not delegate the job of firing someone else unless it is *directly* his or her responsibility, that is, that person is the employee's immediate supervisor; and 2) do not tell anyone except the supervisor beforehand. Word might get back to the employee sooner than it should.

In times of high unemployment, it appears people are more willing to take employers to court over the loss of work. In the following case, the firing, coupled with the surrounding circumstances, resulted in a legal hassle for the foodservice employer when the alleged damages were "emotional distress." (Note the inconsistency in the employer's two "reasons" for the firing.)

Shirley Ann Counce v. M.B.M. Company, Inc.
Court of Appeals of Arkansas
597 S.W.2d 92 (1980)

Facts: Shirley Counce, a waitress, was working with one other employee on the night shift and had responsibility for the cash register, including inventory of funds in the register at the time of closing. The other employee had the key to the establishment and was responsible for locking up. The following day, the manager called Counce at home and told her there was a shortage in the cash register. Later in the day, the manager called Counce again and told her she was laid off because there was too much counter help. There was evidence there was no excess counter help. M.B.M. withheld $33 in excess of normal deductions from Counce's last pay. She took a polygraph test at the request of M.B.M. which indicated she had no connection with the money shortage. Notwithstanding this, M.B.M. did not re-employ her or pay her the $33 withheld from her wages.

When Counce applied for unemployment benefits, M.B.M. sent a report to the Employment Security Division stating that she was terminated "because of numerous customer complaints and because she failed to follow company policy." Counce sued M.B.M.

Jerrell Moss, another supervisor, testified that he had received telephone complaints from two customers about the service of a waitress meeting Counce's description. Moss could not give the names of either of those making complaints.

M.B.M. filed for a summary judgment by the court, claiming there was no cause of action for the plaintiff's complaint. The summary judgment was given in favor of the defendant. The plaintiff appealed. On her appeal, the plaintiff noted that a cause of action existed based on the breach of employment contract and the wrongful acts of the employer, which caused her "emotional distress." The appeals court reversed the summary judgment and ruled in favor of Counce.

Reasoning: The court found that a breach of contract did not exist because there was no infringement of a legal right owed the plaintiff. However, on the allegation of mental distress, the court found differently, citing cases where damages were upheld for such distress. The court found that since there were factual issues in dispute (i.e., the reason for the firing, the problem with the missing money), a summary judgment was improper, and the plaintiff should have been allowed to argue her case fully.

Conclusion: This case illustrates two points: 1) Termination for reasons that are nebulous, false, or simply unlikely to stand up in court can result in costly legal problems for the employer. 2) Follow-up "bad-mouthing" by an employer could precipitate a legal action that might not otherwise occur.

Imperial Diner, Inc. v. State Human Rights Appeal Board
Court of Appeals of New York
417 N.E.2d 525 (1980)

Facts: After being subject to abusive anti-Semitic remarks repeatedly made to her by her employer in the presence of customers, a female waitress demanded an apology. When her employer refused, she quit and complained to the State Human Rights Commission. Upon hearing the case, the Commission ordered her reinstatement with back pay for two years, the payment of compensatory damages, and a written apology. The State Human Rights Board affirmed. The employer sought judicial review. The Court of Appeals affirmed the Commission's order.

Reasoning: The high court decided that this was a blatant example of willful religious discrimination, not justified or condoned by the victim's work record, which was exemplary. Therefore, the Commission was acting fully within its powers in imposing the reinstatement with two years' back pay and the written apology. Three dissenting judges felt the order requiring a written apology was not appropriate because that requirement infringed on the employer's First Amendment right to free speech. The dissenters also argued that the back pay award was excessive, because the victim had secured other employment during the two-year period.

Conclusion: This case represents an extreme example of religious discrimination in the workforce. The high court's affirmance on these facts sends a clear message that the full powers of the appropriate state agency will be brought to bear and that the courts will not hesitate to affirm the penalties imposed in the absence of clear proof of extenuating circumstances. It is fair to say that most courts faced with this issue would reach a similar conclusion. This decision sends an unmistakable message that religious discrimination in the form practiced here will not be tolerated and that courts will not hesitate to deal firmly with violators.

UNIONIZATION: RESPECTIVE RIGHTS AND DUTIES OF EMPLOYER AND UNION

Employees have been unionizing to protect themselves since the beginning of the Industrial Revolution. Still, the foodservice industry is not as unionized as other industries, such as textile mills and car makers. This may be due partly to the high employee turnover typical of the industry. Or it may be attributable to the larger number of small commercial operations in the industry with fewer employees per unit than, say, automobile manufacturers. However, the interest in unions in the industry is growing, and it is important for employers to both their own rights and the rights of employees regarding unionization and collective bargaining.

Under the Labor Management Relations Act of 1947 (popularly known as the Taft-Hartley Act), employees have five basic rights.[34]

1. To form or join a labor organization
2. To select bargaining agents
3. To bargain collectively with the employer
4. To engage in acts of mutual support or protection
5. To refrain from any union activity

Employees have the right to hold union meetings, distribute literature, petition employers and the government, strike, picket, and engage in any legal activities to promote their common interests. But these rights are subject to restrictions. A court may stop any of these activities if carried out unlawfully. Plant or premises seizures, sit-down strikes, and violence are prohibited.

A union or pro-union group cannot force any employee to participate in union activities. However, employees may be required to listen to union solicitors off the premises, to pay union dues, and to join a union—if one is voted in—as a condition of continued employment. Under right-to-work laws in some states, the last requirement is prohibited, but even in these states non-union employees may still be required to pay union dues.

Prohibited Employer Labor Practices

Under the Taft-Hartley Act, the following labor practices by employers may be halted by a court order.

1. Employers may not interfere with employees' rights to unionize and take part in union activities. They may express an anti-union viewpoint if the statements do not contain threats of unfair statements. They may *not* threaten to close the business if a union is formed, or threaten to give a wage increase as a reward for voting against a union.
2. They may not dominate or interfere with the formation or administration of any union. *Company-owned unions are forbidden.*
3. Encouraging or discouraging union membership by discriminating between union and non-union people in hiring, job tenure, and job conditions is prohibited.
4. Employers may not discharge or discriminate against employees because they file unfair labor practice charges or testify against an employer. This provision of the Act prohibits the "blacklisting" of employees for exercising their legal rights. *It does not prevent an employer from disciplining employees who neglect their duties or are absent from work because of union activities.*
5. Discharging employees for legal strikes is forbidden. A legal strike by lawful, nonviolent methods in protest of an unfair labor practice may not result in discharge of the strikers or their replacement. The National Labor Relations Board (NLRB) has the power to order reinstatement of striking workers with reimbursement for wages lost because of an employer's illegal actions.[35]

Can the union and employer agree to a no-strike clause in their collective bargaining agreement and thereby eliminate the threat of strikes? The answer depends on the kind of strike being conducted by the union. A no-strike clause may prohibit an *economic strike*—one concerned with work hours, wages, and working conditions. A no-strike clause *may not* prohibit an *em-*

ployee unfair labor practice strike—a strike against an employer for denying workers the right to bargain collectively, for discriminating against union members, or for interfering with or restraining workers' attempts to bargain collectively.[36] The federal courts, which have the power to hear and determine controversies under the National Labor Relations Acts, will not allow employers to justify violations of the Act by the use of a no-strike clause. Neither employer nor union unfair labor practices are made legal under the guise of a no-strike clause.[37]

What good is a no-strike clause? It gives an employer the right to hold the union financially responsible for business interruptions and other losses resulting from a strike. The no-strike clause is a means of making the union account in dollars for illegal strikes. The clause itself cannot stop the union from striking, but it can make any illegal strike very costly to employees, and hence less likely.

A *wildcat strike* is a local strike, not sanctioned by the parent union or voted on by a majority of the members. *It is illegal*. The NLRB can halt it and will uphold the firing of the striking employees. In case of a wildcat strike, a no-strike clause may permit an employer to recover monetary damages from the union. The no-strike clause serves to caution unions that illegal conduct resulting in economic harm to the company will have to be paid out of the union treasury.

Prohibited Union Labor Practices

To prevent unions from abusing their powers over employers, the Act also contains the following prohibited *union* unfair labor practices. Union employees may not do the following.

1. A union may not restrain or coerce employees. It may not use force, violence, or threats to compel an employee to join a union or to comply with union requirements. It may not blacklist employees for refusing to support a union or try to persuade employees to withdraw unfair labor practice charges.
2. A union may not refuse to bargain with an employer's association or restrain the employer in the selection of the employer bargaining representative.
3. A union may not cause an employer to discriminate against an employee because of refusal to join a union.
4. Refusing to bargain collectively with an employer is illegal. A union cannot bargain with the support of a majority of its members, or for an illegal purpose.
5. Union members may not encourage, threaten, or force any employee to strike or to refuse to handle work or products when the purpose is to compel an employer to join a labor union or employers' association; to bargain with a union not yet certified as such; to bargain with a union other than

the one certified to represent the employees; or to prefer one union over another by assigning work on a preferential basis. Unions are forbidden to engage in a *secondary boycott* of a neutral employer to force that employer to harm another employer with whom the union has a dispute.[38]

6. Requiring employees joining a union to pay excessive fees is prohibited.

7. Causing an employer to pay or agree to pay for services that are not performed or not to be performed is forbidden. This requirement is intended to eliminate *featherbedding*, the practice of paying employees for work that is not done. However, contracts for extra work not actually *needed*, but agreed to by the employer, are allowed.

8. Unions may not picket or threaten to picket any employer to force that employer to accept or to bargain with a union representative, unless the union is properly certified to do so. The purpose of this part of the Act is to prevent the employer from getting caught in a dispute between two or more unions over which union will represent employees.

   Informational picketing outside a restaurant by a union to advise the public that their members are not employed there is permitted. But if actual employees refuse to cross such a picket line, then they violate the law.

9. Unions may not enter into any agreement with an employer where the employer agrees not to handle or deal in any of the products of another employer. Such a refusal to deal with other employers is called a *hot cargo clause*.

Filing a Complaint

Any employee or employer having a complaint can file a *notice* with the NLRB within six months after the activity occurs. If the charge has merit, the NLRB will issue a *complaint*, which will result in a *hearing*. If the NLRB finds that a violation of the Act has occurred, it may restore the parties to the conditions that existed before the start of the unfair practice. Either party may appeal an NLRB decision to the federal appeals court having jurisdiction, and petition the U.S. Supreme Court to review the case.

The Act provides no penalties, either civil or criminal, for unfair labor practices. This means that *no damages can be recovered by victims, nor any fines or jail terms imposed*. However, a finding of a pattern or practice of violations will allow courts to impose heavier penalties when authorized by other provisions of the Act. A practice of violations creates a presumption of *intentional misconduct*.

Penalties are provided, however, for disobeying an *NLRB order* to correct a violation after one has been found. Like any other administrative order having the force and effect of law, an employer's or employee's failure or refusal to comply with an NLRB order may result in punishment by a fine, or, in extreme cases, imprisonment. However, the NLRB cannot alone make a refusal to obey their order of civil violation or crime. The NLRB must go to court to enforce the order if employers or employees refuse to comply with it. If on court review the order is reversed, it is unenforceable. If the reviewing court upholds the order, then either an employer or employees are *in contempt of court* if either fails or refuses to obey the order. It is contempt of a court order that may cause criminal or civil penalties.

EMPLOYEE RELATIONS

Maintaining good relations with employees is essential for all hospitality employers, since many employees are in direct contact with patrons, who are the primary source of income. Any misjudgments by management may result in poor employee morale, which will probably be reflected in indifferent employee attitudes toward patrons. In our highly competitive industry, an operation's profit or loss statement cannot help but suffer without constructive and consistent employee management policies.

The role of the law is to establish and enforce the rights and responsibilities of employers and employees to each other. These laws evolved from a common law, hands-off attitude, giving employers sole discretion to hire, fire, promote, and determine working conditions, to a series of federal and state regulations that restrict some management policies. Employers can no longer do as they please regarding employees. Where, however, the law does not intrude, they remain free to follow the dictates of the marketplace and their own conscience.

To enforce consistent management policies, managers should prepare and circulate an employee manual, even if it is simply a stapled, photocopied handout. It should reflect management policies and promote consistent and fair treatment of all employees. The manual should be shared and discussed with all employees so that any questions can be dealt with early. Topics to include in an employee manual are listed in **Exhibit 8.2**.

All work-related incidents concerning employees should be described in writing, with a copy in the employee's file. This includes promotions, disciplinary actions, terminations, and work appraisals.

Exhibit 8.2   Topics to Include in an Employee Manual

TOPICS TO INCLUDE IN AN EMPLOYEE MANUAL

Employment Policies

- Absence from work
- Schedule substitutions and shift trading
- Vacation and sick days
- Paid holidays
- Overtime
- Tips
- Pay periods
- Shift changes
- Time cards
- Performance appraisals
- Wage/salary reviews
- Breaks

Rules and Procedures

- Dress code
- Illegal activities
- Grievances
- Disciplinary procedures
- Probationary policies
- Causes for dismissal
- Emergencies (injuries, fires, natural disasters, robberies, etc.)
- Safety rules
- Off-duty time at the operation
- Friends visiting the operation
- Personal telephone use

Benefits

- Medical and dental insurance coverage
- Sick leave and disability
- Meals
- Pension, retirement, and/or death benefits
- Profit sharing
- Retirement plans

Other Topics

- Employee and locker areas
- History and mission of the organization
- Organizational and departmental structures (include a graphic organization chart)
- Job description
- Where to enter and leave the facility
- Smoking and nonsmoking areas
- Restrooms
- Breakage
- Parking
- Training opportunities
- Employee assistance program
- Job openings and postings

Reprinted with permission from *Orienting New Employees* skillbook (Management Skills Program). Copyright © 1992 by The Educational Foundation of the National Restaurant Association.

SUMMARY

An employer's first responsibility is to comply with federal and state equal employment opportunity laws in screening and promoting employees. Federal laws, such as the Equal Pay Act of 1963 and Title VII of the Civil Rights Act of 1964, set the pattern, but may not be as inclusive as many state and city laws. State and local laws may cover other areas not regulated by federal acts. In some cities, employees may be protected from discrimination based on their sexual orientation or on the fact that they may be recipients of welfare. Federal employment laws do not cover these areas. Once hired, legal aliens are covered by federal equal employment laws.

Legal defenses for violations of civil rights laws, normally based on proving a *bona fide occupational qualification (BFOQ)* for the job or proving *business necessity*, are narrow and hard to establish under federal and comparable state laws. The costs and bad publicity involved in defending a discrimination case, regardless of the outcome, oblige employers to *prevent* discriminatory practices.

Employer duties also carry over into areas of employee supervision. While it is important to maintain a secure operation, surveillance measures must respect employee rights and privacy.

Child labor laws come from both state and federal governments. The federal government usually sets conditions under which children may work. The state laws usually cover the number of hours an employee may work. Generally, the laws are concerned with the environmental conditions of jobs for minors and may prohibit their being given hazardous jobs handling hazardous equipment, such as meat slicers and dishwashers.

Sexual harassment in the workplace is of growing concern in all industries. This type of harassment by supervisors, co-workers, or patrons should not be tolerated by employers.

QUESTIONS

1. A question on a job application asks, "Where were you born?" What is wrong with this question? Is it legal? If not, what law does it violate? What is a related question that could legally be asked?

2. Marty of Marty's Marvelous Pies gives a baking test to applicants for baking positions. He also gives a grammar and math skills test to all applicants for every job, including dishwashers. Is there anything questionable about either of these tests? Why or why not?

3. When would a state child labor law be applied over a federal child labor law?

4. Maureen Crown is the manager of a posh nightclub with a large staff. What steps would you recommend she take to prevent sexual harassment of employees and to minimize liability?

5. Present an argument in favor of a no-strike clause in a union contract from the *employer's* viewpoint.

NOTES

1. 29 U.S.C. sec. 210 *et seq.* (1976 & Supp. IV 1980 & Supp. V 1981).

2. *Schultz v. Wheaton Glass Co.*, 421 F.2d 259 (3d Cir.), *cert denied*, 398 U.S. 905 (1970).

3. 29 U.S.C. sec. 206(d)(1) (1976).

4. *Corning Glass Works v. Brennan*, 417 U.S. 188 (1974).

5. 42 U.S.C. sec. 701 *et seq.* (1976 & Supp. IV 1980 and Supp. V 1981). Discrimination on the basis of sex, religion, and national origin was prohibited by the Equal Employment opportunity Act amendment to the Civil Rights Act in 1972. Congress amended the Civil Rights Act in 1978 to include pregnancy.

6. B. Schlie and P. Grossman, *Employment Dis-*

crimination Law 249–50 (1976). "Interim earnings or amounts earnable with reasonable diligence by the person or persons discriminated against shall operate to reduce the back pay otherwise allowable." Civil Rights Act of 1964, *amended* 1972, sec. 706 (g), 42 U.S.C. sec. 200e-5(g) (1976). The 1972 Amendments to the Civil Rights Act limit back-pay liability to a date not more than two years prior to filing of a charge with the Equal Employment Opportunity Commission. *Id.*

7. See *Kamberos v. GTE Automatic Elect., Inc.,* 603 F.2d 598 (7th Cir. 1979).

8. *Id.* Statutory authority is found at Civil Rights Act of 1964, *amended* 172 sec. 706(g), 42 U.S.C. sec. 200e-5(g) (1976).

9. Schlie and Grossman, *supra,* note 6, at 1258–59.

10. 42 U.S.C. sec. 1981 (1976). See *Johnson v. Railway Express Agency, Inc.,* 421 U.S. 454 (1975).

11. Civil Rights Act of 1964 sec. 706(k), 42 U.S.C. sec. 2000e-5(k) (1976). See *Newman v. Piggie Park Enterprises, Inc.,* 390 U.S. 400 (1968).

12. 42 U.S.C. sec. 2000e-3(a) (1976).

13. *Id.* sec. 703(e).

14. Schlie and Grossman, *supra,* note 6, at 279. Employment of a French chef in a French restaurant was cited in the Senate Interpretive Memorandum on Title VII. Also see *Diaz v. Pan American World Airways,* 442 F.2d 385 (5th Cir.), cert. denied, 404 U.S. 950 (1971).

15. *Id.* (male airline stewards).

16. *Willingham v. Macon Telegraph Publishing Co.,* 507 F.2d 1084 (5th Cir. 1975).

17. Schlie and Grossman, *supra,* note 6, at 99–100.

18. *Carroll v. Talman Federal Sav. and Loan Ass'n of Chicago,* 604 F.2d 1028 (7th Cir. 1979), *cert. denied,* 445 U.S. 929 (1980).

19. 29 U.S.C. sec. 621+34 *et seq.* (1976 & Supp. IV 1980 & Supp. V 1981).

20. 29 U.S.C. sec. 794a(A)(1) (Supp. IV 1980). See also *id.* secs. 701–94.

21. Human Rights Law, N.Y. Exec. Law sec. 296(2) (McKinney 1982).

22. *Griggs v. Duke Power Co.,* 401 U.S. 424 (1971).

23. J. Sherry, *The Laws of Innkeepers, Third Edition,* sec. 11:1.

24. See the following labor arbitration cases: *Electronic Instrument Co. (Eico),* 44 Lab. Arb. 563 (1965); *EMC Corp.,* 46 Lab. Arb. 335 (1966); and *Colonial Baking Co.,* 62 Lab. Arb. 587 (1974) involving visual surveillance of employees by management. Only in *Eico* did the arbitrator rule in favor of halting management's visual surveillance practices (closed-circuit TV).

25. See *Rakas v. Illinois,* 439 U.S. 128, 140–48 (1978) (examination of privacy doctrine); *Thomas v. General Elec. Co.,* 207 F. Supp. 792 (W.D. Ky. 1962).

26. Waks and Starr, *Sexual Harassment in the Work Place: The Scope of Employer Liability,* 7 Employee Relations L.J. 369 (1981–82).

27. 45 Fed. Reg. 74676 (Nov. 10, 1980).

28. 42 U.S.C. sec. 2000e *et seq.* (1976); *Tomkins v. Public Serv. Elect & Gas Co.,* 568 F.2d 1044 (3d Cir. 1977); *Barnes v. Costle,* 561 F.2d 983 (D.C. Cir. 1977); *Garber v. Saxon Business Prods., Inc.,* 552 F.2d 1032 (4th Cir. 1977) (*per curiam*).

29. 29 C.F.R. sec. 1604.11(a) (1982).

30. *Ludington v. Sambo's Restaurants, Inc.,* 474 F. Supp. 480 (E.D. Wis. 1979).

31. 29 C.F.R. sec. 1604.11(3) (1982).

32. *Id.* sec. 1604.11(f).

33. The 14 states known to limit employers' rights are Arizona, California, Illinois, Indiana, Kentucky, Massachusetts, Michigan, New Hampshire, New Jersey, Ohio, Oregon, Pennsylvania, Washington, and West Virginia.

34. 29 U.S.C. sec. 141 *et seq.* (1976 & Supp. IV 1980 & Supp. V 1981).

35. *Id.* sec. 151 *et seq.*

36. *Id.* sec. 158 defines unfair labor practices. See also *Mastro Plastics Corp. v. NLRB,* 350 U.S. 270 (1956).

37. 29 U.S.C. sec. 158 (1976).

38. *Id.* sec. 158(b)(4)(A).

Employer-Employee Administrative Issues

OBJECTIVES

After reading this chapter, you should be able to:

1. Discuss insurance options and requirements.
2. Explain requirements of the Occupational Safety and Health Administration (OSHA).
3. Outline employer obligations, regarding employee wages, taxation, and tax credits.

Case in Point

A man identifying himself as an Occupational Safety and Health Administration inspector came to Frank's High Street restaurant at noon one day when the business rush had just begun. He asked to be shown around the facility, and for the main oven to be switched off so he could check the interior safety features. After touring the facility, he told Frank that he found three violations and that Frank would hear from the local agency "sometime." What legal options could Frank have used to protect his rights in this situation?

First, Frank could have asked the man for credentials. For all he knew, the man could have been casing the operation to find the safe. Second, Frank should have arranged with the inspector to come at a more convenient time, when the restaurant was not quite so busy. That would have given Frank the opportunity to accompany him to point out safety compliance measures and answer questions. Finally, Frank could have asked the inspector to discuss the specific violations.

ADMINISTRATION AND THE LAW

There are administrative tasks involved in managing hospitality employees. Some of them are a matter of employer choice, such as whether to require certain employees to obtain bonding insurance. Others, such as minimum wages and taxation, may be regulated by both federal and state governments. This end of employee management may have legal implications for operators and requires them to be very astute when it comes to record keeping and supervision.

EMPLOYEE INSURANCE

There are various kinds of employee-related insurance plans. Some are designed to protect employers, from liability for employee actions. Others compensate employees for injuries suffered on the job.

Liability Insurance

Liability insurance provides protection or *indemnification* against losses for damages resulting from injuries to a person or property by employees. (See Chapter 5 on liability.) This insurance is optional and employers must use their own judgment as to whether or not to use it, depending on the size of the operation, the turnover, and the type of job. Operators who serve alcohol, offer flaming dishes, or provide live entertainment may want to consider whether this insurance would be helpful. Others may want to rely only on good employee training to prevent liability.

Surety bonds, issued by a pending company, make the company liable for the payment of an employee's, or another person's, debt. Liability to pay is primary; the creditor need not sue the debtor first. This is important in cases where the debtor (employee) has minimal financial resources. A common type of surety arrangement is *fidelity insurance*.

Fidelity Insurance

Fidelity insurance protects employers against embezzlement—taking property entrusted to one's care for personal use—by an employee. It is normally issued for key people who handle or transport cash receipts, such as a restaurant's night supervisor. This form of insurance, called a *fidelity bond*, is issued by a bonding company.

As with liability insurance, the same exceptions and defenses are available to the bonding company, except that with bonding, criminal misconduct, and carelessness are covered. Employer criminal misconduct, however, is an area from which the carrier is exempted in most policies.

Fidelity coverage is quite expensive. Both the employer and the bonded employee are very carefully screened. It is solely within the discretion of the bonding company to accept or reject an application.

WORKERS' COMPENSATION

Before the development of major industrial growth, society was rural, mostly agricultural, and economically self-contained. Work was simple and the family head was responsible for the few on-the-job accidents that occurred. In the 18th century, the development of machines and the size and complexity of business enterprises created industrial hazards unknown to the pre-industrial period.

During the transition to an industrial economy, increased numbers of job-related injuries required remedial legislation. Two major types of laws were passed to cope with this problem: workers' compensation laws and Occupational Safety and Health Administration (OSHA) statutes.

Under common law, a worker injured on the job had to sue his or her employer to recover compensation. To succeed, the worker had to prove that negligence or intentional misconduct by the employer caused the injuries. The employer could defend on the ground that the employee's own fault or inattention caused the injuries, or that the employee voluntarily assumed the risks of injury by accepting the job. The employer could also say that the injuries were caused by the carelessness or misconduct of another employee. Both defenses could defeat the injured employee's lawsuit.[1] Very few employees were able to recover under these rules of common law, and many lost their jobs as a result either of lawsuits or of the injuries.

The first effort to provide workers' compensation was the federal *Workmen's Compensation Act* of 1908. Today every state has adopted some form of worker compensation. However, federal law still covers federal employees. The state statutes are not uniform, but certain basic similarities do exist. In both situations, the statutory remedies are exclusive, except where the employer intentionally injures an employee. Then the employee can sue the employer for compensatory and punitive damages.

Each statute provides a plan to pay employees or their dependents for death, injuries, or physical disabilities caused directly by their employment. The carelessness of the employee is no longer a defense, nor need the employee establish any fault or misconduct on the employer's part. Further, an injured employee cannot sue the employer for injuries covered by worker compensation.

Payable amounts for injuries are fixed by state or federal law, either at a definite sum for each type of injury or some percentage of the employee's weekly wage or salary. Compensation includes coverage of medical, doctor, hospital, and rehabilitation expenses, as well as payments for diminished earning capacity resulting from the injury. No payment is provided for injury-related pain and suffering, disfigurement, or loss of *consortium* (services) to a spouse.

For an employee to be covered under compensation, the injury must arise in the scope of the employee's duties to the employer. One test of the term "in the scope of" is whether the employee was acting for the benefit of his or her employer at the time of the injury. This time can include commuting time, mealtime, and washup time. It is an elastic term, depending on the facts of each case. The U.S. Supreme Court has rejected any narrow definition and held that it is sufficient if the condition of employment creates a *zone of special danger* out of which the injury arises. A worker injured while moving cases of wine in a storeroom is covered, as is the same employee mugged on the way to deliver nightly profits to the bank.[2]

An injured employee is virtually compelled to claim worker compensation benefits, since there is no legal alternative against the employer. But the employee *may sue a third party* who caused the injury and still recover worker compensation. In such cases, a *lien* or claim by the Workmen's Compensation Board is available to be applied against the third party, to reimburse the Board for any compensation payments made to the employee. Neither the employee nor the employer may initially have a court decide the merits of any claim. This is the function of the state worker compensation hearing board. However, either side may appeal any adverse ruling to an appropriate reviewing court.

The courts recognize certain employer defenses to workers' compensation claims. An employee who deliberately causes an injury in order to collect benefits will be denied payment. Injuries caused by worker intoxication are also causes for denial of benefits.

Workers' compensation insurance *must* be carried by every employer. A failure to carry it, whether or not intentional, is a crime in many states, making violators subject to fines, and even imprisonment. If no insurance exists, the Workmen's Compensation Board may sue the employer for the value of any payments the Board makes to an employee.

As with any insurance, insured parties are subject to an *experience rating* based on the number and frequency of claims filed and paid. To keep this necessary business cost reasonable, it is in every employer's favor to make every effort to *prevent* employee accidents. **Exhibit 9.1** shows administrative guidelines for minimizing workers' compensation claims.

OCCUPATIONAL SAFETY AND HEALTH

Workers' compensation laws can only *compensate* for job-related deaths and injuries. They do not help *prevent* deaths or injuries caused by unsafe working conditions.

Employer Duties

Confronted with the spiraling costs to industry in the form of lost production time, lost employee wages, and the increasing cost of workers' compensation insurance premiums, Congress enacted the federal *Occupational Safety and*

Exhibit 9.1   Workers' Compensation Administrative Measures

1. *Prepare a statement of policy.* Put this in writing and base it on concern for the mutual prosperity of the company and its workers. Cultivate the understanding that "safety is high priority."
2. *Set goals.* After implementing the safety program, establish goals for the reduction of workplace injury and illness. Schedule periodic meetings to discuss achievement of goals and establish new ones.
3. *Keep records.* Develop a comprehensive and accurate record keeping system. Good information can only arise out of skillful accident investigation.
4. *Outline the responsibilities of managers and supervisors.* Designate responsibilities for workplace safety at each level of management and supervision. Include a general statement of safety duties in the job description.
5. *Enforce the rules and policy.* To a limited extent, rules can control worker behavior, which is a causative factor of accidental injuries. Encourage worker input during the rule-making process.
6. *Investigate accidents.* Make sure supervisors seek the facts and get all of the pertinent information about cause and effect. Workers' compensation is a no-fault system, investigation should not be performed to place blame or punish. The goal is to remove accident causes, not people.
7. *Provide job analysis and training.* In many workplaces, safety training consists of a long list of "don'ts." Efforts to design on-the-job training should be guided by three principles.
   - Keep it positive.
   - Integrate safety concerns with production concerns.
   - Specify critical procedures only.

   It is especially important to analyze jobs and prescribe work procedures where there is hazardous equipment or a history of near-miss accidents.
8. *Hold safety meetings.* Reinforce the idea that safety is part of doing business by talking about it along with production, maintenance and marketing concerns. Effective communication is a process of information exchange as it encourages and stimulates worker participation.
9. *Conduct inspection and correction.* Periodic inspection, hazard identification and correction are functions which are traditionally the mainstay of workplace safety programs.
10. *Give managers authority and accountability.* Make sure managers have the authority to take the action necessary to accomplish the tasks for which they are responsible. In some instances, a supervisor is held accountable for situations over which they have no direct control.

    Accountability is critical to safety performance. Supervisor performance evaluations should measure effectiveness in carrying out safety program responsibility. If possible, tie achievement of safety program goals to pay increase and promotion opportunities.
11. *Hire the right person for the job.* One of the most direct means of preventing injury and illness is to hire and assign the right person to the right job. Workers whose mental and physical capabilities are ill-matched to job demands operates at greater risk to their health and safety. Since the passage of the Americans with Disability Act (ADA), extreme caution must be used on hiring practices. One must understand the job and learn about the applicant.
12. *Involve workers in policy development.* Success relies heavily on the cooperation and contribution of workers. A safety committee should include workers in formulating policy and identifying and correcting program deficiencies and physical hazards.
13. *Solicit worker suggestions.* A good way to get workers involved in safety is to solicit and take action on their suggestions, by doing the following:
    - Advertise that management welcomes workers' ideas.
    - Acknowledge the receipt of a suggestion immediately.
    - Evaluate the merit of the suggestion.
    - Implement the idea, or explain to the worker why it cannot be implemented.
    - Involve the area supervisor in implementation.
    - Announce the change that resulted and the benefit to workplace safety.
    - Give credit to the worker who suggested the change.
14. *Establish incentives.* Many kinds of incentives—safety contests, prizes and banquets—can have a substantial impact on an employer's health and safety problems. However, they should supplement, not substitute for other aspects of a comprehensive safety management system.

SOURCE: U.S. Department of Labor

Health Act of 1970.[3] This comprehensive statute encourages both employers and employees to reduce job hazards by voluntary cooperation. To support this joint effort, Congress established responsibilities for employers.

1. *General*—Employers are required to provide employees with a workplace free from hazards that may cause death or serious injury. You must act on your own to remedy safety violations. *This duty is in addition to and not a substitute for state or local safety requirements.*

2. *Safety standards*—Employers must comply with any safety and health standards established by the Occupational Safety and Health Administration (OSHA), the regulatory agency that enforces the Act. The standards are often complex, not to mention costly to implement. Failure to comply, however, may be even more costly.

3. *Records*—Employers should keep a record of every accident involving employees. A report of any accident causing death *or* resulting in injury to five or more persons *must* be filed immediately with the Department of Labor. Employers must post OSHA notices advising employees of their legal rights regarding safety.

   The filing deadline for OSHA is 24 hours and the file must be maintained in log #200. OSHA inspections, to check its own safety standards, involve specific employer duties based on the nature of the industry (safety guards on kitchen equipment) and a general duty to provide a work environment free from recognized hazards that are causing, or likely to cause, death or serious physical harm to employees. The general duty is violated only when the employer knows, or should have known, of the dangerous condition or activity.

   OSHA inspections, whether triggered by an employee or a third-party complaint, must meet constitutional safeguards. All inspectors must present an administrative search warrant before conducting an inspection.

   If a violation is found, OSHA can issue a notice of *de minimis* (not serious) violation, those having no direct or immediate relationship to the health or safety of workers or of the work place. These violations are issued no citations or penalties. For more serious violations, OSHA can issue a written citation that requires the employer to correct the problem within six months. The citation becomes final unless the employer contests it within fifteen days. Contested citations are reviewed by the Occupational Safety and Health Review Commission (a three-member body appointed by the president). Judicial review of Commission decisions is also provided by appeal to the appropriate federal circuit court of appeals. The secretary of labor can seek a temporary restraining order if the hazard cannot be promptly addressed by regular OSHA procedures.

4. *Safety programs*—Employers are encouraged to establish a safety checklist to discover and correct job hazards, and to train employees in on-the-job safety. Training should include all current OSHA compliance procedures and additional procedures recommended by OSHA. The operators

specific safety rules should be posted and periodically reviewed with employees. Continuous supervision and education for safety are management's responsibility.

5. *Inspections*—An OSHA compliance officer is given broad authority to conduct an inspection of any work area at reasonable times and to interview employees. However, a provision of the law giving OSHA inspectors the right to conduct an inspection over an operator's objection—without a search warrant—has been held unconstitutional by the U.S. Supreme Court.[4]

6. *Penalties*—Civil and criminal penalties are provided for violations of OSHA safety standards and failures to comply with OSHA violation notices (*citations*). The maximum penalty is $10,000 for each intentional or repeated violation of OSHA standards. Failure to correct a violation or to prepare required reports, can result in a fine of up to $1,000. Willful violations that result in death to an employee are punishable by a maximum $10,000 fine and/or imprisonment of up to six months. Fines and jail terms double for second offenders. Any person who informs an employer about a planned compliance inspection by an OSHA officer can be fined up to $1,000 and/or imprisoned for up to six months.

These penalties are independent of penalties prescribed by state or local safety laws.

The following factors are considered by OSHA in fixing any penalty: 1) the size of the business, 2) the seriousness of the violation, 3) the good faith of the employer (violation was not intentional), and 4) the history of violations by the employer.

Employer Rights

1. *Exemption from compliance*—OSHA also spells out the rights of employers. If an operator cannot comply with a certain standard immediately, he or she may request a *variance*, or an exemption from the standard. To obtain the variance, it must be shown that the operator is unable to comply at present, is taking all possible steps to keep employees safe from any hazards, and will have an effective safety program eventually. Variances are limited to no more than one year in length, but can be renewed for good cause, especially for delays due to circumstances beyond the operator's control.

2. *Establishment of standards*—Operators have the right to present views on any existing or proposed OSHA standard to the National Advisory Committee on Occupational Safety and Health, which is made up of employer, employee, and professional safety representatives. They also have the right to comment on any proposed standards before they become law.[5] They are entitled to request a public hearing on all proposed standards.

Once any proposed safety standard is adopted by OSHA, it has the force of law, and the courts will enforce it as written, unless it is found to be arbitrary in enforcement or application. The normal rule of court review is whether there is substantial evidence to support the standard. If there is, the standard is binding on the courts. Courts will not typically second-guess OSHA's administrative expertise in safety.

3. *Compliance inspections*—Employers have certain rights regarding inspections that OSHA is required to respect.
    a. The right to review the credentials and identification of the inspector before allowing any inspection
    b. The right to insist that the inspection be conducted at a reasonable hour
    c. The right to require the OSHA inspector to conduct the inspection in a reasonable manner, so as not to disrupt your food and beverage production
    d. The right to present evidence of your good faith in showing compliance, by pointing to existing safety programs and rules
    e. The right to accompany the inspector in order to answer questions and point out compliance steps you have taken
    f. The right to obtain copies of the inspector's findings
    g. In some states, employers may also have the right to request a search warrant for any inspection of records[6]

OSHA inspectors are required to discuss any violations found and to fix a time period in which to correct the problem. OSHA must notify operators within six months of an inspection if a citation or violation notice will be issued. Any OSHA citation must be in writing, stating the nature of the violation, the remedy required, and whether any penalty will be imposed.

Employers have the right to discipline any employee who refuses to comply with any OSHA standard or rule.

Employers do not have the right to fire or discipline an employee for reporting a safety hazard covered by OSHA. Retaliation *of this kind is prohibited*, and violators may be prosecuted and forced to reinstate the employee with all seniority rights, fringe benefits, and back pay.

A foodservice operation can be a hazardous place to work unless employers take steps to prevent accidents. Rather than relying on agencies such as OSHA to dictate safety steps, it is better to come up with a workable safety program. Doing so not only will keep employees safe, but also will keep insurance costs down.

Chemical Safety

The 1988 OSHA Hazardous Chemical Communication Act requires employers to provide free annual access to all employees regarding information surrounding toxic substances or harmful physical agents at the workplace. Records must be maintained that such information was sought and obtained by each employee (a signed sheet). Employers can obtain, from the government, a registry of chemical substances that aid in identifying toxins.

ADMINISTERING WAGES AND BENEFITS

The federal *Fair Labor Standards Act* (FLSA) of 1938, enforced by the Wage and Hour Division of the Department of Labor, establishes requirements concerning minimum wages, work time, overtime pay, equal pay for equal work (the Equal Pay Act of 1963), and regulates child labor.[7] The Age Discrimination Employment Act of 1967 prohibits age discrimination. (The EEOC administers the regulations on equal pay and age discrimination.)

Enterprise and Employment Terms Defined under the Fair Labor Standards Act (FLSA)

An enterprise is the related activities performed either through unified operations or common control by any person or persons for a common business purpose and includes all activities whether performed in one or more establishments or by one or more corporate or legal entities including departments of an establishment operated under lease.

Trainee or training wages no longer can be paid. The cut-off date was March 31, 1993.

An *employer* is any person who directly or indirectly manages, supervises, or acts in the interest of an employer in relation to an employee. Everyone from the president of a foodservice company down to a cafeteria manager is an employer, which means they can all be held responsible for wage and hour violations.

Employee means any individual employed by a covered establishment.

Employ means to permit to work, meaning an employer's knowledge of work done for him or her by another person. Proof that the employer supervises or manages the work of an employee is *not required.*

Trainees (or *students*) who work without any expressed or implied compensation agreement are *not* employees if *all* of the following criteria are met.

1. The training given on the premises is similar to that given at a vocational school.
2. The training is for the benefit of the trainees or students.
3. The trainees or students do not displace regular employees.
4. The employer providing the training derives no immediate benefit from the activities of the trainees.
5. The trainees are not guaranteed jobs at the end of the training period.
6. Both the employer and the trainees or students understand that no compensation is to be paid for the time spent in training.

Who Is Covered under FLSA?

Coverage under the Fair Labor Standards Act is based on two tests: 1) individual employee status, and 2) enterprise status.

Individual employees are covered if their jobs involve interstate commerce or they work in producing goods for transportation in interstate commerce. Working in interstate commerce means: 1) dealing with patrons, goods, or services that cross state lines; 2) an employee travels across state lines either to and from work or in the course of employment; or 3) communicating across state lines is a major part of the job. Food production employees whose products cross state lines, as well as foodservice employees who deal with out-of-state patrons, are covered under the FLSA.

Enterprises are covered if they operate a business, individually or as a chain; engage in interstate commerce, or are in the production of goods for interstate commerce; and meet a *dollar volume of business test*.

The dollar volume of business test for retail or service establishments is met if the business makes yearly gross sales of $500,000, not counting retail excise taxes that are separately stated.

There is an important exception for family operations. The law does not cover these if the only employees are the owner and a spouse, children, and other members of the immediate family.

All *units* of a foodservice chain are covered even though an individual unit of a chain may not meet the dollar volume of business test, as long as the total sales of the chain equal or exceed the dollar volume amount.

Seasonal amusement or recreational establishments are exempt from the Fair Labor Standards Act, if they do not operate for more than seven months in a calendar year *or* if the average receipts for any six months of the preceding calendar year were not more than 33.3 percent of the average receipts for the other six months of that year.

Work Time under FLSA

Work time includes not only all required hours the employee must be on duty or at a designated place to work, but also all times when the employee is permitted to work by the employer. Employees must be paid *at least the federal minimum wage for all hours worked*. Employee meetings and training are treated as compensated work time.

In addition, every employee who works over the maximum number of hours in the employer's work week must be paid at least one and one-half times his or her regular rate of pay for the overtime hours.

The Fair Labor Standards Act does not fix the hours worked in order to determine employee rights to overtime pay. FLSA overtime is any time over 40 hours in a 168 hour work week. However, the labor union contract or state labor laws may stipulate overtime pay rules.

According to the FLSA, permitting work includes voluntary overtime not requested, but not forbidden by the employer. Should an employee challenge an employer for overtime pay, the burden is on the employer to prove that overtime was *expressly* prohibited. Employers must communicate and enforce overtime rules, or they may face very costly back-pay claims if challenged.

Absences caused by illness or disability and time off for holidays and vacations are not considered work time under the Act. It is up to each employer to decide how to pay for such times off the job, or to negotiate times off under any labor agreement with employees.

When employees are required to change clothes and wash up on work premises, time spent doing so is counted as work time and must be compensated.

Minimum Wage Requirements

The federal minimum wage is $4.25 per hour for all restaurants having employees engaged in commerce. The minimum wage fluctuates and is based on a *single work week*. Each work week is defined as a regularly recurring period of 168 hours in seven consecutive 24-hour periods. It need not be a calendar week, and it may begin on any day and any hour of the day.[8]

No state may require a minimum wage payment less than the federal scale. States may, however, require a higher minimum wage.

Tip Credit   Federal law provides for the fact that many foodservice employees receive tips from patrons. Employers may add a *tip credit* to the cash wage due tipped employees in order to meet the federal minimum wage scale.

A *tipped employee* is any employee engaged in an occupation in which he or she regularly receives more than $30 per month in tips.

A tip is a voluntary gratuity for services rendered by the employee, and is made by the patron in cash or by credit card. A compulsory service charge, imposed by the employer, is not a tip, nor is a charge negotiated between employer and patron under a banquet or other standard function contract, nor is any tip-pooling agreement between employer and employee that requires employees to treat such tips as income to the employer. A tip is a tip when employees may keep most or all *of it*. In no other case may an employer claim a tip credit. The tip credit permitted is 50 percent of the minimum wage. To qualify for the tip credit, the law requires employers to do the following.

1. Inform employees about this allowance before using the credit.
2. Permit tipped employees to keep all tips, either on an individual basis or under a valid tip-pooling agreement among the employees. An employer cannot force the employees to agree to an arrangement.
3. Be able to prove that tipped employees received the minimum wage after the credit was applied.
4. Pay every tipped employee at least half the prevailing minimum wage after the credit was applied.

If an employer decides to forego the tip credit and agrees to a pooling arrangement with employees, servers cannot be required to contribute a greater percentage of their tips than is customary and reasonable.

The 1989 amendments to the FLSA raised the credit to 50 percent of the minimum wage. The amount of the credit can never be more than the employee actually receives in tips or more than the prescribed percentage of the minimum wage.

The *Tax Equity and Fiscal Responsibility Act* of 1982 (TEFRA) instituted new tip income rules. First, within the large food and beverage establishments, the tip pool allocation can only be made among those employees regularly and customarily receiving tips. Second, the allocation will be 8 percent of gross receipts among those regularly receiving tips over the aggregate amount reported. The method of allocation will be based upon a good faith agreement by employers and employees.

TEFRA defines "large food and beverage establishments" as those employing more than 10 employees on a typical business day during the preceding year; and whose employees, serving food and beverages, customarily receive tips. TEFRA requires these employers to withhold federal income and social security taxes.

All of the above amendments to the FLSA were upheld against the hospitality industry's challenge in *Foodservice and Lodging Institute v. Regan* (809 F2d 842 DC Circuit 1987.)

The following case involves a foodservice employer who failed to fulfill the FLSA requirements regarding the tip credit.

Barcellona v. Tiffany English Pub, Inc.
Federal Court of Appeals, 5th Circuit
597 F.2d 464 (1979)

Facts: Former waiters of TGI Friday's restaurant, which was doing business as Tiffany English Pub, sued Friday's for using their tips to satisfy the requirements of the Fair Labor Standards Act that every employer pay a minimum wage. The waiters sued to recover back wages, damages, and attorney's fees, all of which are available remedies under the FLSA. The waiters said they were required to surrender their tips to the employer and had not agreed to the tip credit on that basis.

Friday's argued that the waiters had agreed to Friday's withholding of tips to satisfy the minimum wage law. The federal district trial court found that no such agreement existed, and that Friday's had committed a serious violation of the FLSA. The court awarded back wages and attorney's fees, but refused, because of a failure to find that the employer intentionally violated the law, to award damages to the waiters. The district court findings of liability and award of attorney's fees were upheld. The district court finding of no justification for damages was reversed and remanded for a further hearing on that question only.

Reasoning: The appeals court found Friday's contention that it had a valid agreement with the waiters regarding the tip credit false, and if there was no agreement, the tips belonged to the employees. The court upheld the federal trial court's findings.

In a curious maneuver, the employers used ignorance of the law as one of their defenses. This defense was used against the back pay award but not against the compensatory damages. However, the appeals court held that defense was insufficient and that the trial court's earlier finding of a flagrant violation by the employer contradicted its failure to require Friday's to pay damages.

Conclusion: The foodservice operators were clearly in the wrong on two counts: they failed to set up a valid agreement with the waiters in order to qualify for the tip credit, and they failed to make themselves aware of the law regarding wages and tip credits.

Other Credits   The meal credit allowance is restricted to the value of the raw food plus cost of preparation. The credit is further restricted to meals "customarily furnished" to employees. This term means "regularly provided" by the employer and *not* "voluntarily accepted" by employees.

Employers may credit the minimum wage with the reasonable cost of board or lodging, or other benefits customarily furnished to employees, *if* the costs of such arrangements are not excluded from wages they receive under a collective bargaining agreement. These benefits may include meals furnished by foodservice operators to employees, and transportation between home and work when it is not part of the job. However, these benefits must be included as part of the regular wage in figuring overtime pay.

Uniforms and Uniform Maintenance   Employers are required to reimburse employees for the costs of purchasing uniforms necessary for their jobs. They are required to reimburse them for maintenance of the uniforms only if they require that they be washed daily, or if they require special treatment, such as dry cleaning. If the uniforms can be washed in the employee's regular laundry, employers are not required to reimburse employees for laundry costs. You may claim a uniform credit against each employee's wages, as long as each employee's pay is not reduced below the minimum wage.

If special maintenance is required, each employee must be reimbursed one hour's pay per week at the minimum wage rate.

Student Employment   Full-time students are paid a special minimum wage—85 percent of the applicable minimum. The following factors apply.

1. A *full-time student* is defined as a student who receives primarily daytime instruction at a bona fide educational institution.
2. Such a student can be employed for no more than 20 hours a week, except during school vacation when he or she can be employed 40 hours per week at the special rate. For all hours over those limits, the student must be paid the minimum wage.
3. Any employer can employ up to six students by notifying the Department of Labor. Any number over six requires authorization from the Department.

4. The number of students permitted to be employed will be expressed by the Department of Labor in a proportion of total student hours of employment in the establishment to the total hours of employment in the establishment where the student works.
5. Authority to employ students at special rates does not authorize an employer to reduce wages of any existing employees.

Federal Income Tax Laws

Payroll tax law is contained in the Federal Insurance Contributions Act (FICA), the Federal Unemployment Tax Act (FUTA), and the Internal Revenue Service Code.[9] FICA subjects employers to liability for these taxes from the first day an employee is hired. FICA is filed quarterly, FUTA annually. The IRS is empowered to require monthly filings of FICA taxes.[10]

Spouses and minor children (under 21) are exempt from FICA and FUTA taxes if they work in a sole proprietorship. A corporation is not exempt. Employers who pay wages of $1,500 or more in any calendar quarter, or who have one or more employees at any time in each of 20 calendar weeks are liable for FUTA taxes.

Employers must withhold federal income taxes from all employee's wages. This requirement does not extend to certain religious, charitable, educational, or other organizations certified as exempt by the Internal Revenue Service.[10] However, not every nonprofit organization qualifies for this exemption.

Under federal law, both withholding and FICA taxes are combined for purposes of payment. Employers must report all wages paid each calendar quarter. Bank deposits are the method of payment required by law, with the time of deposit determined by the total amount owed. Federal unemployment taxes are reported separately.

Meals   The value of meals provided to employees is not subject to income tax withholding, *if* they are not paid as wages. Two requirements must be met: 1) the meals must be furnished on the business premises; and 2) the meals must be furnished for a substantial business reason other than compensation for the employees—otherwise they are taxable as a benefit. A record must be kept of the value of meals provided to employees. Free meals are classified as benefits and are taxable under both FICA and FUTA regulations.

Meal value for taxation varies with the circumstances. No fixed amount is used. Factors considered are the charges entered in accounting records, the type of meals served, where they are served, and the nature of the meal service.

Taxation of Tipped Employees   Tips are considered wages subject to withholding by the employer. The Internal Revenue Code requires employees to report in writing tips of over $20 in any calendar month. Employers are not held responsible for the truth of such reports.

Income taxes on tips can be computed at a flat 20 percent rate, rather than the graduated withholding rates used for regular wages.

Tip Reporting   Under the 1982 amendments to the Internal Revenue Code, the tip reporting provision requires foodservice operators employing 10 or more employees in establishments in which tipping is customary to report to the IRS: 1) gross food and beverage sales (except carry-out sales and sales that have mandatory service charges of at least 10 percent); 2) total tip income reported by employees; 3) food and beverage credit sales; 4) total tips on credit sales; and 5) service charges of less than 10 percent.

All foodservice operators whose total reported tip income from all sources is less than a minimum percentage of total food and beverage sales must file information returns for the first quarter on Form 8027. Any employee is subject to IRS audit if the minimum threshold is not met.

In case the minimum (called a *safe harbor*) is not met on the first quarter report, the following additional reporting requirements must be met for each calendar year (including the last three quarters of that year).

1. All foodservice operators must report each employee's share of tip income equal to the minimum percentage of total food and beverage sales.

   All large-scale foodservice operators will allocate tips annually among tipped employees to the extent of any actual amount of tips reported that is less than the minimum percentage of such sales.

2. The allocation of tips will be made either by good-faith agreement between employers and employees, or, where no agreement exists, by a mandatory formula, which the Secretary of the Treasury shall fix by regulation. In either case, the tip allocation will not affect FICA (Social Security), FUTA (Unemployment Insurance), or income tax withholding imposed on all employers by the Internal Revenue Code. Moreover, no employer will be held liable in any dispute with employees regarding allocation of tip income. Tip *allocation* is *not* the same thing as tip *pooling*. Under the TEFRA amendments, the employer is now made responsible for hospitality employees reporting a minimum of 8 percent sales as tips.

3. Each tipped employee must receive a statement of the amount of tipped income allocated to that employee. The inclusion of this allocation on the employee's W-2 Form will satisfy this requirement.

To exempt foodservice operators who do not meet the minimum, an appeal procedure will be established by the government in cooperation with the foodservice industry. Relief may be granted, on a case-by-case basis.

Foodservice operators may avoid the tip reporting provision by instituting a mandatory service charge of at least 10 percent on all cash and credit food and beverage sales.

Federal Penalties   The *intentional* failure to pay federal employment taxes, including the intentional failure to file employment tax returns and withhold income taxes, is a felony carrying severe fines and prison terms. Intentional violations authorize not only backpay to affected employees, but also an additional amount for liquidated damages. All of these penalties are mandatory.

Tax authorities, like other government agencies, have powerful weapons to enforce compliance. A failure to withhold taxes from employee wages for income and disability purposes may give the Internal Revenue Service the right to seize and sell a business without a court order.

Unlike the normal rule of law in criminal cases—that you are innocent until proved guilty—tax laws reverse the burden of proof. Citizens are presumed liable until they demonstrate the contrary. The tax laws are complex and require strictest attention.

Common Violations under the FLSA

1. *Failure to pay overtime*—Overtime must be paid unless the employee is exempted under the FLSA. Overtime for a minimum-wage-tipped employee is one and one-half times the minimum wage for all hours over 40 during one week's time, less 50 percent allocated for tip credit. Special categories exempted are executive employees, administrative employees, professional employees and outside salespersons, all of whom are paid $250 or more per week. Merely giving an employee a high salary or a title does *not* automatically exempt him or her from overtime pay. Specific criteria set for each category must be met, and this requires a case-by-case analysis of each employee's job duties. Nonwillful violators must pay two years of lost overtime. Willful violators must make restitution for three years.
2. *Failure to keep records*—As in other violations, the FLSA $10,000 maximum applies.
3. *Normal work periods*—The FLSA requires compensation for all work performed, including work necessary to begin and end the regularly scheduled activities. Compensation must be paid to employees who must change uniforms at the beginning and end of the work day.
4. *Employing minors in hazardous occupations*—This topic is covered separately. One special concern has been power-driven food-slicers, which the Department of Labor has recently prohibited 16-and-17-year-olds from operating. This is in addition to the equipment restrictions previously noted.

State and Local Taxes

For federal income tax purposes, the distinction between employees and independent contractors (IC) is adopted by the courts based upon: degree of control over work performed; opportunity for profit or loss; investment in facilities; permanency of relationship and length of employment; degree of skill required and amount of skill involved; method of payment; and parties view of the relationship. The major factor is whether the person utilizing the services of the alleged IC has the right to control the manner and method of performing the work as well as the result to be accomplished.

Many states and major cities tax employee wages and salaries as well as fringe benefits, and make provisions for payment of unemployment and disability insurance taxes.

Recordkeeping

Accurate, up-to-date wage and tax records are an essential management responsibility for all employers. There is a simple reason for this: The government is presumed right in tax cases, with the burden on citizens to prove otherwise.

Special Arrangements   Whenever persons are employed with special wage and salary arrangements, such as tipped employees, employees who receive pay in the form of board, lodging, or other facilities, and employees who are not covered by FLSA minimum wage and overtime requirements, the employer must maintain separate records verifying the arrangements.

The following FLSA recordkeeping requirements apply generally to all employers.

1. Name, employee identification number, home address, birth date for employees under 19, sex, and job position
2. Time of day and day of week on which employee begins the work week
3. Regular rate of pay per hour for every week of overtime worked and for which excess overtime pay is due; basis on which you pay wages; and amount and nature of each payment excluded from your regular wage rate under the FLSA
4. Number of hours each employee worked each work day (consecutive 24 hours) and total hours worked each work week (consecutive 168 hours or seven consecutive 24-hour periods)
5. Total daily or weekly regular wages, including overtime, but excluding excess overtime pay
6. Total excess overtime compensation for each work week
7. Total additions to or deductions from wages paid during each pay period— each employee must be provided a record of the dates, amounts, and nature of any additions or deductions
8. Total wages paid for each pay period
9. Date of payment and pay period covered by that payment

The following special pay arrangements must be separately categorized, recorded, and maintained.

1. Restaurant employees exempt from the overtime pay required for a greater than 40-hour workweek
2. Tipped employees
3. Bona fide executive, administrative, professional, and outside sales employees

4. Commissioned employees of a foodservice establishment exempt from overtime pay
5. Employees subject to collective bargaining agreements exempt from overtime pay, and employees who are permitted to be paid a constant weekly wage even when different amounts of weekly overtime are worked (these are called "Belo" wage contracts)
6. Learners, apprentices, messengers, students, or disabled persons working under special certificates
7. Board, lodging, and other facilities furnished employees for which wage deductions are made

How long must records be kept? There are two broad categories of records that require different holding periods.

Two-Year Records

1. Basic employment and earnings records
2. Order, shipping, and billing records
3. Records of additions to or deductions from wages paid
4. Explanations of any wage differentials based on sex for employees within the same establishment

Three-Year Records

1. Payroll records
2. All collective bargaining agreements, plans, trusts, and employment contracts affecting wages authorized under the FLSA
3. Sales and purchase records
4. Retirement plans

All records must be kept in a secure place on the premises, preferably in a central office. Wage and Hour representatives are entitled to access to these records for inspection and copying at any time.

SUMMARY

Some form of workers' compensation is required by each state to cover job-related injuries. Workers' compensation protects both the employer and the employee.

Worker safety is enforced by OSHA under the federal act of the same name. Employers must comply with safety standards and record-keeping requirements, and may be inspected by OSHA.

Employers must comply with all federal, state, and local income tax laws for employers. Such laws are broad in coverage and strictly enforced.

QUESTIONS

1. Under what circumstances would you wish to *bond* an employee? How does this type of insurance protect the employer?

2. How does workers' compensation protect the employer if an employee is injured?

3. Why is a safety program for employees important? Be specific; include other issues from this book.

4. Bart runs a club. He wishes to add a tip credit to help meet the minimum wage in his state. He presently pools the employees' tips with them, as the employer. What must he do to use the tip credit?

5. Explain why you must keep good employee records, covering different topics in this chapter.

NOTES

1. H. Lusk, Hewitt, Donnell, and Barnes, *Business Law and the Regulatory Environment* 1108–10 (5th ed. 1982).

2. Thus the death of an employee who drowned while trying to rescue an endangered swimmer at a recreation area owned and operated by the employer for the employees was held to fall within the definition of the term. *O'Leary v. Brown-Pacific-Maxon, Inc.*, 340 U.S. 504 (1951).

3. 29 U.S.C. secs. 651–78 (1976 & Supp. IV 1980-federal level).

4. *Marshall v. Barlow's, Inc.*, 436 U.S. 307 (1978).

5. Standards must be published in the Federal Register before becoming effective.

6. 29 U.S.C. secs. 651–78 (1976 & Supp. IV 1980-state level).

7. *Id.* sec. 201 *et seq.* and 26 U.S.C. sec. 3101 *et seq.* (Supp. V 1981).

8. Executive, administrative, and professional employees are exempt from both the minimum wage and overtime requirements of the FLSA if all of the following requirements are met.

   a. Primary duty involves management of the business or of a recognized subdivision or department.

   b. Directs the work of at least two other employees.

   c. Has authority to hire or fire or has clearly defined power to recommend hiring and firing of employees.

   d. Regularly exercises discretionary authority over assigned tasks.

   e. Nonmanagerial duties do not exceed 40 percent of his or her time (not applicable to an owner of at least 20 percent of the business or executive employee who is solely in charge of the business).

   f. Is paid at least $155 per week on a regular salary basis. All employees who earn a salary of over $250 per week, manage as a primary duty, and direct work of two or more employees qualify.

9. 26 U.S.C. sec. 3301 *et seq.* (Supp. V 1981).

10. *Id.* sec. 3121 (b) (8) (B).

10 Foodservice and Hotel Contracts

OBJECTIVES

After reading this chapter, you should be able to:

1. Define the elements of a contract.
2. Discuss various foodservice and hotel contracts within the framework of contract requirements.
3. Outline tests for legality, validity, and enforceability of contracts and the remedies available when one party fails to perform.

> ## Case in Point
>
> *Martin and Muriel Glick owned and operated the Chef's Delight Restaurant, a popular, well-established operation in the suburbs. They had water seepage problems in their basement. Martin hired a plumber, Ron Shepard, to install a sewer line into the basement to correct the problem. There was no written contract. After Shepard installed the new line, but before payment of the agreed-upon price, Martin died. Muriel refused to pay, claiming that the oral agreement was only between her late husband and Shepard. Muriel did not dispute the quality of the work performed or the contract price. Shepard sued for payment of his services, arguing that he had performed according to a valid contract entered into by both Martin and Muriel.*
>
> *Was Shepard right? Was Muriel a contracting party? Yes, on the basis of the following rule of law: A contract exists if implied from the conduct, situations, or mutual relations of parties, and enforced by the law on the grounds of reason and justice. An oral contract is presumed valid when there is a meeting of the minds of the parties and the contract meets the legal requirements for contracts. There was no dispute that a definite oral contract was entered into by Martin. In this case, Muriel's refusal to pay would have resulted in Shepard being the loser, even though he had entered into the contract in good faith. Why was Muriel a contracting party according to the law? She was a co-owner of the restaurant, including the land and the building; she knew the work was necessary and wanted it done; and she co-signed an advance deposit to Shepard. Moreover, Muriel did not object to the terms and conditions when Martin negotiated with Shepard in her presence and guaranteed him payment.[1]*

TYPICAL HOSPITALITY CONTRACTS

Foodservice managers must negotiate and deal with a variety of agreements in order to operate their business. The life of any foodservice business requires that agreements—with manufacturers and suppliers, professionals such as lawyers, florists, entertainers, architects, and engineers, and government bodies at all levels—be enforceable. These agreements, or *contracts*, provide the foundation for business relations.

A *contract* is any legally enforceable agreement, in which each person *agrees to perform or not to* perform some act. The law gives each party to a contract reciprocal rights and responsibilities. In any contract each party has the *right* to receive the performance agreed to by the other party and the *responsibility* to perform any duty agreed to under the terms of the contract.[2] If one party fails to receive the performance agreed to by the other, he or she may seek a legal remedy, usually money damages or the performance of the act.

There are three forms of foodservice contracts: 1) contracts used to negotiate purchases of goods and services from foodservice manufacturers, growers, suppliers, and professionals; 2) contracts used to negotiate sales of finished products and services—foods, beverages, or the use of space or function rooms—to patrons; and 3) *management contracts* between establishments such as hotels and hospitals with foodservice firms for management of their foodservice operations.

In the first form, most operations buy goods, such as meats, poultry, and vegetables, from a seller (*vendor*) using a *purchase order*, a printed form provided by either the seller or buyer. Terms and conditions of the order are contained on the form, which is considered a simple contract.

Hospitality operators also negotiate *service contracts* with professionals who furnish entertainment, decorations, and special lighting, as well as legal or accounting services. In the first category are repetitive service contracts for which a standard form may be used. In the second category, the uniqueness of the services (law or accounting) usually requires that the buyer negotiate from a contract provided by the professional.

Almost every commercial foodservice operator will negotiate *reservation* and *function contracts* with patrons. A reservation contract involves the use of tables, generally available for walk-in patrons, except that a specific table, tables, or space is reserved *in advance* for the patron for a specific date and time. A standard menu is used and no charge or deposit is required to hold the reservation.

A function contract is a form of reservation contract. The operator agrees to reserve private space in the establishment for a patron in the future, rather than merely regular table space. A function is a special event, such as a wedding, banquet, or awards dinner, for which the patron wishes to buy the exclusive use of space. Usually a function involves preplanned food and beverage service, at an agreed-upon price per portion, rather than the regular menu selection. An advance deposit is usually paid by the patron to assure performance by the operator.

Management contracts are common among institutions, especially those whose primary purpose is not to serve food but to offer other services, such as education, mental health care, and shelter. Hotels, schools, prisons, and recreational parks may contract with foodservice firms to manage all or most of their foodservice operations. The firm may either be paid outright or may split the profits. The firm is usually in charge of staffing, as well as purchasing, preparing, and serving food.

Each of the agreements has elements that make it an enforceable, legal contract. When an element is missing or the terms of the agreement are misunderstood by one or both parties, a court may have to correct the problem. An understanding of the elements of a contract should prevent hospitality operators from having these kinds of legal problems.

CONTRACT LAW

Contract law assures the enforcement of agreements without resort to physical violence or economic warfare. Most people voluntarily live up to business agreements as a moral duty. However, when adverse economic conditions make it costly or disadvantageous to do so, moral duty may not be enough to compel compliance from everyone. Here contract law steps in to provide an impartial method of contract enforcement. In this way the wronged party in a contract is given relief and the party who broke the contract is held responsible. Without contract law, business transactions would constantly suffer from uncertainty. Some understanding of contract law will help foodservice operators avoid unnecessary and costly mistakes.

TYPES OF CONTRACTS

Contracts are classified in various ways: how they are stated, which of the parties must perform, how they are formed, and to what extent they are legally binding.

How Contracts Are Stated

An *express contract* is one in which all of the terms and conditions are fully stated, either orally or in writing. An *oral express contract* was made in the Case in Point. An *implied contract* is one in which the terms and conditions are implied from the conduct of the parties, rather than their words.[3] Muriel's conduct as a co-signer of the deposit check created an implied contract between her and Shepard to pay him for the sewer line.

Who Must Perform

In a *bilateral contract*, a promise is exchanged for a promise, and both parties are bound to perform. In a *unilateral contract* one party promises to perform only in exchange for an *act* of the other party.[4] In this case only the party promising to perform the act is bound. A unilateral contract existed between Shepard and the Glicks, but once Shepard performed, the Glicks were bound to pay for his work. Until the agreed-upon act is completed, the other party need do nothing, and may in fact withdraw from the promise without liability. For example, if Charles Teeter, the owner of a motel, agrees to obtain a 15 percent discount from Carol Cayce for 500 styrofoam cups in exchange for his agreement to buy 600 cups next month, they exchange promises to perform, and have a bilateral contract. If Carol tells Charles she'll deliver 500 cups to him next Tuesday, only a unilateral contract exists. Only Carol has agreed to perform. Charles has agreed to nothing until Tuesday, and he may change his mind. However, in such cases the courts normally prevent injustice to the party performing the act by holding the other party liable for the reasonable value of the benefits received.[5] Thus, Charles may have to pay Carol the reasonable value of the cups.

How Contracts Are Formed

A *formal contract* is actually a *negotiable instrument*, a promise to pay for goods or services. Checks, notes, drafts, and certificates of deposit are examples of negotiable instruments. The source for the law of formal contracts is the Uniform Commercial Code. All other contracts are *informal contracts*. A purchase contract is an example. No special form is required, except in the case of real estate transactions.[6] The service contract between Shepard and the Glicks was an informal contract.

Quasi-contracts are a legal classification. They are not really contracts at all, since they are created by the courts to avoid injustice. No agreement or consent between the parties exists. Rather, the courts impose duties on one party to prevent unjust enrichment, or one party getting something for nothing without legal justification.[7] Under this doctrine, Muriel might have been required to pay Shepard for the fair value of his services even in the absence of any written contract.

Stage of Performance

A contract fully performed by both parties is an *executed contract*. A contract not yet completed by one is an *executory contract*.[8] Shepard had executed his responsibilities under the sewer installation contract. Muriel had not yet paid for the services Shepard rendered. Her part of the contract remained executory.

Extent to which Contracts Are Legally Binding

Contracts are either *valid, void, voidable*, or *unenforceable*. This classification determines which, if any, legal remedies are available in the event that one of the parties violates the contract.

A *valid contract* is one that contains all the essential elements of a contract and is fully enforceable. A *void contract* is not a contract at all, meaning that it will not be recognized as one that is legally enforceable. Neither party is required to perform and no legal obligations exist.

A *voidable contract* gives one of the parties the option to avoid performance without liability or to perform at his or her option.[9] Performance of a voidable contract is called a *ratification*.[10] If one party chooses not to perform, then both parties are released from performance. If one party performs, the other party is obliged to perform. Until the contract is avoided by the party choosing to do so, it is fully enforceable. Contracts made by a minor are avoidable by the minor, but not by the adult with whom the minor contracts. Other examples are contracts induced by fraud, duress, or undue influence, which may be avoided by the innocent party.[11]

An *unenforceable contract* is a contract in all respects except one, which makes it legally unenforceable. For example, if a party to a contract fails to sue to enforce it within a time period (the statute of limitations) required by law, that person is prevented from enforcing any claim based on that contract, even though all necessary elements of a contract exist.[12] A statute of limitations is the maximum amount of time after something happens for a remedy to be sought in court. If Shepard had failed to sue within the proper time period, and Muriel had raised that as a defense, the court would have dismissed Shepard's claim even though it was otherwise valid.

PURCHASE ORDERS

Purchase orders are actually informal express contracts to buy foods and nonperishable items such as silverware, tables, and chairs.

Purchase orders must contain all essential contract elements to be enforceable. In these contracts a promise to deliver goods is made by the seller and the buyer promises to pay for them when received. These are typically unilateral contracts with the buyer promising to pay only if the seller delivers the goods. **Exhibit 10.1** is a typical purchase order between a foodservice operator and a supplier.

By seeing how this simple contract meets the legal elements, the law of contracts should be clearer. Foodservice operators should know their legal rights and responsibilities for all the different contracts they must use to perform their jobs and maintain their operations.

Exhibit 10.1   Purchase Order

Purchase
Order

ORIGINAL ....W
ACCOUNTING........Y
RECEIVING ....P
DEPARTMENT......G

TO

    Maxi-Fish & Suppliers, Inc.
    255 Pier Street
    Boston, Mass.

DATE ____ April 2, 19XX _____

PURCHASE ORDER NUMBER ____ 732 _____

hereby orders from you the goods or services specified below. The
Purchase Order Number must appear on all packages, papers and invoices
pertaining to this order. Invoices must be submitted in duplicate and a
packing slip must accompany each shipment.

Company Telephone No.

QUANTITY	DESCRIPTION	PRICE EA.	AMOUNT
50	Fresh Grade A filleted codfish	$ 1.00	$ 50.00
2 cases	cocktail sauce	20.00	40.00
1 case	tartar sauce	15.00	15.00
25	lemons	.25	6.25
			$ 161.25
			+ 6%

Delivery needed by April 10.

SHIP TO	CHARGE ACCT. NO.	AMOUNT
Dover Seafood Shoppe 10 Massachusetts Street Boston, Mass.	Kitchen 12	

AUTHORIZED BY ___ Billy Dover, Jr. _____

ORDERED BY ____ cb _____

RECEIVED BY _____

DATE RECEIVED _____

ESSENTIAL ELEMENTS OF A CONTRACT[13]

Agreement

Contract law requires that one party make an offer and the other party accept the offer on the terms contained in it. This is an *agreement*. Each of these elements includes other factors that are essential to enforcing an agreement. The offer must identify the parties, the subject matter of the contract, the time for performance, and the price. If all of these terms are expressed in the contract, the contract is enforceable.

A purchase order is really nothing more than an offer to purchase goods or services from a seller, and does not become a contract until accepted by the seller. The purchase order in Exhibit 10.1 is enforceable. It contains the essential contract terms. (Sometimes the court will imply reasonable terms as long as the intentions of the parties are not violated.[14] If the purchase order did not state an exact price, but rather "cost plus 15 percent" for the items, the court might imply a reasonable cost figure.)

The Offer The offer itself must also contain three elements: 1) an intention that the court can ascertain objectively from the conduct of the parties if it is not otherwise expressed; 2) definiteness; and 3) communication.

Intention by the seller to make an offer is illustrated by the seller leaving the blank purchase order form with a potential buyer, or by notifying buyers that the seller wishes to obtain business. This means that the courts will ignore the secret, unspoken intentions of the parties.

At common law the offer must be *definite* on all essential terms, meaning identification of the parties, subject matter, time for performance, and price. The Uniform Commercial Code covers otherwise indefinite terms. The purchase order may be enforced by the establishment of a reasonable price, places of delivery, time for shipment or delivery, and time for payment. The common law they adopted to determine whether an offer is definite is as follows: are its terms *specific* and firm *enough* for the court to measure the damages directly related to the violation or breach of that contract? The purchase terms in the Dover Inn order are specific. Let's take a less obvious example, so you can determine the difference.

Suppose that the Glicks wish to hire you to manage their restaurant. The written employment contract provides a fixed monthly salary, but the Glicks orally promise you "a reasonable share of the profits" as an additional inducement. After 10 years you voluntarily end your employment. Then you request an accounting of the profits. The Glicks refuse.

You sue to recover your share of the profits; the total profits on which the share would be based add up to $2 million over the 10-year period. You would probably lose your case.[15] A *reasonable share* of the profits is too indefinite a term on which to base any damage award. Had the Glicks agreed to pay you a fixed salary plus 10 percent of the annual profits, this promise would meet the test of definiteness and permit you to recover your share.

For the foodservice industry, the Uniform Commercial Code also permits omitted or indefinite terms to be supplied by *"custom and usage"* in trade and by prior dealings. This means that if practices such as a blank purchase order or special delivery terms are widely accepted in the foodservice industry, and are legal, then they may be enforced as long as both parties understand the terms. If terms customarily have one meaning in the foodservice industry, they are acceptable, even if they have other meanings in other types of business. It is up to the party claiming custom and usage to prove such existence and application to any issue in dispute. Suppose you ordered frozen chicken from a poultry seller for resale to restaurant patrons as a weekly chicken feature. When the chickens are delivered, you discover that they are "stewing chickens" rather than "young chickens" suitable for broiling and frying. The sales contract did not define the word "chicken," and you failed to prove that the generic word "chicken" in the foodservice business always means "young chickens." There was no clear definition of what you wanted in the contract.[16] To the courts, the word "chicken" means any chicken suitable for human consumption. There was no breach or violation of any warranty of fitness by the seller.

The offer must be *communicated* by the person making the offer, the *offerer*, to the person intended to receive the offer, called the *offeree*.[17] Every person who offers goods or services to another has the right to limit the offer to certain specific persons.[18] Once an offeree accepts an effective, valid offer, the offer and acceptance together make up a legally binding agreement or contract.[19] Assuming that all other elements of a contract exist, only those persons may accept the offer and thereby create a contract.[20] This means that a total stranger cannot accept the offer.

Termination of Offers   Offers can be terminated in a number of ways, as follows.

1. Just as a person can limit the offer to meet his or her requirements, the offerer can *revoke* the offer before it is accepted. For example, Maxi-Fish could state that a fish special is no longer a special and the Dovers have to pay full price. A withdrawal of the offer must be specific and communicated to the offeree before he or she accepts.[21] Otherwise the acceptance will create a valid contract. The rule of revocation permits offerers to revoke any time *before* acceptance, even though they may have promised to hold the offer open for a specific time period.[22]

   The only exception to the right of revocation is if the offeree enters into an *option contract* by paying the offerer for the promise to hold the offer open for a specific time period. Payment for the promise to keep the offer open makes the option right binding upon the offerer.[23] Although contained in the offer, it is a separate, enforceable agreement. This is why it is called an option contract. The offerer remains free to refuse to provide an option, but once such a bargain is made, is bound to honor it for the time specified.

The Uniform Commercial Code does away with payment or other consideration in the case of a *firm offer*. The firm offer rule applies as follows: If a merchant dealing in goods sold in the regular course of business signs a written offer stating that it will be open for a specified period, it is firm for that time period, not to exceed three months.[24]

Sometimes courts prevent a legal revocation when the offeree changes position in *reasonable reliance on* the offer.[25] (If Dover had turned down a lucrative deal with Just Fishin' to buy the special with Maxi-Fish, and suffered losses because of Maxi-Fish's withdrawal of the offer, Dover could sue.)

2. Another method of termination of an offer is *rejection by the offeree*, the person or party to whom the offer is made. (Mr. Dover can simply decide he can get a better deal elsewhere.) As in the case of revocation, a rejection of the offer must be received by the offerer to be effective.

3. Offers are also terminated by *operation of law*, or circumstances beyond the control of the parties but recognized as legal grounds for termination.[26] Four general situations fall into this category of termination.

   • The *lapse of time period specified* to accept an outstanding offer is a failure to accept terms within a reasonable time period. This constitutes a rejection of the offer, unless unusual circumstances justify late acceptance. For example, a reservation contract for a business luncheon may contain a provision requiring a deposit before the contract is accepted. Failure to make the required deposit within a certain time period will automatically terminate the offer. This frees the operation from any duty to perform and allows the manager to negotiate a new reservation, either with the same client or another group.

   When no time period for acceptance is specified, the courts will create a reasonable time period for termination. This concept of "*reasonableness*" will vary with circumstances, price fluctuations, and market conditions, among other factors.[27] Sales of perishable commodities, such as fresh meats and vegetables, will require a shorter time period for acceptance than the sale of tables and chairs. Fuel to heat your operation might also require a shorter time period for termination, not because the commodity is perishable, but because a short supply may cause sharp changes in the market price.

   • *Destruction of the subject matter* is another ground for termination. In this case the circumstances causing the destruction must occur before acceptance of the offer.[28] If a supplier offers to sell dressed poultry for delivery to your restaurant, and you show interest but do not commit yourself by signing or accepting a purchase order, the destruction of the supplier's poultry warehouse automatically prevents you from accepting the offer.

- *Death or incompetency of either the offerer or the offeree* will also terminate any offer.[29] (If the Maxi-Fish supplier died after making the offer, and he had no business partner, the offer would have terminated.)
- *An offer legal when made but later declared illegal* by a court or legislation before acceptance also terminates the offer.[30] Suppose that under existing laws you are able to buy imported alcoholic beverages at wholesale from private warehouses for resale to your customers. You receive an offer from the wholesaler for imported vodka. Before you are able to accept the offer, the federal government places an embargo on all shipments from that country and prohibits sales of products from that country already on hand. Such government intervention makes the vodka sale to you illegal, preventing you from accepting the offer.

Any of these situations could short-circuit the formation of a legal contract based on a purchase order. However, the Uniform Commercial Code does permit merchants to vary the terms of an acceptance or add new terms that automatically become part of the contract unless: 1) the buyer requires acceptance of the original terms of your offer; 2) the new terms materially alter the contract; or 3) the buyer rejects the new terms.

Non-offers   Not all statements made by one party to another are considered offers. To know what an offer is, you should also know what it is *not*.

1. When a party expresses an opinion about a future happening, that expression is not an offer that, if accepted, makes the opinion legally binding.[31] (If the Maxi-Fish salesman said the cod would probably increase dinner sales, any drop in those sales would not be a breach of contract by Maxi-Fish. The supplier was not guaranteeing anything, but merely expressing an opinion.)
2. *Preliminary negotiations*, or an expressed willingness to negotiate, are not considered offers.[32] (If Mr. Dover merely asked Maxi-Fish to submit a bid for the items, the invitation to bid is not an offer, nor does Maxi-Fish bind Dover by submitting the requested bid. Once the bid is submitted, it is an offer Dover is free to accept or reject.)
3. *Statements of intention*, such as promises to make a future contract, are also non-offers.[33]
4. *Advertisements, circulars,* and *catalogue promotions*, made to the general public, are non-offers because the seller (offerer) does not have an unlimited supply of merchandise, and everyone who accepted could sue the seller for breach of contract. However, if the advertisement contains all the essential elements, it can create contract liability, if accepted.[34] Thus, sellers may protect themselves by including the phrase "subject to prior sale" in their promotion material.

5. *Non-contracts* are non-binding agreements. A contract is an agreement, but not all agreements are contracts.

The law excludes *social transactions* from the field of offers.[35] If the Dovers invited the Maxi-Fish supplier to a free meal at the restaurant as a friendly gesture, and not as an occasion to discuss future business, a social obligation would be created, involving a *moral duty* on the Dovers to provide the promised meal.[36] If he refused to do so when the supplier arrived at the appointed date and time, the supplier could not sue for breach of contract, even though he may have spent money to get there. A social obligation is not a legal obligation. Had the invitation been business-related, the supplier would have a better chance of creating a contract by accepting the invitation.

Acceptance must normally meet buyer requirements. This means that the seller may not attach new or different conditions. The UCC creates some flexibility on the terms of acceptance between merchants.

Acceptance of the purchase order must be *communicated* by the seller to be effective. The normal rule is that an offer by mail is accepted when the signed purchase order is mailed, not when it is received. *A purchase contract is created when a valid purchase order is accepted.*

Acceptance How does one accept an offer? Communication is the key to acceptance of an offer.[37] This means by words or conduct communicated to the offerer that shows the offeree's agreement to the terms of the offer. Under common law, the acceptance must normally be made in the manner stated in the offer.[38] Any changes will create a *counteroffer*, or rejection of the original offer.[39] A counteroffer, if accepted, creates a new contract.

One party may express dissatisfaction with the offer without creating a counteroffer. (The Dovers, in responding to Maxi-Fish's offer, could agree to the terms, but add, "we accept, but want credit instead of a cash deal." That condition is a counteroffer, and not acceptance of the original contract.)

The Uniform Commercial Code changes this common law acceptance rule. An acceptance of an offer to sell goods that change the terms of the offer *slightly* is enforceable because the changes are treated as *additions* to the contract. The requirements of the UCC to meet this criteria are as follows: 1) both parties must be merchants; and 2) the proposed changes are enforced unless the offer specifically states that the offer must be accepted without any changes, or the changes specifically alter the contract, or the offerer objects to the changes within a reasonable time. If neither party is a merchant, then the offerer must agree to any changes.[40]

Silence (neither verbally accepting nor rejecting the offer) usually does not mean acceptance of an offer. There are three exceptions to this rule, however: 1) when the offeree accepts the benefits of the offer with an opportunity to reject goods or services and understands that the goods or services are not intended as gifts; 2) where the recipient must notify the offerer only if rejecting the goods or services; and 3) when the recipient exercises possession and control over the goods. (If the Dovers cooked the codfish, it is a safe bet that the court will find they had accepted the fish.) This use establishes an intention to pay for the reasonable value of the goods.[41]

Generally, any reasonable and appropriate acceptance is effective unless the offer requires a specific kind of acceptance. Normally the acceptance must be transmitted by the same method used in transmitting the original offer. Where an offer is mailed, mailing the acceptance would be reasonable.[42]

Consideration

A second major element of every contract is *consideration*.[43] This term means something, usually money, given by one party to the contract in exchange for something given or to be given by the other party. (In the purchase order, Dover exchanged the promise of money for food he would receive.) The giving may take the form of an act, or a promise to act. It may also consist of a promise to not act.[44]

Consideration does not have to be economically valuable to be legal.[45] *Detriment* is necessary to prove that the consideration for the bargain made is legally sufficient.[46] Both parties to a contract suffer detriments and gain benefits at the same time in any legal bargain.[47] (Dover has to give up his money and Maxi-Fish has to give Dover the goods—*on time*.) Negotiations can help minimize detriments and maximize benefits.

Certain promises can be enforced without consideration. For example, if you are persuaded by a quick-service owner to give up your independent restaurant to run a franchise and rely on assurances that the franchise will be granted, and the franchise owner does not grant it, you can recover damages for the losses suffered when you gave up the restaurant. The franchiser is prevented from raising the issue of lack of consideration for the promised franchise. The courts can only avoid an injustice by requiring the franchiser to compensate your losses.[48]

The Uniform Commercial Code eliminates the question of consideration when a *firm offer* is made.[49]

Capacity of Contracting Parties

The third major element of an enforceable contract is that both parties must have the *capacity* to enter into a contract. Capacity means the emotional maturity to understand the consequences of a contract.[50]

Non compos mentis, the legal phrase used to describe the state of mind of the person trying to avoid a contract, means that person is unable to understand the legal consequences of his or her conduct. The legal test of *non compos mentis* is objective, meaning that a person's conduct under the circumstances is examined to determine his or her ability to understand legal duties.[51]

In this area, transactions with *minors* represent the most difficulty. (In most states, a minor is any person who has not attained the age of 18.)[52] In general, contracts entered into by a minor are voidable if the minor chooses to avoid them, *but only by the minor*. The adult who contracted with the minor cannot avoid the contract. Unless the minor chooses to avoid a contract, it is presumed valid and enforceable.[53]

The exercise of a minor's rights to avoid a contract obligation is called a *right to disaffirm*, that is, set aside the contract. Such disaffirmance may be expressed by words or conduct.[54] A minor may disaffirm at any time while under 18.[55]

Let us assume that Billy Dover, Jr., was only 16 years old. Might the purchase order still be enforceable? Probably, since it was on the regular form used by Dover. However, if Billy went to Maxi-Fish, said he wanted the cod for "his food service," and then changed his mind, the supplier has a problem.

Operators or suppliers dealing with minors should know exactly whose signature is valid when it comes to purchase orders. Operators in the catering business especially have to be on the lookout for minors who want to contract for a party. Find an adult to sign agreements.

When a contract has been fully performed by the seller, the minor must *restore* the goods or services received to the seller, as long as they are in his or her control. Such property may be returned "as is," without compensating the seller for any damage caused, in the majority of states.[56] A few states compel the minor to provide *restitution*, meaning returning the seller to his original position.[57] A minor cannot disaffirm a part and retain the benefits of the rest of the contract. *The disaffirmance must be total.*[58] Given the perishability of items in the foodservice industry, this is small comfort.

What happens when a minor misrepresents his or her age to a seller? Statutes in some states prohibit disaffirmance by a minor who lies about age. Other state courts require the minor to repay the full value of the consideration received in such cases.[59] A third group of state courts allow the minor to disaffirm, but permit the seller to recover damages on the basis of fraud.[60]

Parents are not usually liable for contracts made by their children. Therefore, to protect yourself, it is wise to obtain the co-signature of a parent or other responsible adult for any contracts involving a substantial value; for example, the sale of a foodservice van. That signature will make the parent a *guarantor of payment*, meaning that the parent is personally responsible for payment, even if the minor disaffirms.

Minors present a serious dilemma to all business persons. Foodservice managers are not as vulnerable as some sellers of high-cost durable goods (business vehicles, restaurant equipment), since food and beverages are usually paid for at the time of sale. The contract, completely performed by both the minor and the operator, is over or *discharged*. There is nothing for the minor to disaffirm. However, the minor may disaffirm food and beverages bought on credit with the restaurant's own charge system. Again, caterers should be wary of minors contracting for a party. Find a parent to sign the actual agreement. Since this can run into large sums of money, credit transactions with minors should be reviewed carefully.

Intoxicated persons are another class of people who can avoid their contracts. To do so, the intoxicated person must prove that he or she was so drunk as not to know what he or she was doing. (If the salesman was drunk when he said he had 50 codfish and actually had 50 halibut steaks, Bill may have to change his menu—and his supplier.) Intoxication does not *automatically* void a contract. The contract is enforceable until the intoxicated person chooses to avoid it.

Incompetent persons are also able to avoid otherwise binding contracts. Incompetence is defined as *lack of mental capacity* to understand the legal consequences of one's actions. *Physical incapacity* is *not* a ground on which someone may disaffirm or avoid contracts, unless physical incapacity impairs one's mental capacity to make contracts.

Insanity recognized by a judicial body makes all contracts of the declared incompetent *void*. If insanity is not judicially declared, that person's contracts cannot automatically be avoided. Rather, the issue is whether the person was incompetent at the time the particular contract was executed. If the contract of an otherwise incompetent person was executed by that person in a *lucid interval* (period of mental competence), it will not be voided by the courts.

Legal aliens who are not residing in the United States have the same power to contract, sue, and be sued as U. S. citizens. Illegal aliens are denied the right to assert contract claims in the courts. Some states deny all aliens the right to own real property. Enemy aliens from a country with whom the United States is at war will be denied the right to enforce contract claims, if such claims would afford aid to the enemy country.

The *death* of one of the contracting parties automatically terminates an agreement. In our Case in Point, however, it was determined that Muriel, as co-owner of the business, was obligated as a full contracting party to pay Shepard.

Legality

Let us assume that neither Billy nor the supplier was underage, drunk, incompetent, or an illegal alien, and that the purchase order meets these requirements. The contract must also be *legal*. The purchase order also meets the contract requirements for legality.[61] A contract that is not legal will not be enforced by the courts. Illegal contracts include business loans at interest rates higher than permitted by law and contracts that illegally restrain trade if they tend to create a monopoly or restrict competition.[62] (Chapter 13 deals with franchising and antitrust pitfalls.)

Unconscionable contract clauses are also illegal since they violate public policy, additional grounds for legality.[63] Unconscionable clauses are clauses that are so unfair that no court could enforce them. A foodservice operator persuaded to contract with the local police for "extra protection" would be an example. Both at common law and under the Uniform Commercial Code, courts are empowered to strike out unconscionable clauses.[64]

Another type of illegal, unenforceable contract clause is an *exculpatory* clause. Here one party conditions the offer on a waiver by the other party of all legal rights arising out of the first person's negligence or willful misconduct in the performance of the promises.[65] (If Maxi-Fish had conditioned the purchase order on Dover's promise not to take legal action if the supplier failed to deliver, such a clause would not hold up in court.) This is to prevent the law from being used as an instrument of injustice by permitting people to benefit by wrongdoing.[66] Using illegality as an excuse to get out of or terminate a contract is not always a sure solution. Courts are quick to terminate contracts that are obviously illegal but not those that tend to involve hazy areas where even the law is not clear, and especially when the contract does not address an illegal issue or require illegal action as performance. Courts tend to view with distaste attempts to escape contracts by various ploys.

Genuineness of Assent

A final requirement necessary to create a contract is *genuineness of assent*, a legal phrase meaning that a contract must be *voluntarily* negotiated by both parties. A contract by definition is an agreement. When either party induces the other to contract by *mistake, fraud*, or any other conduct other than normal business risks, the innocent party may cancel the contract and avoid the duties the contract creates.[67] (The purchase order meets this final requirement. If the Maxi-Fish had misrepresented the quality of the items, Mr. Dover might have a legal right to bow out of the contract.)

Mistakes   There are two kinds of mistakes. The first involves a mutual mistake, made by both parties, for which neither party is held responsible.[68] Suppose the Dovers wish to employ a new chef. They contact a reputable foodservice recruiter, and the recruiter gives them the name of John Snyder. The Dovers believe the identity of the person to be famous chef John Snyder, Sr., whereas the recruiter believes they want to hire John Snyder, Jr. John Snyder, Jr., is sent to apply, and when the Dovers do not hire him, the recruiter sues to recover the agreed-upon finder's fee. The lawsuit will be dismissed. No contract existed between the parties, because both were *innocently mistaken* as to the identity of the subject matter of the contract. This rule requires that the mistake be serious, or *material*; that is, neither party would have made the contract had they known the true facts.

When only one of the contracting parties makes a mistake, the law defines it as a *unilateral mistake*, one not wrongfully caused by the other party, and that party cannot normally cancel the contract.[69] This means that if only the Dovers were mistaken as to the identity of the Snyder she wished to employ, the recruiter could win his lawsuit and recover the finder's fee.

Fraud   Fraud requires the innocent party to show: 1) a misrepresentation of a material fact, meaning a false statement or conduct, which leads the person to believe something that is not true; 2) that the false statement or conduct is known to be false by the party making the statement or performing the act, as a dealer concealing a dangerous defect in a van being sold to the Dovers for business purposes; 3) that the deception or false statement is intentional; 4) that the innocent party is justified in relying on the false statement or deception; and 5) that the innocent party suffered some damage or injury as a result of the false statement or deception.[70]

When all of these requirements are met, the innocent party has two choices. One is to cancel the contract and be restored to one's original position, such as getting back a down payment or deposit given to the seller at the time the contract was signed.[71] The other is to enforce the contract and sue to recover for any injury suffered as a result of the fraud.[72]

Previously we noted that silence normally does not mean an acceptance of a contract. *Likewise, silence normally does not mean fraud.* The fact that not all pertinent information is volunteered by one party is not a legal reason for the other to cancel a contract. However, when the other party is under a duty to speak, his or her silence is not legally justified, thus permitting the innocent party to cancel. For example, if a condition is dangerous and cannot be discovered by reasonable inspection, the seller must inform the buyer of it.[73] When a restaurant owner buys equipment that contains a concealed defect (called in law a *latent defect* as opposed to a *patent* or *obvious defect*) or a potential risk that a reasonable person could not detect, such as pressure cooking equipment that may explode without warning unless precautions are taken, *the seller has a duty to warn the buyer.* This is a requirement imposed by the Uniform Commercial Code, which requires a product to have an implied warranty of fitness for its intended purpose. Fraud is also a common law concept applicable to situations where the UCC does not apply, such as in real estate sales, employment contracts, and contracts for personal services.

When a misrepresentation is *innocent*, meaning there was no intention to deceive, the innocent party can cancel but cannot recover damages for any losses suffered.[74]

All five elements must be present to form a contract. In any major agreement, operators should keep all of these elements in mind. If any of the elements seems to be missing, it may be wise to contact a lawyer either to fill in the gaps or to help renegotiate the agreement.

SERVICE CONTRACTS

Service contracts are another important contract used in the foodservice industry. Service contracts are used to buy live or recorded entertainment, interior decorating, and renovation or repair of the pieces, and for the hiring of extra staff for special occasions from an independent employment service.

Service contracts also are used to purchase accounting, consulting, computing, and legal services.

Service contracts are treated differently in law from purchase orders. Here the strict, sometimes uncertain, rules of the common law apply and the more consistent and flexible provisions of the UCC do not. This is because the UCC governs sales of goods, not services. A court decides whether a sale of goods or services is involved, before determining which law to apply.

At common law, an acceptance of a service contract requires compliance with the conditions or terms determined by the buyer or seller. In other words, both the buyer and the provider of services are free to negotiate the contract. Normally the service provider will fix the method of acceptance of the offer. Contracts for services such as legal and accounting consultants are usually standard, although there may be some room for negotiation.

The following four common law rules must be applied to guide a court in deciding whether or not a contract exists.

1. The offer must be communicated to the person or organization providing the service. The buyer has the right to limit the offer to as few or as many persons or parties as he or she sees fit. The only restriction is the legal capacity of any person to accept, and any laws prohibiting businesspersons from refusing to deal with certain individuals, such as civil rights laws.
2. The method chosen to communicate the offer determines the method of communication acceptance, unless a contract clause says otherwise. If the buyer telephones an offer, the provider must normally telephone the acceptance.
3. An acceptance is effective to form a contract when sent to the provider. This means that a mailed acceptance, in response to an offer by mail, is effective *when mailed, not when received.*
4. The offer must be accepted without any new conditions or terms that change the original offer. Any such changes will operate as a counteroffer, which legally is a rejection of the original offer.

These rules apply *unless there are specific contrary provisions in the contract*. Negotiation is usually possible in contracts involving entertainment, interior decorating, or even repair services. Contracts with attorneys and accountants may be more rigid and less open to negotiation. Operators seeking any of these services, however, must explore what areas are open to negotiation. Fees, modifications in services, dates, times, and additional services may all be negotiable.

FUNCTION CONTRACTS

Another type of hospitality contract is the function contract that operators negotiate with patrons, individually or in groups, for special parties.

Unlike the purchase order contract, usually prepared by the supplier, the operator has control over the terms and conditions of the function contract. The operator has a choice of standard clauses for function contracts, which must be chosen carefully to suit business needs. Some terms may develop into standard clauses used for each function contract. Others may be used by the operator as negotiable terms to attract business. **Exhibit 10.2** is a typical function contract.

The terms and conditions in Exhibit 10.2 are similar in purpose to those contained in the purchase order. The menu is usually filled in before the contract is legal. Note the following aspects typical to a function contract:

1. *Definitions*. All parties to the agreement are defined, as is the event, to include any private function forming the subject of the agreement.
2. *Taxes*. All taxes applicable to the event are a separate charge payable to the operator in addition to all other charges.
3. *Guarantee*. The minimum number attending the event must be communicated in writing by the patron to the operation 48 hours beforehand. This figure represents the number of covers for which the patron must pay, regardless of how many actually attend.
4. *Deposit*. This assures that the patron will perform or the party will show up. It is similar to a security deposit. Such a deposit should be large enough to cover operational expenses (extra food ingredients, beverages, dining staff salaries, etc.), yet remain competitive with other establishments.
5. *Cancellation*. The patron's cancellation or breach of the agreement gives the operator the right to recover a minimum cancellation fee of $100 when the date can be rebooked and, in the case of a refusal to perform, recover actual losses together with compensation for attorneys' fees to collect those damages.

   When the date cannot be rebooked, the cancellation fee should be the difference between the contract price (which includes the profit) and normal function expenses, plus any other reasonable expenses, such as flowers, special foods, and beverages, that cannot be used on another date.

   Notice of cancellation by the patron via registered or certified mail is essential to enable the operator to rebook the date. This protects the operator and the patron against unreasonable losses. The operator minimizes his or her losses. The patron, by giving the restaurant the chance to rebook, reduces his or her potential liability to the operator.
6. *Payment in advance*. This clause assures full payment by the patron who has not established credit in advance. The operator is protected against any failure or refusal to pay after the event. A certified or bank check prevents the patron from stopping payment and makes the bank that certifies the check or issues its own bank check a guarantor of payment. This guarantee does not affect the operator's right to recover for breach of contract against the patron.

Exhibit 10.2   Function Contract

St. James on-the-Park, Inc.
70 Park North
New York, NY 10017
212/555–9999

NAME   Judy Horton

ADDRESS   200 E. Clark St.

Eldin, NY 10122

ORGANIZATION

FUNCTION   Wedding Reception

DATE   May 28, 19XX

TIME   6:00 pm–11:30 pm

TELEPHONE   555-4322

ROOM   Walnut

GUARANTEE $500

This AGREEMENT includes the Menu and Arrangement Proposals set forth below
together with the CONDITIONS on the reverse side which are made part of this Agreement.

MENU	ARRANGEMENT PROPOSALS

Assorted hors d'oeuvres
   (see separate sheet)
Cold poached salmon
Pasta (tortellini, choice of marinara,
   al fredo)
Tenderloin, horseradish sauce
Desserts: cherry cheesecake
            blueberry flan
            assorted cookies
	Up to 200
	25 8-top tables
	2 service bars
	dance floor

This contract must be returned within ten days of receipt.

The acceptance of this agreement will constitute a contract between the signatories hereto.

St. James on-the-Park, Inc.
By ....

....
Accepted

THIS CONTRACT IS SUBJECT TO THE TERMS AND CONDITIONS PRINTED ON THE REVERSE SIDE AND MADE A PART HEREOF.

7. *For banquet personnel.* This clause covers tips for service workers.

8. *Price increases.* This clause protects operators against supplier price increases that must be recovered. It does not give the operator the right to increase prices arbitrarily or without justification. In case of any dispute, the operator bears the burden of proving that any price increases added were justified and that any menu substitutions, which the patron agreed to accept, were reasonable.

9. *Extras.* Since the patron requests extras, the patron is expected to pay for them.

10. *Excused nonperformance.* The operator's only responsibility is to terminate the agreement and return the patron's deposit. No *consequential damages*—loss of profits—are recoverable by the patron.

11. *Displays and decorations; patron's property.* This clause gives operators the right to permit or refuse to allow patrons to use certain displays or decorations. Further, operators assume no responsibility for property brought onto the premises by the guests. The specific language concerning combustible material is to place liability for any claims arising out of fires caused by such material on the patron. Fire codes prohibit the use of some combustible materials; they must be obeyed regardless of patrons' special desires.

12. *Provisions of beverages.* Operations that provide alcoholic beverages usually want to prevent other liquor coming into the establishment. This allows the operator to control the types and amounts of liquor served, prevent liability, and enhance the profits of the operation.

13. *Conduct of event.* This clause places the responsibility of control of the event on the patron; makes him or her liable for any overtime labor costs caused by the patron's failure to vacate the function space on time; and makes the patron liable to make good any claims or lawsuits against the operator arising out of the negligence of guests, agents, or independent contractors during the event.

14. *Security.* This clause requires the patron to provide suitable security personnel if in the operator's judgment such personnel are necessary to protect property, other patrons, and employees. These security guards, supervisors, and ushers should limit their activities to the space assigned to that function, for the main purpose of keeping out uninvited guests.

The uncaptioned clause concerns the responsibility of the person signing the agreement when he or she signs in a representative capacity, that is, as an agent for a partnership, corporation, association, club, or society. The signer becomes *personally liable* for payment in the event he or she lacked legal authority to bind the organization.

The waiver of trial by jury is to eliminate delays in the trial of any claim or lawsuit brought against the patron, by allowing a judge to hear and determine all questions of fact and law. Usually the jury must decide questions of fact, and it is more likely that a jury would decide those questions in favor of the patron. Juries are often more sympathetic to an individual being sued in a commercial case, where the other side represents a large corporation. No matter how faulty that view is, juries unconsciously tend to favor the underdog, whereas a judge may not be as easily swayed.

Letter of Intent

Letters of intent may be used by both buyers and sellers of function space to announce either the function or available space for a specific function. The question often arises as to the legal effect of a letter of function issued by the function host or foodservice manager concerning a forthcoming event, such as a wedding or business luncheon. Is such a letter a function contract, or merely a promise to make a contract in the future? The difference is critical. If the letter is merely a promise to make a contract in the future, it is not enforceable as a contract. The function host is free to walk away, find a better deal, and leave the manager empty-handed. Of course, the operator also has that option, but takes the risks that he or she may be unable to reschedule a new function in time to avoid an economic loss.

The common law of contracts covers this problem. Generally oral or written statements of intent to do something *are not offers and may not be accepted to form a contract.*

For example, a printed form letter sent to the business community reads: "I have function rooms available for business luncheons. I am asking $15.00 per cover for a fixed menu, which I have enclosed. Beverages are extra, as set forth on the menu. Minimum number of covers: 15 per room. I look forward to serving you. The Metropolitan Restaurant, signed, *Bill Morgan*, proprietor."

Does such a letter of intent create a contract if the function host replies in writing: "I accept your offer. Reserve 15 covers for September 15, 19__, at $15. List of attendees to follow. The Utopia Chamber of Commerce, signed, *Harry Hargraves*, President."?

The answer is no, because the intent to contract must be demonstrated by the language used by Mr. Morgan. The courts must find intent in the oral statements or the writing. The test of intention is an objective one, not what was in the mind of Bill Morgan. The fact that he said he wanted function business is not enough, since wanting to do business at some future time and contracting to do business at a specific date are not the same. Courts will also support this conclusion, pointing out that a form letter is usually understood by businesspersons to be sent indiscriminately (not intended for any particular person or party), and that the seller can only sell to a limited number of function buyers.

Nonetheless, any spoken or written communication that contains all the essential terms of contract is a contract, and will be enforced. Assume that the letter has been addressed to Mr. Hargraves personally, and said: "I offer the Utopia Chamber of Commerce the Red Room, seating 25 persons, at $15 per cover, on September 15, 19__, at 1:00 p.m., plus 15 percent gratuities to be added to the bill. Closed bar. Menu items to be as follows: fruit cocktail, roast ham, candied yams, green beans, rum raisin ice cream, coffee or tea. Signed The Metropolitan Restaurant, by *Bill Morgan*, proprietor." This is an offer, meaning a promise to perform.

The practical difference is this: In the first case Morgan merely made a statement of future intention. The intention to perform is not a promise to perform. In the second case he made a definite promise to perform in the future.

Because each spoken or written communication stands on its own, there is no one right answer in every case. Each communication must be examined to determine what the parties intended by their exchange of words.

RESERVATION CONTRACTS

When a patron calls up and wants to reserve a table for a party of four at a certain time, or a hotel room for a night, and you agree to hold them, you and the patron have created a *reservation contract*. A reservation contract need not be in writing (and usually is not), but it is enforceable, especially by the patron.

A reservation contract is another form of service contract, but differs in that the service is rendered by you, the operator, directly to a patron, rather than by a provider of a service to you.

MANAGEMENT CONTRACTS

Management contracts between institutions and foodservice management firms usually call for considerable negotiation. Often institutions are legally bound to accept some liability and some functions cannot be delegated. However, courts tend to frown on the use of illegality as a contract out, especially in vague areas.

Thacher Hotel, Inc. v. Economos
Supreme Judicial Court of Maine
197 A.2d 59 (1964)

Facts: Thacher, a corporate hotel owner, entered into a foodservice management contract with Economos relating to the operation of dining rooms previously licensed to serve liquor in Thacher's hotel. Thacher sued Economos to recover minimum payments due under the contract for sales, social security, and state and federal unemployment and withholding taxes, which Economos admitted were due. Economos did not dispute Thachers's lawsuit for payment. Rather, Economos argued that Thacher's management contract was illegal, and therefore unenforceable.

The defense of legality raised by Economos rested on the claim that the Maine liquor control statute prohibited Thacher from operating his dining rooms other than under the hotel management. Under the management contract Economos argued that she was an independent operator, and that the statute had been violated. This being so, her contract with Thacher was illegal and not enforceable.

The Maine Superior Court, which heard the case, entered judgment in full for Thacher. Economos appealed to the Supreme Judicial Court to reverse judgment for Thacher and to dismiss the lawsuit. The judgment for Thacher was upheld, and Economos' appeal denied.

Reasoning: The following sections of the management contract are relevant to the case:

1. The Owner shall employ the Food Manager for the term of five (5) years from [January 3, 1955] until the close of business on the Saturday following [January 3, 1960], as manager of the Coffee Shoppe and Dining-Cocktail Room located in the premises of the Owner known as Thacher Hotel.

    (a) The Food Manager will have exclusive direction and responsibility of purchasing, storage, preparation, and service of all food within the Thacher Hotel, specifically set forth as the kitchen, Coffee Shoppe, Dining-Cocktail Room, and in the rooms of the Hotel whenever such is required by guests of the Hotel.

    (c) The service of beer, wines, and liquors will be made by the waiters or waitresses, and such sales of beer, wines, and liquors shall be accurately recorded on a separate check from the food check, and such separate checks shall be identified as Cocktail Bar checks, and payment of such Cocktail Bar checks shall be made after each serving in a manner prescribed by the Owner, and shall not constitute in any manner a part of food income.

    (e) The Food Manager will have the exclusive responsibility for the food operation without undue interference by the Owner, except in such cases specifically referred to herein. However, at times mutually agreeable, the Food Manager and the Owner or the Owner's representatives shall discuss and consider matters which may be of mutual interest in maintaining efficient and profitable operation . . .

7. The Food Manager shall have the authority and responsibility to operate the food business without interference from the Owner. It is understood, however, that the Food Manager will conduct the operation in a high grade and orderly manner, and will not permit questionable conduct or entertainment on the premises; that the Food Manager will pay all food and other expenditures of said operation from the income thereof and within a time consistent with good business practices, and not through negligence impair the good credit of the Owner . . .

12. It is understood by the Owner and the Food Manager, that wherever the word *food* or *food operation* is used herein, beers, wines and liquor are specifically excluded therefrom and that the Food Manager has no connection therewith or responsibility therefor.

1. The court found nothing to indicate that the hotel had given up management of the dining room; in fact, the hotel retained management due to licensing law laws. The court said:

> The "management contract" was unquestionably entered into with the operations of dining rooms to meet the requirements of liquor laws in the minds of the parties. In short, the contract was in furtherance of the operation of the hotel with a liquor license.
>
> The Thacher Hotel, that is the plaintiff corporation, met without challenge on this record the strict requirement that it is a "reputable place operated by responsible persons of good reputation." It chose to give broad authority to a "food manager." It did not, however, give up or transfer, or lose its "management" of the dining rooms and thereby fail to qualify for the license so obviously a vital part of the business enterprise . . .

2. To the defendant's claim that "public policy" denies recoveries to the hotel because the contract was illegal, the court decided public policy did not protect the defendant from escaping a contractual obligation. The court found no wrong on the part of the hotel to cause them to terminate the contract.

Conclusion: This case illustrates that courts are not likely to upset contracts on the grounds of illegality *unless no other reasonable conclusion is possible*. Here there was no proof of evil motive or intent of the hotel owner to operate outside the scope of the Maine liquor control law. The purpose of the Maine law, requiring that a hotel liquor license for public dining facilities be under the same management, is to prevent control over the service of alcohol from being transferred by the license holder, the hotel owner in this case, to an unlicensed foodservice manager. This is a universal requirement under virtually all liquor control laws. The usual way for a hotel operator to transfer *complete* control of the foodservice operation to an independent operator is to lease the restaurant facility and require the operator to obtain his or her own liquor license.

PERFORMANCE OF CONTRACTS

Every party to a contract is obligated to perform the duties promised. Not to do so is a breach of contract. Standards of contractual performance are either: 1) complete or satisfactory; 2) substantial; or 3) unsubstantial enough to constitute a material breach of the contract.[75]

The common law of contracts speaks of performance in terms of *discharge*. To the degree that the promised performance measures up to the standards of the contract, the party performing is relieved or discharged of any further duties to perform.[76]

Some contracts, such as those requiring the transfer of a deed for title to real property, can be performed completely or satisfactorily without any need to determine how completely the promise was performed. Building and construction contracts, however, do not require complete performance. *Substantial performance* is enough to permit the builder to recover payment, with an allowance to the owner for remedying minor defects.[77] For example, a substantial performance may occur when the operator performs all the requirements for a banquet contract, except that blue table cloths are used instead of blue and white as specified in the contract.

Most businesses will fulfill the terms of contracts. There are situations where complete fulfillment might have to be delayed, as when a number of items on a purchase order are back-ordered or equipment in a certain color has to be brought from out of town. Foodservice operators should recognize whether the other person is making an honest effort to fulfill the contract or is simply trying to get out of it by cheating.

A material breach of contract means that the party in question has failed to meet the standards of performance required by the contract. If workmanship on kitchen plumbing is shoddy enough that the operator has to have someone else do it over, this is a *material breach*. In such a case, the other contracting party is relieved of any responsibility to pay or perform.[78]

Partial performance is not to be confused with *substantial performance*. Partial or unsubstantial performance is incomplete performance. An example would be if a painter painted only half of your dining room when the contract included the entire room. If one party accepts partial performance, then that party is obligated to pay the fair market value of the services provided, but not the contract price.[79] If partial performance is not accepted, then the job must be completed.

Contracts to Personal Satisfaction

A contract to the *personal satisfaction* of one or the other parties means that the party's performance must be to the personal satisfaction of the other party. Performance is tested by the *subjective* reactions of the party, and not by an *objective* standard of a reasonable person.[80] For this reason, such promises should be avoided.

WRITTEN AND ORAL CONTRACTS

Some contracts must be in writing to be enforceable. In case of doubt, a written contract is preferable to an oral contract. This is true not only to satisfy the *form of contract*, but to avoid possible memory lapses and fraud where the contract must be proved by oral testimony. The most obvious example of a contract that must be in writing is one dealing with real estate.[81]

Oral contracts are enforceable, but disputes more readily lead to trouble. As the following case illustrates, whether the contract is or is not a contract, it is going to be harder to prove the facts.

Jones v. Hartmann
Court of Appeals of Colorado
514 P.2d 123 (1975)

Facts: Jones, an architect, was hired by Hartmann and his business associate to build a lodge and restaurant building. Hartmann was to perform according to a letter of agreement containing fees for various aspects of work proposed by Jones and sent to Hartmann. This letter was to include a clause giving Jones the right to charge extra fees for work beyond making the schematic drawings. The agreement was never signed by Hartmann nor returned to Jones. Jones began work on the drawings. He then sent a standard architects' contract to Hartmann, which Hartmann never signed. This contract included a 9 percent fee for additional work. Jones then sent two bills, one of which was paid by Hartmann's wife. Jones sent Hartmann a series of letters on the work progress, and advised him to obtain a building permit. Finally, Hartmann appeared at a hearing for a liquor permit.

A dispute over the fees charged by Jones and billed to Hartmann arose. Jones, not being entirely paid, filed a mechanic's *lien* on the lodge and restaurant building, a legal claim ultimately authorizing him to seize and sell the building to liquidate his bill. Jones then sued both Hartmann and his associate for the reasonable value of the services (called a suit *in quantum meruit*). The trial court found for Jones against both Hartmann and his associate individually and awarded damages. The trial court also said that Jones could recover 75 percent of the original contract price even though the work was unfinished. Hartmann appealed. The appeals court upheld the trial court's decision for Jones.

Reasoning: The appeals court upheld the trial court's ruling that there was an oral agreement. The court said that the agreement was in keeping with Hartmann's schedule to obtain a building permit and a liquor license.

The appeals court also upheld the trial court's finding that a contract existed, saying that this was a question of fact for the lower court to decide.

On the 75 percent contract price finding, the appeals court said that this partial payment was justified by the work done under the legal theory of *quantum meruit*, or compensation for work done.

Conclusion: The decision makes it clear that an oral contract for services is enforceable, when there is sufficient proof to establish the contract.

BREACH OF CONTRACT: PURCHASE ORDERS, CONTRACTS OF SALE, AND SERVICE CONTRACTS

A *breach of contract* exists whenever a party to a contract fails to perform those duties he or she promised. A breach, unlike a discharge of contractual duty, requires the party violating the duty to respond in damages or perform the duty. A discharge ends a contractual duty without imposing liability on either party. A breach, if serious or material, ends the contractual duty of the innocent party but imposes liability on the guilty party.

Contract law, including the common law and sales of goods under the UCC, provides a variety of remedies to the innocent party.

Damages

This is usually an award of monetary compensation, known as compensatory *damages*. Damages enable innocent parties to recover the loss of the bargain. This means that they are to be placed in the position they would have been in had the contract been fully performed. Damages to compensate an innocent party must rest on some form of economic loss. If the innocent party suffers no loss, then the court can only award *nominal damages*, a trivial award for the technical breach of contract.

In sales of goods, the measure of damages is an amount equal to the difference between the contract price and the market price of the goods at the time and place of performance, usually delivery of the goods to destination.

In a typical case of a purchase contract for food, the buyer would obtain the difference between the contract price of a refrigerated carload of dressed beef, costing $15,000, and any rise in price, say to $20,000, caused by the supplier's failure to deliver on time; the damages would be $5,000.

The innocent party may also recover *consequential* or *special damages* that arises only from the results of the original breach of contract. Say that the buyer had contracted to resell the carload of dressed beef as individual portions to a patron for an office party. Consequential damages would consist of the operator's *loss of profits* from the party contract. Consequential damages are available in addition to compensatory damages. To recover these special damages, the buyer would have to prove that the meat supplier knew of the resale of the dressed beef for the party.

Liquidated damages are usually included in a contract by agreement of the parties. This provision specifies a certain amount to be paid in the event of a failure to perform or breach of contract. A foodservice operator might negotiate with a supplier to receive $100 for every day beyond the delivery date the shipper fails to deliver the meat under the purchase contract. Since such a provision is negotiable, the buyer needs sufficient economic clout to include this provision in a contract.

In contrast, a *penalty clause* requires a payment designed to penalize the party who violates the contract.

Liquidated damage clauses are enforced by the courts; penalty clauses are not. The courts use the following two tests to determine whether or not a provision is a penalty clause: 1) the difficulty of estimating damages under the contract when executed; and 2) whether the amount set is reasonable. If the damages are difficult to estimate, and the amount set is reasonable; the provision is enforceable.

Rescission and Restitution

Rescission is a legal term meaning the right to cancel a contract. *Restitution* is the right of each party to be returned to the position each occupied prior to the contract. Fraud, duress, mistake, or failure of consideration will justify rescission. Restitution may be required by each party as a condition of canceling the contract.

Specific Performance

In certain cases courts will require one party to perform the duties that person agreed to perform in the contract, rather than award money damages to the innocent party. This remedy may be applied in service contracts. In sale contracts this remedy is normally not available, because the law requires proof by the innocent party that there were no adequate money damages available. In the dressed beef example, money damages would be considered adequate, since identical goods can be purchased in the open market, although at a higher price, from either the original supplier or someone else. However, if the goods or services are unique and unobtainable from other suppliers, then a court might order specific performance.

Reformation

The common law permits revision of a contract by a court to express the real intentions of the parties. Fraud and mutual mistake are the grounds most often used to reform a written contract, but both grounds must be established by clear and convincing evidence. For example, if both the buyer and the meat supplier in the example were mistaken as to the kind of meat to be delivered, then the court could correct the error and allow the parties to perform the contract they truly intended to perform.

Quasi-contracts

A court may create a *quasi-contract* when no contract exists, to avoid unjust enrichment of one party. Let us assume that the meat contract is unenforceable. Let us further assume that the buyer partially performed by making an advance payment on the contract price. Here the supplier received a benefit. In such a case the court would probably permit the buyer to recover the fair market value of the meat.

BREACH OF A FUNCTION CONTRACT

A function contract might be violated by either the foodservice operator or the patron or host organization. A breach by the operator would occur when he or she failed completely to perform, by not holding the promised function at all.

Another example of an operator breach would be a *partial breach*, such as holding the function but not providing the food and beverages agreed to, using substitutes without the consent of the patron.

Another partial breach would be to furnish less than the standard of service agreed upon by not providing enough servers, providing stainless steel instead of silver flatware, or using water glasses instead of wine goblets.

Another breach would be to cancel the scheduled function without notice or justification.

An unjustified breach would consist of any delay, cancellation, or failure to perform a material or important contractual promise.

A *justified breach* would excuse performance without any liability to the patron. One example is a contract clause allowing the operator to cancel if the patron fails to make required payments under the contract. This is called a *discharge by agreement of the parties.*

A *discharge by operation of law* would occur in the event of damage to or destruction of an operation's premises. Although it is always desirable to cover this type of occurrence in the contract, it is likely that a court would excuse performance even if the contract were silent on this point.

An example of a patron's breach would be a failure to perform at the appointed date and time.

A *partial breach* would consist of a failure or refusal to pay the full price agreed to without any legal justification. Such a lack of justification is illustrated by unproven claims that the soup was "cold," the meat entree was "raw," the bread was "stale," and so on. On the other hand, serving non-kosher food instead of a contracted kosher dinner would justify a total or partial patron refusal to pay the promised price.

Another patron breach of contract is a failure to disclose an illegal purpose of the function or the identity of the host, which might change the operator's mind about holding the function.

Yet another example of a patron breach or contract violation is when the patron submits false credit information.

If either the breach or the justification for a breach is legally insufficient, the contract will be enforced, and will be applied to aid the innocent party.

BREACH OF
A RESERVATION CONTRACT

The normal remedy for a breach by the operator is money damages to compensate the patron for the cost of substituted services. For an individual patron, this is often too small an amount to justify a lawsuit, and is settled on the spot or at a future time and date. However, when the failure to honor a group reservation is established, then the individual loss of the benefit of the bargain is multiplied by the number of disappointed patrons and can be much more expensive. Some operators may impose a fee for broken reservations.

Note that a small claims court action can be filed if the patron wishes to pursue a matter to indicate hurt feelings. In one reported case in New York City, a disgruntled patron sued when he claimed that the table provided him was not up to the standard he deserved. The issue was not the lack of accommodation, but rather, in his eyes, the provision of an inferior accommodation. Since the restaurant had not agreed to provide a specific table, the court found for the restaurant.

However, where a patron requests a specific table location and you agree to provide that table and no other, the case could very well be decided against you. The best solution is to inform the patron booking a reservation that you will do your best to comply with his wishes, but cannot guarantee to do so.

In case of a partial breach of contract by the operator, the measure of damages is the difference between the value of the reservation contracted for and the market value of the reservation, determined by comparing other local establishments. This is usually a token amount. If you were told by the patron that the reservation was made to conclude a business deal, and the patron would lose money on that deal if the reservation was not honored, and you in turn guaranteed to honor the reservation on that basis, *consequential damages* for lost profits could be recovered by the patron.

Damages for *emotional harm* and *mental distress* are recoverable in cases where there is a total failure to honor the reservation, or where the broken reservation was deliberate, rather than merely careless.

Specific performance is not generally authorized for your breach of contract. This is because the compensatory money damage remedy is adequate, since a reasonable substitute for the reservation is generally available elsewhere, from you or from another operator.

Rescission and restitution, or cancellation and return of any sums paid to hold the reservation, are generally available in those rare cases where a deposit is paid. Since a typical phone reservation contract is not made with a deposit, this is not generally used.

NEGOTIATION

Negotiation is the key to obtaining good deals in business contracts. Negotiation is the process of discussing contract arrangements to come up with the most workable contract for both parties. Negotiation may involve compromise—trading desires for critical concessions. More often it involves a step-by-step discussion of the key factors of a contract. Many sales contracts are standard, but usually there are flexible clauses and it is up to the buyer to find out what they are, since the seller may not volunteer that information.

The law is not set up to enforce wishful thinking, and both parties to any contract must know what they want. Negotiate from this point and in this order: What do I want? What can I get? What does the other person want? What can I give up?

In any real estate deal you should have a lawyer. Knowing what you want and knowing your own rights will help the lawyer negotiate a good deal for you. Take advantage of the opportunity to negotiate. It is a good business practice.

CONTRACT GUIDELINES

As a manager you will be making business decisions involving contracts with vendors, suppliers, employees, and patrons on a daily basis. Most of these transactions will be carried out without the need for lawyers or court action. However, there is always the chance of conflict. To ensure the successful defense or prosecution of a lawsuit, as well as to minimize the likelihood of litigation, you must know what the contract says, what contract rights you have, and what you need to do to enforce them.

The following is a reference for the six broad contract elements explained in this chapter. Use it as a checklist for contracting in any part of your business.

1. Every contract must contain an *offer* and an *acceptance*. One party must offer to make a contract, and the other party must agree to the terms and conditions by communicating an acceptance of the offer in the manner required.
2. The promises or acts agreed upon by the parties must be supported by legally sufficient *consideration*, an exchange of something that the law recognizes as sufficient to bind the parties. A *binding* agreement is one in which each party has legal rights and duties that can be enforced by law.
3. Every contracting party must have the *capacity* to contract. This means competence not only to understand the terms and conditions, but also the consequences of any contract made by the parties. An adult is presumed to have both types of competence. A minor is presumed to lack understanding of consequences, and because of this may elect to avoid contracts.

4. Every contract must have a *legal objective*. No illegal contract will be enforced.
5. Both of the parties must legally *consent* to the contract.
6. To be truly enforceable, the contract must be in proper legal form. *Form* means *in writing*, not necessarily the style of writing.

As a manager you must know *the effect* of a failure of a contract to contain any of the above requirements.

SUMMARY

Hospitality operators will find themselves dealing with a variety of contracts to buy and sell goods and services. Contracts are the foundation of business relations with suppliers, patrons, and others. Contracts with suppliers include purchase orders, contracts of sale, and service contracts. Contracts with patrons include reservation and function contracts.

Contracts are classified according to how they are stated (*expressed* or *implied*), who must perform, how they are formed, the stage of performance, and the extent to which they are legally binding.

All the elements of a contract must be present in order for a contract to be enforceable. The elements are agreement, consideration, capacity of the parties to contract, legality, and genuineness of consent. These elements themselves contain requirements which may help a court determine whether the agreement is enforceable.

All contracts require some sort of performance by one or both parties. Standards of performance are the guides the law uses to determine whether a party has performed or not. These standards are complete or satisfactory, substantial or unsubstantial.

In sales contracts, legal rights to goods may be transferred from one owner to another by one of two methods—agreement between the parties or delivery. If goods are damaged or lost, either the buyer or seller may bear the risk. This is one area that should be covered in sales contracts during negotiation.

Sales on approval and *sales or return* are special sales that enable buyers to receive durable goods on a trial basis.

A *breach of contract* occurs whenever one party fails to perform according to the terms of the contract. The remedies for breach of contract are usually imposed by the court in a dispute and include damages, rescission and restitution, specific performance, reformation, and creation of a quasi-contract.

Most contracts contain negotiable elements. It is up to each party to use negotiating skills to obtain the best deal.

QUESTIONS

1. A purchase order and a contract of sale are both legal contracts. How are they similar? How are they different? Give an example of each.

2. Harry and Mike Alvin are brothers. Mike owns a restaurant with a bar. They draw up a contract to transfer Mike's liquor license to Harry when Harry takes over the restaurant. The contract is written and appears to contain the essential elements of a contract. What is wrong with it?

3. You receive an offer from a wholesaler to accept a special price on two convection ovens. The federal government orders manufactur-

ers to stop the production and distribution of the ovens and to recall any held by distributors. Can you accept the wholesaler's offer? Explain.

4. Some junior high schoolers want to give a birthday party for a friend, and use Ethel's Catering for food. What should Ethel do to protect herself legally? Why is such protection necessary?

5. Under what circumstances might *specific performance* be required to compensate for breach of contract?

NOTES

1. *Shepard v. Glick*, 404 S.W.2d 441 (Mo. App. 1966) (sale of home).

2. Simpson, *Contracts*, Second Edition—1 (1965) note 1.

3. *Id.* at 5.

4. *Id.* at 5–6. *Cook v. Johnson*, 37 Wash. 2d 19, 221 P.2d 525 (1950).

5. *Dyer Constr. Co. v. Ellas Constr. Co.*, 153 Ind. App. 304, 287 N.E.2d 262 (1972).

6. Simpson, *supra*, note 2, at 4–5.

7. *Puttmaker v. Minth*, 83 Wis. 2d 686, 266 N.W.2d 361 (1978).

8. Lusk, Hewitt, Donnell, Barnes, *Business Law and the Regulatory Environment* (5th ed. 1982).

9. *Id.* at 89.

10. A *ratification* is the right of a party to create an enforceable agreement even though the agreement is otherwise avoidable at his or her election. You may ratify such an agreement by your conduct as well as by your words. The concept of ratification most often deals with agency law.

11. Lusk *et al.*, *supra*, note 8, at 156.

12. Under Uniform Commercial Code, U.C.C. sec. 2–725(1), contracts for the sale of goods are governed by a four-year limitations period in which the party suing must commence his or her lawsuit. Both parties to such a contract may reduce the limitations period to not less than one year but may not

extend it beyond the four-year statutory period. The period begins from the date the breach of contract occurs, regardless of the lack of knowledge of the breach by the party suing.

13. Simpson, *supra*, note 2, at 7–8.

14. *Savoca Masonry Co. v. Homes & Son Constr. Co.*, 112 Ariz. 392, 542 P.2d 817 (1975).

15. *Petersen v. Pilgrim Village*, 256 Wis. 621, 42 N.W.2d 273 (1950).

16. *Frigaliment Importing Co. v. B.N.S. International Sales Corp.*, 190 F. Supp. 116 (S.D.N.Y. 1960).

17. *Barnes v. Treece*, 15 Wash. App. 437, 549 P.2d 1152 (1976); *Lucy v. Zehmer*, 196 Va. 493, 84 S.W.2d 516 (1954); *McAfee v. Brewer*, 214 Va. 579, 203 S.E.2d 129 (1974); and *Glover v. Jewish War Veterans of the United States, Post No. 58*, 68 A.2d 233 (D.C. 1949).

18. See generally *Belger Cartage Service, Inc. v. Holland Const. Co.*, 224 Kan. 320, 582 P.2d 1111 (1978); *Goldstein v. Rhode Island Hospital Trust National Bank*, 110 R.I. 580, 296 A.2d 112 (1972).

19. *Savoca Masonry Co.*, *supra*, note 14.

20. *Glover v. Jewish War Veterans*, *supra*, note 17.

21. *Cushing v. Thomson*, 118 N.H. 292, 386 A.2d 805 (1978).

22. Lusk *et al.*, *supra*, note 8, at 102–3

23. Clarkson, Miller, Blaire, *West's Business Law* (1980) at 96–97.

24. Lusk *et al.*, *supra*, note 8, at 105–6.

25. *Hoffman v. Red Owl Stores, Inc.*, 26 Wis. 2d 683, 133 N.W.2d 267 (1965).

26. Lusk *et al.*, *supra*, note 8, at 103–4.

27. *Id.*

28. *Id.*

29. Simpson, *supra*, note 2, at 40.

30. *Id.*

31. *Hawkins v. McGee*, 84 N.H. 114, 146 A. 641 (1929).

32. *Mellen v. Johnson*, 322 Mass. 236, 76 N.E.2d 658 (1948).

33. *Id.*

34. *O'Keefe v. Lee Calan Imports, Inc.*, 128 Ill. App. 2d 410, 262 N.E.2d 758 (1970).

35. Simpson, *supra*, note 2, at 1.

36. *Osborn v. Old National Bank of Wash.*, 10 Wash. App. 169, 516 P.2d 795 (1973).

37. *Rothenbuecher v. Tockstein*, 88 Ill. App. 3d 968, 411 N.E.2d 92 (1980).

38. *David J. Tierney, Jr., Inc. v. T. Wellington Carpets, Inc.*, 8 Mass. App. Ct. 237, 392 N.E.2d 1066 (1979).

39. *Zeller v. First National Bank and Trust of Evanston*, 79 Ill. App. 3d 170, 398 N.E.2d 148 (1979); *City of Roslyn v. Paul E. Hughes Const. Co.*, 19 Wash. App. 59, 573 P.2d 385 (1978).

40. U.C.C. sec. 2–207.

41. Simpson, *supra*, note 2, at 57–59; *Corbin-Dykes Electric Co. v. Burr*, 18 Ariz. App. 101, 500 P.2d 632 (1972); *Hobbs v. Massasoit Whip Co.*, 158 Mass. 194, 33 N.E. 495 (1893). However, the Postal Reorganization Act of 1970 permits any receiver of unordered merchandise to keep and use it without being required to pay for it. Under common law, a mailed acceptance is effective when mailed, not when received by the offeror, unless the offeror makes receipt of the acceptance a condition of the offer. The *mailbox rule*, as this form of acceptance is called, requires proper posting and mailing.

42. *Morrison v. Thoelke*, 155 So.2d 889 (Fla. App. 1963), *hearing denied*, 172 So.2d 604 (1965).

43. Simpson, *supra*, note 2, at 80–106.

44. For example, if an uncle promises to give you his restaurant business if you in turn refrain from smoking and drinking until your 25th birthday, this exchange of promises for your performance is sufficient consideration to create a contract.

45. *Dedeaux v. Young*, 251 Miss. 604, 170 So.2d 561 (1965).

46. The giving up of your right to smoke may have little monetary value, but it is nonetheless a *detriment*, or a loss to you. You were induced by your uncle to surrender your legal right to indulge yourself.

47. Simpson, *supra*, note 2, at 80–84.

48. *Hoffman v. Red Owl Stores, Inc.*, *supra*, note 2.

49. U.C.C. sec. 2–205. The firm offer rule applies as follows: If a merchant dealing in goods sold in the regular course of business signs a written offer and states that the offer is not revocable for a specified period, the offer is firm for that period of time, not to exceed three months.

50. *G.A.S. v. S.I.S.*, 407 A.2d 253 (Del. Fam. Ct. 1978) (mental incompetence).

51. *Williamson v. Matthews*, 379 So.2d 1245 (Ala. 1980); *Lucy v. Zehmer, supra*, note 17.

52. *Gastonia Personnel Corp. v. Rogers*, 276 N.C.279, 172 S.E.2d 19 (1970).

53. Simpson, *supra*, note 2, at 215–18.

54. *Id*. at 288.

55. Lusk *et al.*, *supra*, note 8, at 156–58.

56. *Quality Motors, Inc. v. Hays*, 216 Ark. 264, 225 S.W.2d 326 (1949).

57. *Terrace Co. v. Calhoun*, 37 Ill. App. 3d 757, 347 N.E.2d 315 (1976) (restitution not applicable to an intangible service).

58. *Langstraat v. Midwest Mutual Ins. Co.*, 217 N.W.2d 570 (Iowa 1974). *Gastonia Personnel Corp. v. Rogers, supra*, note 52. An *exception* to the rule allowing minors to disaffirm or avoid contracts exists in the case of purchases of "necessaries." Here the minor may disaffirm the contract, but must pay the reasonable value of the goods or services. Necessaries include such basic needs as food, clothing, shelter, medicines, hospital costs, and the costs of an education. A minor's station in life and social status are used by the courts to determine whether a particular item meets this definition.

59. *Mechanics Finance Co. v. Paolino*, 29 N.J. Super. 449, 102 A.2d 784 (1954).

60. *Rice v. Boyer*, 108 Ind. 472, 9 N.E. 420 (1886). Certain contracts cannot be avoided by a minor. Statutes in some states compel minors to honor loans for education or medical care. However, loans of money are often avoidable. This means that you extend credit to minors at your own risk.

61. Lusk *et al.*, *supra*, note 8, at 170–71; *Abramowitz v. Barnett Bank of West Orlando*, 394 So.2d 1033 (Fla. App. 1981). Decision quashed 419 So.2d 627 (Fla. 1982).

62. *Id.* at 175.

63. *Id.* at 175–76.

64. U.C.C. sec. 2–302; *Brokers Title Co. v. St. Paul Fire and Marine Ins. Co.*, 610 F.2d 1174 (3d Cir. 1979). But see *Oliver B. Cannon and Son, Inc. v. Fidelity and Cas. Col of New York*, 519 F.Supp 668 (D. Del. 1981).

65. *Hy-Grade Oil Co. v. New Jersey Bank*, 138 N.J. Super. 112, 350 A.2d 297 (1975); *cert. denied*, 70 N.J. 518, 361 A.2d 532 (N.J. 1976). *Henrioulle v. Marin Ventures, Inc.*, 20 Cal. 3d 512, 573 P.2d 465, 143 Cal. Rptr. 247 (1978).

66. *Williams v. Walker-Thomas Furniture Co.*, 350 F.2d 445 (D.C. Cir. 1965).

67. *Boyd v. Aetna Life Ins. Co.*, 310 Ill. App. 547, 35 N.E.2d 99 (1941).

68. *Ohio Co. v. Rosemeier*, 32 Ohio App. 2d 116, 288 N.E.2d 326 (1972).

69. *Clover Park School Dist. No. 400 v. Consolidated Dairy Products Co.*, 15 Wash. App. 429, 550 P.2d 47 (1976).

70. Lusk *et al.*, *supra*, note 8, at 125–26.

71. *Roberts v. Sears, Roebuck & Co.* 573 F.2d 976 (7th Cir.) *cert. denied*, 439 U.S. 860 (1978); *Gardner v. Meiling*, 280 Or. 665, 572 P.2d 1012 (1977).

72. *Griffith v. Byers Constr. Co. of Kansas*, 212 Kan. 65, 510 P.2d 198 (1973).

73. *Janinda v. Lanning*, 87 Idaho 91, 390 P.2d 826 (1964); *Walsh v. Edwards*, 233 Md. 552, 197 A.2d 424 (1964).

74. *Yorke v. Taylor*, 332 Mass. 368, 124 N.E.2d 912 (1955).

75. *Id.* at 204–5.

76. Leete, *Business Law*, 191–92, 2d ed. (1982).

77. *Hunter v. Andrews*, 570 S.W.2d 590 (Tex. Civ App. 1978).

78. *New Era Homes Corp. v. Forster*, 299 N.Y. 303, 86 N.E.2d 757 (1949).

79. Leete, *supra*, note 81, at 1197.

80. *Kichler's Inc. v. Persinger*, 24 Ohio App. 2d 124, 265 N.E.2d 319 (1970); *Columbia Christian College, Inc. v. Commonwealth Properties, Inc.*, reh'g denied, 286 Or. 669, 596 P.2d 554 (1979).

81. Lusk *et al.*, *supra*, note 8, at 190–93.

Property Rights

OBJECTIVES

After reading this chapter, you should be able to:

1. Define various kinds of property and the legal rights and responsibilities created by each.

2. Explain how property is acquired and transferred.

3. Outline types of zoning restrictions that may be imposed by local governments.

4. Discuss contracting and working with real estate agents to buy property.

5. Discuss various ways of financing a purchase of property.

Case in Point

Sharon Yancy, a landlord, leased first-floor restaurant premises to Maureen Gill, who later sold the restaurant business to Peter Roberts and his son. A five-year lease was drawn up between Yancy and Roberts. Several months before that lease was to expire, Roberts and his son moved their restaurant business to a new building.

When they left, they removed restaurant stools, sinks, dishwashers, refrigerators, and other items of restaurant equipment used by them in their restaurant business. They also removed the restaurant lighting fixtures, paneling, sheet rock nailed to the walls, and false ceilings above the dining booths and above the bar. Unused portions of the lighting fixtures and paneling were thrown away.

The removal of the stools, sinks, dishwashers, refrigerators, and other restaurant equipment was not disputed by Yancy. These were conceded to be trade fixtures properly removable by Roberts.

Yancy sued to recover damages for the removal of the lighting fixtures, paneling, sheet rock, and false ceiling, and the damage done to the remaining premises by Robert's methods of removal of these items.

The trial court awarded Yancy $2,500 for all damages suffered by Roberts. Roberts and his son appealed, arguing that they were legally entitled to remove all of these trade fixtures as tenants occupying the premises for the conduct of their restaurant business. The judgement for Yancy was upheld by a higher court.[1]

What was the difference between the two types of property? Movable equipment, such as stools, dishwashers, and refrigerators, are trade fixtures, and Roberts was entitled to take them. However, equipment such as lighting, paneling, and sheet rock are considered immovable if fixed to real estate, especially if their removal could damage existing property. Both landlords and renters have legal rights under the law.

PROPERTY: LEGAL RIGHTS
AND RESPONSIBILITIES

Foodservice operators at some point may own, rent, buy, or sell property. *Property* consists of legally protected rights in and ownership of anything having recognized value. Legal rights refer to the right to use, sell, and protect property against trespass or theft.

Property includes real estate and equipment used for business as well as personal property. A refrigerator in a foodservice operation and a refrigerator in a private home both provide the owners with legal rights. A buyer of a restaurant and a buyer of a home both have legal rights as to the property and responsibilities to local government and their neighbors for maintenance of the property and payment of taxes.

Hospitality operators have the same rights to own, maintain, and sell property as other businesspersons, but their *responsibilities* may differ. Responsibility for maintaining proper conditions on hospitality property is enforced strictly because members of the public are invited on that property, and in most cases operators will be subject to tight regulation.

Property is safeguarded by the Bill of Rights and by the Fourteenth Amendment to the Constitution, which prohibit persons and states from depriving individuals of private property without due process of law.[2] The Bill of Rights also prohibits any person or government from depriving another of private property for public use without just compensation.[3]

Classification of Property

Personal property is movable, such as a regular-size refrigerator. *Real property* or *real estate* is immovable, such as a restaurant building.

Tangible personal property has physical substance. A delivery truck is tangible personal property, as is a refrigerator. *Intangible* personal property has no physical existence, but consists of rights represented by tangible property. For example, a stock certificate is tangible paper, but the rights it represents are intangible. That is, the certificate represents ownership rights in the assets of the corporation whose name appears on the certificate. The owner of the certificate who sells the stock does not sell the assets, but only intangible rights in these assets.

Services such as telephone, gas, and electricity, and rights represented by credit cards are also treated as intangible property. When a person intentionally uses such services in an unauthorized manner, for example, by using a phony credit card to obtain a room at your hotel, that person commits a theft of services—a crime in most states.[4]

Obtaining Ownership of Property

Ownership and *title*, the legal right to own, use, and transfer personal property, may be established in a variety of ways.

Possession   Simple ownership or possession is the most obvious way to control property. The right of possession means the right to control property and to keep others from controlling it. This includes the right to recover the property from someone who wrongfully takes it or attempts to take control of it. Unless the right to the property is challenged by the true owner, the possessor has the right to retain and control the property against the claims of others.

Purchase   Hospitality operators, like other businesspersons, usually obtain property by buying it from a seller. Personal property transactions are governed by the Uniform Commercial Code (UCC), which regulates the transfer of tangible goods by sales contracts. The Code governs the rights of buyers and sellers. It creates separate rules for merchants and consumers. Purchases of food supplies, paper napkins, and heavy equipment such as dishwashers are all covered by the Uniform Commercial Code.

Gifts   A gift is another way of acquiring real and personal property. A gift is a voluntary act whereby property is transferred by the giver, the *donor*, to the receiver, the *donee*, free of any requirement that the receiver give something of value in return. However, a gift must be completed to have this effect. A promise merely to make a gift in the future requires some exchange of value by the receiver.

The requirements to make a gift legally effective or binding are 1) delivery, 2) the intention to make a gift, and 3) acceptance by the receiver.[5]

Delivery usually means physical delivery. However, a *constructive delivery* is sufficient, that is, delivery of a tangible representation of the gift itself. Moreover, a delivery requires the giver to give up complete control over the gift.

Intent is determined from the language and actions of the gift giver and the surrounding circumstances. Courts look with suspicion at large gifts to a competitor or personal enemy. Likewise, courts closely examine a very large gift to someone outside the family or business circle, to ascertain whether fraud or *duress* (physical or mental compulsion) were used to obtain it. If so, the gift will be returned to the donor or donor's estate, since the essential requirement that the gift be voluntary does not exist.

Acceptance by the donee, or receiver, usually presents no problem.

Accession   *Accession* means adding on to some form of property. When you employ someone to paint the interior of your restaurant with paint you provide or pay for, you own the *improvements* but must pay the *contract price* of the job. Problems in this area come up when large sums are spent by someone renovating or making repairs without the owner's permission. Normally the owner is entitled to recover *damages* to a business caused by "improvements" that the owner did not request or approve. However, if the operator had the opportunity to stop the accession, but failed to do so, he or she may be liable to the accessor for the fair market value of the improvements made.

Inheritance   Property may be inherited on an owner's death by a last will and testament, or by the laws covering the estate of a person who dies without a will.

Under a transfer by will, the owner may name the person or party (corporation, partnership, government body, charity) to whom ownership of property will pass on to after his or her death. Unlike a gift, title and ownership do not pass to the person or party named *until death*. Only death makes the will effective.

Under a transfer by *intestate succession* in which an owner dies without leaving a will, state laws determine how and to whom the property is to be transferred. In these types of cases, property is transferred to the living heirs (those persons in the direct line of descent, such as a surviving spouse or children) of the deceased person. No provision is made for transfers of property to other than living persons. Transfers of property to a charity, government body, corporation, or partnership must be made by a will and are not imposed by state laws.

Confusion   Confusion means a mixing or commingling of the property of two people, so that the individual property of each person can no longer be distinguished. For example, if another restaurateur deliberately, without your consent, mixes his potatoes in a bin shipment with yours so as to cause confusion, you, as the innocent party, may be entitled to the entire shipment.

This principle of law does not apply where the mixing occurs by agreement, an honest mistake, or the act of a third party. In such cases, the owners would own the property equally. If co-owned property is accidentally damaged or destroyed, all losses are spread equally unless greater ownership is proved.

Lost and Found Property   The common law recognizes three types of property rights for finders of property, depending on how and where the property was found.

1. When property is mislaid, meaning the owner or rightful possessor voluntarily put it aside and then forgot where he or she put it, the finder holds the property for the true owner as a caretaker, until that person returns to claim it. For example, if a patron of your restaurant accidentally left her purse, coat, or parcel in the booth where she ate, and it is brought to you by another patron, you must take reasonable care (that care which a person of reasonable intelligence and maturity would use under the circumstances) of the article until the patron returns to claim it. If you or an employee negligently delivers it to an imposter, you may be liable to the owner for its value. If the true owner does not reclaim the article within a reasonable time and the property owner is not located, then you as the restaurant owner have the right to keep it.[6]

   What if an employee finds the article? Your rights as owner and employer prevail over the employee's rights as finder. If the article was discovered by the employee in the course of his or her employment.[7] A practical solution is to require all employees to turn over all found property.

2. *Lost property* is property that is inadvertently put aside and forgotten. As a general rule, every finder of lost property is entitled to keep it from everyone but the true owner.[8] However, the finder is required by law in some states to make a reasonably diligent search to locate the true owner and return the property to that person. A deliberate, intentional retention of property that does not belong to the finder is considered *larceny* in many states.[9]

   Some states determine a finder's rights to lost property on the basis of where the lost property is found.[10] A finder of property in a public space, such as a hotel's or restaurant's dining area, foyer, lobby, or any other area to which the public is given access, takes precedence over the owner of the premises.[11]

   However, an exception exists in the case of an employee who finds lost property in the course of his or her employment. In that case, the employer or owner of the premises has a superior right to possession, unless he or she agrees to turn the property over to the employee as a reward for honesty.[12]

   Some states have passed statutes intended to promote the return of found property and, in the event the owner does not claim an article within a certain time period, to reward the finder by giving him or her full ownership rights. These statutes change the common law rule that always permitted the true owner the right to recover lost property at any time.

3. *Abandoned property* is property deliberately discarded by the true owner, who has no intention of ever reclaiming it. Abandoned property becomes the property of the finder in most states. There is no duty to hold, protect, or return abandoned property, or to turn it over to the police, as is true of mislaid and lost property.

Protection of Intangible Property

The law protects intangible property rights in a variety of ways.

1. A *trademark* is an identifiable mark, symbol, or device that is fixed, stamped, or printed on tangible goods; such as the Coca-Cola legend fixed to its containers. The federal Lanham Act of 1946 permits a trademark to be registered by the owner or user.[13]

   Protection depends on the use to which the goods carrying the mark are put. If use is continuous, no other person may use the mark without prior consent. Such unauthorized use is called *infringement*, and the original owner may sue to stop the infringement and recover for damages suffered as a result of the unauthorized use.[14]

   Use by consent takes the form of a licensing or franchise agreement where the owner transfers the use of a trademark to another for a fee, often called a royalty.

2. A *service mark* performs a function similar to a trademark. Such a mark identifies and distinguishes the services of one owner or user from another. A service mark need not be attached to goods, but may be registered and will provide the protection given property under trademarks.

3. A *patent* is a government grant that provides an inventor the exclusive right to make, use, and sell his or her discovery for a period of 17 years. Subjects that are unpatentable are laws of nature, physical phenomena, and abstract ideas.

   The term patent includes not only tangible items, but a process separate from the article created by that process. The process, to be eligible for a patent award, must consist of *new and useful* acts or a series of acts performed on a particular subject whereby the subject is made into a different thing.

   Patent law is intended to create a reasonable balance between two competing economic interests: 1) the right to promote free enterprise or competition in property transactions; and 2) the right to encourage invention by rewarding inventors with temporary exclusive rights over the fruits of their inventive efforts.[15]

4. A *copyright* is a means of protecting the *form* of original literary, musical, dramatic, pictorial, audio, and audiovisual works. Some foodservice or hotel operators copyright especially creative menus or advertising materials. The ideas, concepts, or methods of operation themselves are not subject to copyright. To obtain protection, registration is required by application to the Register of Copyrights, together with payment of a nominal fee and with one or two copies of the work. No public distribution of the work must be made by the applicant prior to the effective registration date, since prior public distribution makes the copyright unavailable. The period of life of the copyright is 50 years beyond the death of the last surviving author. No right to renew the copyright exists.[16]

   Once granted, the copyright gives the holder an exclusive right to reproduce, perform, or display the work. An exemption is granted for fair use by the public. Fair use means a use without the consent of the copyright holder. In broad terms, fair use means very limited noncommercial use of copyrighted work for teaching, research, criticism, comment, or news reporting.

   An unauthorized use is called an *infringement* of the copyrighted work. *Innocent infringement*, if proven, makes the infringer liable for any actual damages the copyright owner can establish. Independently, a court may award not less than $250 and not more than $10,000 in punitive damages if no actual damages are suffered. Willful or deliberate infringement permits heavier fines to be imposed together with a one-year prison term.

   Fair use requires a case-by-case analysis of the facts by the courts. Four factors are usually examined: 1) the purpose and character of the use; 2) the nature of the work; 3) the amount of the material used in relation to the copyrighted work as a whole; and 4) the effect of the use on the copyright holder's potential market for the work.[17]

Foodservice and hotel operators are most likely to encounter copyright laws in the use of music in their facilities. A concern in the industry is the controversy over whether operators must pay royalties on recorded copyrighted music played by means of radios and loudspeakers. The issue has not been fully resolved.[18] Music played by means of coin-operated jukeboxes or radios that are rented is currently exempt from copyright laws, unless the restaurant operator owns the machine or refuses, on request, to identify its owner.

The general rule for live music is that songs played by a group or individual performer on a song-by-song basis are not a problem. Song writers, however, may sue for royalties if their songs or arrangements are played exclusively in your operation; for example, if you play country music in your western-type bar and grill, chances are you may be playing a few favorites by specific songwriters, and you may be required to pay them royalties for use of their songs.

Two songwriting societies own most of the music rights to recorded songs and represent performers and songwriters. You may have to pay one or both of these societies royalties for rights to use their songs as well as copyrighted music performed by their members. They are the American Society of Composers, Authors, and Publishers (ASCAP) and Broadcast Music, Inc. (BMI). The best method is to choose songs from one society and you may only be required to pay royalties to it, thereby limiting this business costs.

Another way to sidestep conflicting claims and contracts when playing recorded music is to contract directly with those companies offering "piped-in" music, such as Muzak. They handle payments of royalties to performers.

Purchases of Property from Nonowners[19]

What happens when you purchase property from someone who does not own it or have a legitimate interest in it—for example, if you unknowingly purchase goods from a thief?

The Uniform Commercial Code makes a distinction between a *void* and a *voidable* title. A void title to property means that the seller has nothing to sell. A thief does not obtain title to property through a criminal act. Since the thief does not have a title to transfer, you as the buyer cannot receive title. Even if you act in good faith without knowledge that the goods were stolen and pay for the goods, their transfer to you does not affect the rights of the true owner to recover them. An innocent buyer's only right to recover losses is through the thief.

Title is *voidable* when a seller obtains goods by fraud or trickery and resells them for value without giving the buyer knowledge of the fraud. Here the buyer can rightfully keep the property free of any claims by the true owner. However, the law does not permit a wrongdoer to benefit entirely from wrongdoing. It permits the true owner to recover from the fraudulent party. But since the seller had some title, the seller may transfer the property to a buyer without notice to the owner. If the purchaser knows about the fraud, then he or she must either return the property or pay the true owner for the goods.

You should exercise care when buying goods that seem too good a bargain to be true. The goods may be stolen and without title, and you may be forced to return them or make payment to the true owner.

Financing Purchases of Equipment, Fixtures, and Inventory

Hospitality operators may wish to finance the purchase of foodservice equipment, food and beverage inventory, and fixtures, rather than use cash.

To protect their interest, creditors or lenders will want to obtain a *security interest* in the property and to make sure that their interests have priority over claims of other creditors. A security interest is the right of the lender or seller to take back property if the debtor defaults on the payment for the item. Article 9 of the Uniform Commercial Code applies to these transactions.

Article 9 defines *equipment* as "goods used or bought for use primarily in business"; *inventory* as "goods held for sale or lease or for use under contracts of service as well as raw materials, work in process, and materials used or consumed in a business"; and *fixtures* as "goods so affixed to real property as to be considered a part of it." Equipment, in this sense, would include refrigerators, dishwashers, cash registers, phones, televisions, and smaller items such as knives and plates. Inventory would include food and beverages prepared for resale. Hospitality fixtures include a hotel's front desk and a restaurant's walk-in refrigerator, which could be sold along with the building.

Every creditor wants his or her security interest to be good against three parties: 1) the debtor; 2) other creditors; and 3) a person purchasing the property from the debtor. They accomplish this by *attaching* the property and, by doing so, *perfecting* their security interest.

Attachment is a legal word that requires: 1) a security interest in the property given by the debtor to the creditor, and 2) the creditor to give something of value in return to the debtor. For example, the creditor will give you the dishwasher you want in return for the collateral you put up as security if you don't pay for the dishwasher. A written security agreement is required unless the goods to be financed, the collateral, are already in the hands of the creditor. This agreement may cover *future advances* to be made by the creditor to the debtor. This means that you may obtain additional credit on the strength of the *collateral* already in existence, and need not provide additional collateral to do so. This agreement may also create a security interest in *proceeds of the collateral* (cash received from your use of the collateral) and in *after-acquired* property you purchase (new equipment, inventory, or fixtures).

The security agreement, to be effective, must be signed by you and must contain an accurate description of the property for which the interest is created. A failure to provide an adequate description may cause the security interest to be insufficient, meaning that the creditor's rights will be jeopardized.[20]

The second requirement, *perfection*, is the legal claim that a creditor or seller needs to take action if the bill goes unpaid. Perfection requires use of one of the following methods: 1) public filing of a *financing statement*; 2) taking possession of the property (collateral); and, in more limited cases, 3) attaching the security interest in the article sold.

A *financing statement* is filed at either the secretary of state's office (state capital) or at the local recorder of deeds office (county seat), depending on the type of collateral involved. The filing is good for five years in the absence of a maturity date. This initial period can be extended for an additional five-year period by filing a *continuation statement*. When you have paid off the loan, you are entitled to receive a *termination statement* that clears your financing statement from the records.

As for *priority of payment*, the first security interest to be perfected has priority over any interests perfected later or any that are not perfected. If none of the security interests are perfected, then the first security interest to attach has priority.

Where fixtures are collateral, it is important to note that special care is required of a creditor who wishes to maintain priority over other creditors who obtain security on the land and building. Otherwise, once the fixtures are permanently attached to the real estate, they will lose their identity as personal property, and the creditor will find his or her claim inferior to that of a real estate creditor.

What happens to the property if you default? Unless you agree otherwise, the creditor (secured party) has the right to take physical possession of the property (collateral). If the property is bulky or hard to move, the creditor may sell it at a public or private auction. However, the creditor must act in good faith and the sale must be commercially reasonable. The proceeds of the sale are usually distributed as follows: attorney's fees, satisfaction of indebtedness, junior creditors, and the balance to you as the debtor. In most cases (unless otherwise agreed), the creditor may obtain a deficiency judgment against you personally if the sale proceeds are insufficient to cover the loan balance. Where you have paid 60 percent of the debt, and the property secured is consumer goods, the creditor must sell the property, and the debt is canceled if the sale makes up for the debt. If you have paid less than 60 percent of the debt, the creditor, after notifying you, may keep the property and cancel the debt.

REAL ESTATE

Real estate is all immovable property, consisting of land, buildings, trees, crops, and other natural or artificial improvements permanently attached to the land, including mineral and water rights under or on its subsurface and air rights above it.

Real estate is treated differently in law from other forms of property—more rights and responsibilities go with ownership of real estate. Foodservice operators will encounter more restrictions on ownership of restaurant real estate than on personal real estate or property used in other business ventures.

Personal property attached to real estate and associated with the land is known as a *fixture*. Fixtures are treated as real estate and are transferred to a new owner with the land unless otherwise provided in the sales contract. The *intention* of the person affixing it to make it a fixture and the actual *affixtion* of attachment determine whether personal property will be treated as a fixture by the law. For example, the lighting and paneling installed by Peter Roberts and his son in the Case in Point were permanent fixtures. Floors, ceiling, plumbing, central air-conditioning equipment, walk-in refrigerators, custom-built cabinets, and storm windows all qualify as fixtures. The test of intention is *objective*, meaning that the court will examine the words and actions of the owner and not attempt to determine unexpressed thoughts.[21]

Trade fixtures are items of personal property installed by an occupier of land under a lease contract permitting the use of land owned by another for a specific purpose. Trade fixtures also apply to commercial leases. If fixtures, such as window air conditioners, shelves, and partitions, are removable without materially altering or changing the premises, and the lease does not treat them as part of the real property, then they are removable at the end of the lease by the tenant. The tables, chairs, tablecloths, flatware, phones and computers purchased by the tenant are movable.[22]

Land Ownership

Land ownership is classified according to its type, duration, and the amount of interest a person holds in the property. *Freehold* estates are held indefinitely; *non-freehold* estates are restricted and may only be held for a specified time.

Types of Ownership   If you obtain land for an operation or facility, you should be aware of the type of ownership conveyed to you. Some types are restrictive of owner rights.

1. The most complete land ownership is ownership in *fee simple absolute*, or in *fee simple*. A fee simple estate is an estate without restriction on its disposal or use. This estate gives the owner the largest bundle of rights and powers the law can provide. As long as the owner does not interfere with the rights of others, he or she may use the land unconditionally, sell it, or convey it by inheritance to heirs through a will. The give it away. The owner may *commit waste*, meaning deplete the land of value without any requirement that trees, crops, herds, and other renewable items be replaced or restored; nor account for natural resources such as minerals, or set up any replenishment reserve.[23] Short of any local zoning and fire restrictions, the land is the owner's to use or misuse.

2. A *life estate* is non-freehold (meaning restricted in duration), and is a less absolute form of ownership. Ownership rights are conveyed for life only by the owner to a *life tenant*.

   A life tenant is a non-owner, an occupier of land, and has fewer rights than an owner in fee simple. This tenant may not commit waste without the owner's express permission, since doing so would adversely affect the value of the land to the future owner, who obtains the land when the life tenant dies. The rights of the future owner are called *future interests*.[24]

   A life tenant can mortgage, or *encumber*, the land, meaning it can be put up as security to finance improvements.

   The following example illustrates the rights and restrictions of a life estate.

   Jerry Jamesway, a wealthy restaurant executive, conveys valuable real estate to his daughter Emily for life, and after her death to his nephew George Boomer. The land has potential value for development of a restaurant. In the conveyance to Emily there are no restrictions against development. Lacking the money needed for development and not wishing to commit her own money, Emily mortgages the land and obtains construction financing from a group of local banks. The banks require a 99-year mortgage commitment from Emily as a condition of the financing. Emily is now 35 years old and in all likelihood will die prior to the end of the stated period.

   Can she mortgage the land? Yes, but only with the consent of her cousin, George. Her authority to mortgage the land ends with her death, and the mortgage cannot extend beyond her lifetime, because George's rights are affected as the holder in the future interest in the land. By consenting to the long-term mortgage, George has voluntarily given up his right to object to the mortgage. In this way, the interests of Emily, George, and the banks are protected.

3. Landholding by *possession* is another right in land. Only non-freehold rights are created. No interest in ownership exists, but merely the right to *possess* and use the land for a specified period of time. This form of land rights is most commonly known by its paper contract, the *lease*. The owner is normally called *landlord* and the occupier a *tenant*. A lease gives the tenant the right to occupy and use land for a specific period of time. Those starting a restaurant may want to consider this form of land use.

Leasing Real Estate

Non-freehold estates include *tenancy for years, tenancy from period to period, tenancy at will*, and *tenancy at sufferance*. In each case the owner (landlord) leases use and possession to an occupier (tenant) but qualifies the tenant's right to possession by reserving the right to evict the tenant for failure to comply with the lease. Nonpayment of rent, holding the property over the date specified for surrender, committing waste, and improper use are examples of situations that permit landlords to remove tenants.

Tenancy for years is the most common form of tenancy, and its duration is specifically stated in the lease. When the lease term or duration is over, possession returns to the landlord, unless the lease contains a renewal clause or extension provision.

A *tenancy from period to period* is created when the lease does not specify any duration, but requires that rent be paid at certain times; for example, on the first day of each month. Such language creates a tenancy form month to month. Since the duration of the occupancy is not specified, the landlord must notify the tenant of termination of the lease. This notice provision is fixed by state statutes, with 30 days the normal requirement when a month-to-month tenancy is created.

A *tenancy at will* is one created by the landlord for as long as he or she desires. Either the landlord or the tenant can terminate a tenancy at will without notice. It usually comes into being when a tenant remains in possession of the premises after a tenancy-for-years lease ends—the legal term is *holds over*—with the consent of the landlord. The death of either party terminates this form of tenancy.

A *tenancy by sufferance* or mere occupation of the land ("squatting") is not recognized by the law.[25] The tenant is a trespasser who has no right to possession. The landlord may create a lawful tenancy by accepting rent from such a squatter, but is not required to do so.

Non-freehold estates create tenancies and possessory rights that limit use, with the owner always retaining the right to recover possession should the tenant fail to comply with requirements for use. We will examine the rights and responsibilities of the landlord and tenant.

Landlord Responsibilities   In each lease of premises, whether for residential or for business purposes, the law requires a *warranty* or guarantee by the landlord that the tenant will have the sole right to possession, and a *covenant* or promise that the tenant will have *quiet enjoyment* of the premises. Quiet enjoyment means that the landlord will not evict the tenant except for misconduct.

However, a tenant may *want* to leave under certain circumstances. For example, Jill Watson, a landlord, leases a restaurant to Michael Bailey. Bailey takes possession, not knowing that part of the premises is not available for occupancy because of Watson's failure to exterminate vermin and rodents after he promised to do so. The infestation is a constructive eviction, giving Bailey the right to terminate his tenancy without liability for rent. A *constructive eviction* refers to conditions brought about by the landlord, either through negligence or activity, that make the property untenable. Under such conditions, the tenant may be forced to leave and may not be held liable for rent.

Under common law, a landlord was not under any obligation to repair premises rented to a tenant or to guarantee that they were suitable for the particular purposes for which they were rented. Only if the landlord voluntarily agreed to do so under the lease was he or she responsible for repairs or liability. Under most state statutes today, residential tenants are protected by an *implied* promise, one imposed by the law, that premises are fit to live in. A landlord cannot renounce this guarantee unless both parties agree to do so. Most recent decisions also place duty upon the landlord to repair and maintain the structure and all its *common areas* and *permanent fixtures*, such as corridors, public sidewalks, and stairways.

These legal duties are generally limited to protecting residential tenants; they have not yet been widely extended to commercial tenants. This means that if you are seeking to rent space for your foodservice operation, you cannot rely on the protection given apartment dwellers. Your only protection rests in your making a *careful inspection* of the premises before signing any lease and by insisting that the lease contain promises by the landlord to keep the premises in repair.

An exception in the law exists, however. When a landlord rents premises for purposes that involve admission of the public—such as patrons of a bar, tavern or restaurant—and at the time the lease is signed or renewed, the premises are in a dangerous condition, *the landlord may be held responsible for patrons' injuries caused by the dangerous conditions.*[26]

Tenant Responsibilities   The most basic responsibility of a tenant is to pay rent to the landlord, in either money, labor, or services. A lease not only transfers possession, but is a contract requiring the parties to exchange something of value to make their agreement legally enforceable. Usually the rent amount and when it is due are spelled out in the contract. When the rent amount is not contained in the lease, the duty to pay is limited to reasonable rent, due only at the end of the lease period.[27]

Lease Terms   The lease is a contract, and must be negotiated in the same way as any contract. Some states will enforce an oral lease of residential premises if the lease does not exceed one year—but not for commercial property.[28] A written record is always desirable, in either case, to avoid disputes. In a dispute, oral leases must be proved in court by oral testimony, and the credibility of your testimony is usually left for a judge or jury to decide. A clever liar may make a better witness than a person who is truthful, but nervous.

A lease creates far more restrictive rights for the tenant than in the case of outright sale. This is so because with a lease, ownership of the property remains with the landlord. A lease may include an option to purchase, but that is not usually exercisable until the lease period is over.

A landlord is always free to sell the real estate to anyone else. Does the sale terminate the tenant's rights under a preexisting lease? No. The buyer takes the real estate with existing leases in effect.[29] Nor can a landlord shift liability for his or her negligence to the tenant.[30]

Oppressive commercial lease terms imposed due to unequal economic bargaining power can be eliminated by a court. This doctrine is not limited to residential leases. One example of unconscionable conduct is a landlord's unjustifiable refusal to renew a restaurant's lease after the tenant has spent large sums of money to improve the premises and build up the business. Generally courts look with disfavor on such refusals.[31]

When leased premises are destroyed by fire before the tenant takes possession, the tenant may terminate the lease without liability to the landlord.[32]

What happens if destruction of the premises—by, say, a restaurant fire— occurs after the tenant moves in? Leases usually provide that the landlord must rebuild or restore the premises within a reasonable time, with a further provision that the lease automatically terminates in the event the landlord fails to do so.[33]

In most leases the landlord reserves the right to cancel the lease in the event of sale of the property by giving notice to the tenant. Since such a right can drastically affect the use and occupancy of the premises, the courts construe the exercise of that right strictly. Only the landlord may exercise this right, not the prospective purchaser. Unless the lease provides otherwise, notice of cancellation must be served personally or by registered mail.[34]

The duty to repair the premises usually rests with the landlord if the repairs are structural, or out of the ordinary, such as the replacement of built-in refrigeration units or other immovable fixtures. The tenant is responsible for repairing the interior of the premises, trade fixtures, windows, and items that are nonstructural. The parties should define their respective duties in the lease.[35]

What about substantial structural improvements made by the tenants? Unless specifically permitted, or to be undertaken at the landlord's cost and expense, such improvements become part of the real estate and belong to the landlord when the lease expires. This means that the tenant has no claim for reimbursement against the landlord, even though such improvements may enhance the value of the real estate.[36] *Neither party should make substantial improvements without a clear understanding of who is to pay for them.*

The Case in Point at the beginning of the chapter illustrated this issue. The difference is between trade fixtures and property permanently attached to the building. Unless the lease provides otherwise, a tenant may remove trade fixtures when leaving. However, fixtures that cannot be removed without damaging the building are not treated as fixtures according to law *and belong to the landlord.*

Trade fixtures are usually items of property installed by a tenant at his or her expense to operate a foodservice business, with the understanding that they are to be removed by the tenant when the lease ends, unless both the landlord and tenant agree otherwise. A tenant who removes items of property that are not legally trade fixtures may be liable not only for the difference between the value of the premises with the items left intact and the value with them removed, but also for damages caused by the improper manner in which the items were removed.

When a landlord is to construct a new building for you, there may or may not be an implied warranty that the property is habitable. Some courts imply such a guarantee for property constructed for commercial purposes, but it is up to you to insert a clause assuring habitability where state laws do not automatically require it.[37]

Previous mention was made regarding a landlord's liability to patrons of a tenant's business, when the landlord knowingly rents premises to a tenant for a business open to the public.[38] A tenant is liable to business patrons for negligence or willful misconduct occurring on that portion of the premises exclusively occupied and controlled by the tenant.[39] A landlord is responsible for common areas used by other people and not exclusively by your patrons and employees.

Can a landlord grant, through a lease provision, a tenant's exclusive right to operate a business within a specified locality? This matter is of critical importance, since the value of the tenancy may be diminished if the landlord is free to lease to a competing business later on. Generally, such a provision is legal as long as it is not inserted for the sole purpose of enhancing revenue for the landlord, and is necessary to protect the legitimate economic interests of the tenant as well as the landlord.[40] Care must be taken to determine whether such exclusive arrangements violate state or federal antitrust laws. Usually, such a provision will be upheld if the duration of the lease is reasonable.[41]

You must know whether you have a *gross lease* or a *net lease*. In a gross lease the tenant's sole responsibility is to pay rent. All other expenses, such as real estate taxes, insurance, and special assessments, are paid by the landlord. In a net lease, the tenant assumes all of these costs. This means that any increases in these costs during the lease become liabilities that the tenant must bear.

Lease Assignment   An *assignment* is a transfer of the entire interest in the leased premises, in which the tenant/assignor gives up any right to the property. When a tenant under a lease assigns the lease to a third person, that tenant does not terminate the obligation to pay rent to the landlord. An assignment merely transfers the obligation to pay to another, but does not terminate or extinguish the original tenant's obligation to the landlord. If the *assignee*, the person to whom the tenant assigned the lease, fails or refuses to pay the rent, the landlord may sue the *assignor*, the original tenant, for the rent due.

Generally, leases contain a clause requiring the landlord's written consent to any assignment, with a further requirement that such consent will not be unreasonably withheld or refused. However, the law does not imply any duty on the landlord's part to assign a lease. This is a matter of negotiation between landlord and tenant.

Subletting   A *sublease* differs from an *assignment*, in that a sublease gives the original tenant the right to reenter if the rent is not paid and thus involves a transfer of less than the tenant's total interest in the lease. The act of subletting even when consented to by the landlord, does not relieve the tenant of the obligation to pay rent under the original lease. Leases often prohibit subletting without the landlord's consent.

Destruction of Premises   Under the common law, destruction of the premises through no fault of a tenant did not relieve the tenant of the obligation to pay rent. Nor did it permit the tenant to terminate the lease. Today residential statutes qualify this rule by suspending rent payments until the premises are restored by the landlord. Commercial tenants are not afforded relief as an implied legal right and *must protect themselves by inserting such a provision in the lease*. To obtain suspension of rent, the destruction must make the premises totally unsuitable for use.[42]

Abandonment   A common leasing problem arises when the tenant unjustifiably, without the landlord's consent, abandons the premises. This does not release the tenant from liability for rent for the balance of the lease term. The landlord must minimize or *mitigate* damages and make a good-faith effort to relet the premises and thus lessen the original tenant's liability—at least in residential tenancies.[43] There is a dispute among the courts as to whether the landlord's duty to attempt to lessen damages for tenant liability applies to commercial tenancies in the absence of language in the lease. It is best for a tenant to insert a provision regarding abandonment in the lease.

Acquiring or Transferring Ownership of Real Estate

Transfers of title to real estate may be accomplished in a number of ways.

1. As with personal property, an owner may transfer real property by *will* or *inheritance*.
2. The most common way of obtaining real property is through *purchase*. Buying real estate for a hospitality operation requires a lawyer and knowledge of the local zoning regulations.

Deeds   Typically, legal transfers of real estate, including hospitality property, are accomplished through either sale, inheritance, or gift. In all cases of transfer, a *deed* must pass between the previous owner and the new owner. A deed is a written instrument by which legal rights to own as well as use property are established. Unlike a lease, which is a contract and requires an exchange of value between the parties to make it enforceable, a deed does not require *consideration* or exchange of value. This is so because a deed is not a contract but merely a *document of title*, a piece of paper legally accepted as proof of a right to own and use property.

A valid deed requires the names of the buyer and seller, words showing the intent to transfer the property, an adequate legal description of the property, the signature of the seller, and date of delivery to the buyer.[44]

Contract of Sale

The *contract of sale* for land is also critical to the buyer of land for commercial purposes. Most real estate transactions must be in writing to be valid.[45]

This document defines the kind of title the buyer will receive; how, when, and where the seller will prove that the title is good; and what will be done and by whom if the title is found to be defective. Title insurance should be obtained to resolve these problems by allocating these risks to a title insurer. Other important matters that should be resolved in the contract are:

1. If the land on which a hospitality business is to be operated is vacant, do any zoning or environmental ordinances hamper, limit, or prohibit the use of the property? For example, an area zoned for residential use only is a serious limitation if you wish to operate a restaurant. If an existing building is to be renovated, are there any special restrictions, such as off-street parking requirements?

2. Will the structure comply with height, depth from the street, and other requirements of the local building code? If an existing building is to be renovated, will the building meet current electrical, water, and fire requirements? A building built prior to the enactment of new codes is often exempt from the newer requirements; this is called a *grandfather clause*. However, when an existing building is renovated or a new form of business is established, then the building must be brought up to the new standards.

3. Does the intended business require the issuance of a permit or license? For example, many restaurants wish to sell alcoholic beverages. This usually requires a license issued at the discretion of a state licensing body. The licensing law restricts the distance of the business from churches, schools, and banks. The personal character of the applicant, and freedom from conviction for felonies or crimes involving moral turpitude, are also of interest when licenses to sell liquor are reviewed. Without the license, sale of alcoholic beverages included in profit projections would be forbidden.

4. Is there any likelihood that the property will be subject to a government takeover for a highway or for other public purposes?

5. If the purchase requires outside financing, what happens in case the financing does not materialize, or the cost of financing becomes too high?

6. Is the business planned on the property insurable? If so, are the rates affordable? Who is responsible for fire or other catastrophes that destroy the premises before title passes?

These and other questions may be handled by inserting language in the contract of sale that would terminate the contract automatically if any problems arose that could not be corrected, removed, or modified.

The contract, once signed, is binding on the parties regardless of what happens later. The fact that the property is destroyed before the buyer obtains title and possession does not terminate the contract unless the seller has agreed in the contract to do so.

Hospitality operators must anticipate what may be required in terms of zoning, building and environmental ordinances, licenses, permits, insurance, and cost and availability of financing, and negotiate the contract accordingly.

The old Latin phrase *caveat emptor*, "let the buyer beware", has even more meaning for sales of real property than of personal property. A mistake in real estate is usually very costly.

In residential sales, there is a trend to protect buyers through specific rules of law.[46] *Potential buyers of property for commercial use, however, must protect themselves in the contract.*[47]

Subsurface and Air Rights

Subsurface and air rights that go with land may be extremely valuable, and may be transferred or sold to third persons for development and use separate from the land itself. Any unjustified entry on privately owned subsurface or air space is a trespass or intrusion that may be halted by a court order and for which damages may be recovered.[48]

The Decision to Purchase or Lease

Whether to buy or rent is as much an issue for foodservice operators seeking space for a foodservice operation as it is for people seeking a place to live. There are advantages and disadvantages to each. **Exhibit 11.1** gives the factors restaurateurs should be concerned with when making a decision.

In times of high interest rates and a tight real estate market, buying property may not be the wisest business choice. On the other hand, leases generally do not allow the tenant to build up any equity in the property, even though an operator may have to spruce up or make additions to a building to make it into a restaurant or to make it attractive to potential patrons.

The best guide is to use long- and short-term business plans when making a decision. For those starting out in the business, leasing with an option to buy may be the best choice.

Long-term plans should allow for some expansion, not just in the restaurant building itself, but in the parking facilities and the surrounding area. For example, if you are in a high-density area, and the zoning laws require off-street parking, you should choose an "expandable" chunk of land for your restaurant. Parking is one of the biggest sources of headaches when it comes time for restaurateurs to expand.

Financing the Purchase of Real Property

In most cases a buyer of real estate does not have enough money to buy vacant land or land with a building. The buyer must seek a loan from a bank or other financial institution. To persuade the bank to approve a loan, the purchaser usually gives the bank a *mortgage*. This is a form of *security interest* or *lien* on the property, giving the bank the right to *foreclose* or take possession of and sell the property if the borrower does not repay the loan. A security interest is interest in personal property, which secures payment or performance of an obligation.

Exhibit 11.1 Purchase versus Lease

	Purchase	**Lease**
Interest Transferred	Title; ownership complete	Use and occupation (possession)
Method of Transfer	Sale, gift, inheritance	Contract of lease. No gift or inheritance without landlord's consent
Rights	Unlimited	Limited by contract
Financing	More likely	Less likely
Warranties	a. *Title*—full or partial b. *Premises* if building new, otherwise none implied in commercial sales c. Quiet enjoyment	a. *Title*—none b. *Premises* if residential; otherwise none implied
Taxation	Cost of premises depreciable for federal and state income tax purposes	Rent payments deductible as business expenses for federal and state income tax purposes
Assessments	Payable by buyer	Payable by landlord unless net lease negotiated
Liability	Buyer primarily liable for third party injuries	Landlord primarily liable for structural defects—both landlord and tenant liable for injuries to third parties
Fixtures	Buyer entitled if attached to real estate and seller intended to transfer	Landlord retains all fixtures except trade fixtures
Termination	Fixed by nature of estate or interest conveyed	Fixed by contract—never to exceed owner/landlord's estate or interest
Remedies—Default	*Seller's default* a. Rescission or cancellation of contract b. Lien as security for repayment of downpayment c. Compel seller to perform d. Damages *Buyer's default* a. Retention of buyer's downpayment b. Forfeiture of installment contract c. Foreclosure d. Rescission or cancellation e. Damages f. Compel buyer to perform	*Landlord's default* a. Surrender premises without liability for rent b. Damages c. Compel repair and restoration d. Compel performance e. Enjoin or halt violations *Tenant's default* a. Evict b. Accelerate lease and recover total rent due c. Compel performance d. Enjoin or halt violations

When the cost of the loan is too high, the seller may agree to lend the buyer the purchase price in exchange for a mortgage. This is a *purchase money mortgage*. Sometimes a bank will agree to make a first mortgage loan for part of the total purchase price, and the seller will lend the balance, using a second purchase money mortgage. A first mortgage means that the bank's security interest—its right to sell the land and liquidate its loan—is superior to that of the seller. Only after the banks loan is repaid does the seller have the right to claim any remaining proceeds under the second mortgage.[49]

A mortgage is also a *conveyance of land*, but its main use is to provide security for payment of a debt. Typically the lender, the *mortgagee*, seeks a regular return on a safe investment and does not wish to assume any management responsibilities. The lender does not want to have to take possession in the event of a default in mortgage payments by the buyer/mortgager. The lender may require the borrower to execute a mortgage note making the borrower personally liable if the mortgaged land and building do not yield sufficient funds from foreclosure to satisfy the mortgage debt.

Existing Mortgage   Often a buyer will want to take over an existing mortgage to obtain property since the interest rate on such a mortgage may be lower than the current rate. The bank or other lender must agree to the takeover. In such cases it is important to understand the difference between a buyer who *assumes* an existing mortgage and one who *takes subject to* a mortgage.[50] The buyer who assumes a mortgage obligates not only the mortgaged land and improvements, but also personal assets, to possible foreclosure. The personal assets may make up the difference between what the land and improvements yield and the balance due on the mortgage. The buyer who takes subject to a mortgage is usually not personally liable for payment of any deficiency. The difference is critical to a buyer who does not wish to risk tying up personal assets for a mortgage loan on a restaurant operation. If you have a choice between assuming and taking subject to an existing mortgage, your natural choice would be to take subject to the mortgage.

Foreclosure   If a mortgager defaults on a mortgage loan, the lender has the right to take the property through foreclosure.

The normal method of foreclosure is the *foreclosure lawsuit*. In such cases, the court will order a public auction of the property. Up to the time of sale, the borrower, the borrower's spouse, or even a tenant of the mortgager, may step in, pay off the mortgage, and halt the sale. This right to prevent foreclosure is called the *mortgager's equity of redemption*. The period in which these parties have the right to redeem the property is usually fixed by state statute. Notice of default must usually be recorded and a stated period of time must elapse before the sale is held, in order to enable the mortgager to pay up.[51]

The *doctrine of unconscionable conduct* is applied to foreclosure sales to protect the mortgager. *This means that foreclosure will be denied unless the borrower is three or four payments behind and a reasonable effort has been made to settle without foreclosure.*[52]

If you buy property through an auction, you may get a real bargain for the money, but you must be cautious. If you become an owner of property through *foreclosure*, you must see to it that all current insurance policies are endorsed to you or that new policies are issued. Such coverage includes workers' compensation, public liability, dramshop, fire, and other insurance. This is especially important if a restaurant is to be operated on the property. If there are existing leases, they should be examined to see if they have been voided after foreclosure. If so, new leases should be prepared. You may want to prepare new leases just to add new provisions to protect yourself. The real estate, federal, and state tax ledger should be reviewed. The building itself should be checked for compliance with all federal, state, and local laws governing protection of tenants, patrons, and employees; and for any condition that could reduce the marketability of the property.

In the case of a purchase money mortgage, the seller and the lender are *one and the same*. Any default in payment of the mortgage should trigger recovery of the property through normal foreclosure proceedings.[53]

Financing a Lease

Lease Mortgage   A tenant with a lease may mortgage his or her leasehold interest *unless the lease specifically prohibits it*.[54] This is because the leasehold may be of enough value to cause a bank to make a loan to the tenant with the leasehold as collateral. However, a lender may not approve a loan without assurance of the continued existence of the tenancy. For example, you may want to use your lease as collateral to finance improvements on your restaurant. If you have a five-year lease, the bank might approve a loan. With only a two-year lease, it is doubtful. If you violate the terms of the lease by defaulting on one or more rent payments, the landlord may declare the lease *forfeited*, or given up. If default and forfeiture occur, the mortgage interest of the lender is ended.

Sale and Leaseback   *Sale and leaseback* is a real estate transaction where the seller sells land for its full value and the buyer simultaneously grants the seller an option to repurchase the land. A new restaurant operator might not want the full obligations of ownership, and may want to consider this. However, he or she also will not have equity either, as the transaction will build up equity for the seller/landlord.

Often the buyer/tenant leases back the land to the seller/landlord for a period of years. Here the seller gets full value instead of the smaller amount he or she might obtain from a mortgage loan, and is not required to repay this amount as long as he or she does not exercise the repurchase option. Under this arrangement, the landlord benefits by not having to foreclose if the buyer/tenant defaults in rent payments. No mortgage is created; merely a sale and lease.

For the tenant, there are two major disadvantages to this arrangement. First, he or she will not benefit from any increase in the value of the land or improvements. Only the landlord who leases back the premises gains in such a case. Second, the tenant has all the burdens of ownership without the usual ownership benefits. The tenant must pay all taxes, assessments, and insurance charges as an owner would, but must get the buyer's consent to sell, demolish, or remodel the building. In other words, the tenant's hands are tied. The landlord regains possession if the tenant defaults or otherwise violates the lease.[55]

When the landowner leases land and space for a restaurant, the landlord and the lending institution are especially concerned with a prospective tenant's credit standing, since good standing is their basic assurance that the rent will be paid.

Because the tenant selected is critical to the lender as well as the landlord, the lease usually will not permit any assignment or sublease without the landlord's consent. The situation is different with a *ground lease* where the landlord rents vacant land and the tenant agrees to erect a building. In this case, the landlord looks to the building for security, and to the tenant to protect the investment by timely rent payments and by tenant financing for construction of the building. This requires the landlord to permit the tenant to mortgage the leasehold interest, and the lease should provide this. Finally, the lease should say that special notice be given to the lender of a tenant's default so that the lender can cure the default and prevent cancellation of the mortgage.

The right of unlimited assignment of leases is basically intended to protect the landowner and the lender. However, it is also of importance to tenants. A tenant's ability to obtain institutional financing for a restaurant may depend on whether he or she can assign a lease to a lender. Assignment gives those tenants the right to transfer the lease to another operator in the event of illness or disability.

Real Estate Agents

It is customary for the sellers and buyers of restaurant property to use the services of a real estate broker or agent, although the law does not require it.

Many states require that a real estate broker's agreement be in writing to be enforceable.[56] Such a contract is called a *listing contract*, and usually permits the broker to recover a *commission*, or a percentage of the selling price, by producing a buyer who is ready, willing, and able to purchase real property on the terms and conditions agreed to with the seller. The refusal of the seller to complete the sale, or the destruction of the premises prior to the sale, will not deprive the broker of his or her commission, unless the listing contract provides otherwise.[57] In a few states, this basic rule of agency law has been modified by the courts to require that the sale be concluded and title transferred from the seller to the buyer before the broker's commission must be paid. The only exception is where the failure to conclude the sale is caused by the seller.[58]

Buyers of property must be sure that the seller is either the actual owner or an authorized party for the owner. Otherwise the buyer may not have any legal rights in the property and may have to surrender it to the real owner.

In any purchase of real property, you will very likely deal with a seller through a real estate or sales agent, the latter often used in negotiating restaurant franchises. To avoid grief, you must thoroughly understand the authority of the agent to negotiate, who is responsible for the agent's commission or fee, when the agent's commission becomes due and payable, and when and under what circumstances you may hold the seller liable for the agent's false statements or other wrongdoing.

What legal responsibilities do you have to your real estate agent once a valid contract is drawn up? You must not do anything to prevent the agent from performing the objectives of the contract. You may not terminate the agency contract without just cause.[59] Both the seller and the agent must perform the contract in good faith. As a seller you are liable for any substantial damages you cause which result in wrongful interference in or termination of the contract.[60]

Your agent owes you a *fiduciary duty* (duty of trust) not to create any conflict of interest in representing you by disclosing any potential conflict to you beforehand and receiving your consent to continue.[61] This duty is called a *duty of absolute loyalty*. The violation of this duty is illustrated by an agent who, instead of representing your interests in selling property, tries to make a secret profit for his or her benefit. Another example is that of an agent who represents both a buyer and seller in a real estate sale without informing either party of this dual agency. This conduct is so frowned upon by the law that the agent must return any sales commission obtained from either the buyer or seller, even though the agreement may have resulted in an otherwise proper and profitable deal.[62]

The following case is an extreme example of an agent's dishonesty.

Henderson v. Hassur
Supreme Court of Kansas
594 P.2d 650 (1979)

Facts: Henderson, a real estate developer, was hired by Hassur to locate possible sites for Pizza Hut restaurants in Mexico. Hassur told Henderson that he owned the Mexican franchises. Along with his partner Perry, Henderson agreed to locate both the sites and landlords for the Mexican properties. Hassur agreed to pay Henderson and Perry $4,000 per site location plus 1 percent of the gross revenue derived from each location for the lease period.

While in Mexico, Henderson met a building contractor named Vorhauer, who worked with Henderson in buying one site, required by Mexican law to lease other sites, for Hassur. The cost to Henderson and Vorhauer of that site, called the Satellite

City site, was $56,000 but it was turned over to Hassur for $88,000. The profit of $32,000 was kept by Henderson and Vorhauer, who split the amount equally and drew up an agreement to split future profits from the Hassur deal. This profit was not disclosed to Hassur.

Ultimately Henderson was additionally paid $16,000 by Hassur for locating four Pizza Hut sites, including the Satellite City site.

Later, Henderson sued to recover the 1 percent of the gross revenues for each site he secured, for which Hassur never paid him. Hassur counterclaimed for breach, or violation of the fiduciary duty, owed him by Henderson. The trial court found for Hassur on his counterclaim, and entered a judgment in his favor for actual damages of $48,000 and punitive damages of $215,000, based on a court verdict for punitive damages. The judgments entered by the trial court for actual and punitive damages were upheld.

Reasoning: The Supreme Court pointed to the facts of the case and current law in upholding the lower court's finding that Henderson had breached his duty. The court then went on to examine the damages paid. The amount in dispute was the actual damages. In applying the damage award, the trial court had not only awarded to Hassur the profit amount which Henderson and Vorhauser had pocketed, but the entire commission paid for the sites. In upholding this award, the court said:

> An unfaithful servant forfeits the compensation he would otherwise have earned but for his unfaithfulness. This court considered this legal principle in *Bessman v. Bessman*, 214 Kan. 510, 520 P.2d 1210 (1974), where it is held: "As a general rule an agent who realizes a secret profit through his dealings on behalf of his principal not only must disgorge the profit, but also forfeits the compensation he would otherwise have earned."

The Kansas Supreme Court agreed with the trial court that the second agreement to share profits permeated the entire transaction, whether profits were actually shared or not.

Conclusion: This Kansas Supreme Court decision is a clear warning of how strictly the common law views the duty of loyalty and good faith required of a real estate agent or broker to a client. Actual damages may be recovered from an agent or broker for any secret profit the broker obtained at the expense of the client.

Business Tax Incentives for Property

The Economic Recovery Tax Act of 1981, among other things, provides *recovery deductions* replacing previous depreciation deductions. Under this Accelerated Cost Recovery System (ACRS), purchases of buildings may be written off in 15 years; purchases of machinery and heavy equipment may be written off in 5 years; and purchases of automobiles, light-duty trucks, machinery, and equipment used for research and development may be written off in 3 years.

In addition, leasing transactions qualify for ACRS treatment, meaning that lessees who rent buildings and equipment can shift these benefits to their lessors, should these landlords or building and equipment owners be in a better position to use them. This may be used to obtain a lower rent or better lease terms from a landlord.

Investment tax credits for building rehabilitation and renovation have been liberalized. The tax credit for structures at least 30 years old is 15 percent; the credit for structures at least 40 years old is 20 percent; and the credit for certified historic structures is 25 percent.

BULK PURCHASE OF BUSINESS ASSETS

At some point you may wish to acquire the major portion of the assets of another business to add on to your own operation or to establish a new business in another location. A *bulk transfer* is any large transfer of a major part of the material, supplies, merchandise, or inventory not made in the ordinary course of the transferer's business, such as food and beverages sold to patrons. For example, a bulk transfer would consist of the food and beverage inventory, tables, chairs, china, silverware, and other items you agree to buy from a restaurant that is closing. Article 6 of the Uniform Commercial Code covers this type of transaction.

You must be cautious. The seller of the business may have numerous creditors whose claims have not been paid. An honest seller would pay off these obligations with the proceeds of the sale. But sometimes a dishonest debtor would take the purchase money and depart, leaving the creditors empty-handed, and leaving the buyer to answer to creditors.

Under the Uniform Commercial Code Article 6 on bulk transfer, a potential buyer of a business must exercise care to see to it that the seller provides a list of existing creditors, that both parties prepare a list of property to be sold; and that the purchaser notify the seller's creditors of the proposed sale of all or a majority of the assets no later than 10 days before the physical transfer of the assets or the payment for such assets, whichever comes first.

Failure to comply with the UCC Article 6 on bulk transfer means that the assets purchased may be subject to the claims of the seller's creditors. In short, the buyer purchases the seller's obligations, as well as the assets. Only third parties who buy the assets without receiving any notice of existing creditor claims take the assets free and clear. This requirement is critically important to you if you purchase a business. You do not wish to assume someone else's obligations in the process.

Bulk transfers involve major physical assets that are included as part of the sale of a business. They do not involve transfers of intangible items, such as health and liquor licenses. You must apply for these on your own.

ZONING LAWS
REGULATING LAND USE

Once you buy or lease property for a hospitality operation, the law will continue to affect your business. No owner or user of real property has such absolute control as to keep the government from regulating some aspect of the owner-ship—least of all businesses—and of those, least of all hospitality businesses.

Unless you immediately adjoin or lease property from the federal gov-ernment, most land-use regulations come from state and local authorities. Usually the only time you will encounter problems with the federal govern-ment is if you own land in which the government is interested. Then you may find yourself in a heated battle over just compensation for your property.

There are two methods of regulating land use. One is by *private controls*, meaning that a seller of land has the right to sell on terms satisfactory to the seller, including the right to restrict the use of the land as the seller sees fit. The other method is by *public controls to protect the public interest*. This form of control is exercised for the most part by state and local governments.[63]

Zoning

Land-use regulation by public control usually takes the form of zoning. Zon-ing means allocating land into areas reserved for a specific purpose. Indus-trial, commercial, and residential uses are the most common.

As a hospitality operator you will be dealing with *commercial zoning*, or *mixed commercial and residential zoning*. Zoning laws require that a rea-sonable balance be struck between private ownership and public benefit. In a dispute, the courts will examine the character of the community, and the strain of your proposed use on utilities, transportation, and public services, as well as traffic congestion, off-street parking, noise, and other factors.[64] When the regulatory agency seeks to limit or prohibit a particular use, such as por-nographic bookstores, and that prohibition is challenged, the general test is whether the prohibited use bears a reasonable relation to public health, safety, and welfare.[65]

State and local land-use regulations are subject to review by the courts as a check against possible violations of state and federal constitutional pro-tections. A state may not regulate land use arbitrarily or unreasonably, since this conduct would be a denial of due process under the Fourteenth Amend-ment of the United States Constitution.[66] Likewise, local or urban land-use regulations may not violate similar protective provisions in state constitu-tions. A state or locality may not discriminate against classes or individuals in buying or leasing land on the basis of race, religion, or nationality.[67] Dis-crimination on other grounds, such as prohibiting entry of low-income groups, may be justified if there is a rational or reasonable basis for doing so.[68] This means that if the zoning regulation is not intended to keep out minorities, it will not violate constitutional requirements.

Usually zoning land regulation takes two forms. One form is by *eminent domain*: taking private property for a governmental purpose, such as for public school, highway, or hospital construction.[69] Regulation which totally deprives the landowner of any beneficial use of his or her property is a *confiscation of property*, meaning taking property for public use for which the landowner must be paid the fair market value of the land.[70] The other form is by the use of *police power*,[71] for example, forbidding a food and liquor operation in a family residential zone. The preferred method is by police power, because eminent domain requires the payment of reasonable compensation to the private land-owner, whereas the exercise of police power does not.[72]

Nonconforming Use

If the area around your business becomes residential and is zoned for that purpose, you can usually apply for your land to be designated for *noncon-forming use*. This will allow you to continue operating your restaurant in the area until you sell the property, at which time its use must comply with cur-rent zoning requirements.[73]

Accessory Use

Accessory use occurs when another type of business is incidental to the main type of business for which an area is zoned. Accessory use is usually upheld if it is *commonly* incidental to the main business. For example, the builder of a bowling alley, under existing zoning regulations, would probably be able to provide refreshment facilities.[74]

Variance

What if you wish to change the use of your land? Suppose you want to con-vert some existing property, a small office building, into a restaurant. Your present-use terms do not include a foodservice operation. Are you "locked in" without any alternative except to sell and build a restaurant in another area? No. A variance issued by the city will authorize you to convert your property. The following tests are used by the courts to determine whether you are le-gally entitled to a *variance*, a special use authorized on an individual basis:

1. The landowner must prove it would be impossible to obtain a reasonable return on the land under the existing regulation.
2. The negative effect on the use must be peculiar to the individual land-owner, and not also apply to others in the area.
3. The approval of the variance must not cause the remaining area to change substantially. This is the most important consideration.[75]

Landmark Designation

Zoning regulations protecting *historic areas* and *landmarks* have been gen-erally approved as well within the concept of public welfare or benefit. This court-developed doctrine is especially favored where the economy of the area depends on tourism attracted by the area's historical qualities.[76] This means that if you and other hospitality operators in the area show a positive correlation between the historic area and your tourism business, the restaurants may be designated for preservation zoning. However, an individual landmark not located within a historic district may not be "frozen" in the midst of high-density economic development.[77]

A number of restaurants and hotels throughout the United States have been designated landmarks and zoned accordingly.

Landmark status may result in problems. Usually, you may not make any structural changes in the premises without obtaining the prior approval of the local landmark preservation commission. Approved changes must meet strict aesthetic and historical requirements as to materials and designs used. Compliance with these requirements may be expensive as well as time-consuming. Someone wishing to purchase a landmark building should understand what is involved *before, not after*, purchasing.

You can appeal a landmark designation, if you can prove it would bring hardship to you and your business. The fact that landmark status might diminish the cash value of the property and that a "higher" or "more beneficial use" of the property might exist has been held insufficient proof of hardship, however.[78]

Sometimes a landowner will seek a change in the zoning law to apply only to a single property, to increase its value. Such zoning, called *spot zoning*, is illegal when the change in zoning does not benefit the public and injures the surrounding zoning.[79] For example, a firehouse in a residential area at a main traffic artery may be permitted, since they are justified by the benefit to the surrounding area.[80] A tavern, however, may not.

Incentive Zoning

Incentive zoning is a relatively new concept of particular importance to commercial land users. In this type of zoning, a city or village will provide additional floor space to a high-density commercial developer to induce the landowner to allocate certain space for "desirable" uses. This innovative land-use plan has the advantage of stimulating voluntary inclusion of public improvements by using a financial incentive, rather than by compelled use. *Compelled use* means that the developer or landowner may be discouraged from investing, since it may be unprofitable to include the required improvements. Compelled use may be adverse to the city because a new use could mean a corresponding loss of tax revenue. As a result, some cities now see incentive zoning as the preferred method of regulating land use, with the traditional compelled land-use regulation held out as a last resort.[81]

Problems with Zoning Laws

In some cases private landowner controls may conflict with public ones. Suppose you wish to purchase a lot on which to build a foodservice operation in a shopping center. At the time the lease or sales contract is prepared, the seller restricts the use of the premises to a restaurant, and the location is zoned for this form of use. Later the area is zoned for professional office use. You decide to subdivide your premises and lease a portion for use as a dental or law office. Can you do this? Probably. More recent cases allow such use, on the theory that where a change in the character of the neighborhood is involved, the zoning ordinance overrides the former owner's sale restrictions. Normally, however, the courts take a hands-off attitude, allowing the lease or sale restrictions to be enforced notwithstanding the rezoning of the area.[82]

Suppose you wish to build two restaurants in a shopping center. The zoning board rejects your application on the grounds that a moratorium exists on new restaurant construction. You sue to test the legality of the board's decision. The board then argues that your intention to build two restaurants on one lot violates the one-building/one-lot ordinance. In court you argue that the ambiguous ordinance does not apply to shopping centers. The trial court agrees with you. On the board's appeal, the reviewing court upholds the trial court's decision, saying that the court was right in accepting your argument that the ordinance is ambiguous, and having created the ambiguity, the zoning board must accept the consequences.[83] You can build your restaurants.

Or suppose you own a delicatessen exempt from city off-street parking requirements for establishments in that zone. You decide to convert the premises to a restaurant an cocktail lounge and apply for a variance to permit increased seating capacity *without* providing off-street parking facilities. The planning commission denies your application. You sue to reverse the board decision. Are you entitled to the variance? No. The variance you sought is a *use variance*, and genuine hardship must be demonstrated before the courts can overturn the board's decision. No real hardship on your part is shown to exist.[84]

Suppose you own and operate a restaurant under a nonconforming use status. You apply to the zoning board for permission to convert a lot near your premises into a parking area. You submit proof that the best use of the land would be for parking. Residential owners across the street object that such a change would reduce their property values. Permission is denied by the board. You seek to overturn the decision in court. Will you succeed? No. The adverse impact on residential properties outweighs the economic gain to you in this situation.[85]

Suppose that under existing ordinances construction of a high-rise residential hotel is permitted. No specific provision is made in the ordinance for the installation of a full-service restaurant, or of food and beverage vending machines. Are such uses authorized? Yes. The courts uniformly treat these functions as *accessory uses* which are incidental to, even though different from, the main or permitted use.[86]

Finally, suppose you are the owner of a small restaurant in an upstate New York community. From reliable sources, you learn that a large shopping mall is to be constructed on land near your restaurant. After a year's notice, you do not bring legal action to stop construction of the mall until three months after the commencement of construction. You go to court to stop the construction of the mall, arguing that the large size of the mall would have an adverse impact on the downtown area, causing economic blight and deterioration of that section in which your restaurant is located. Will the court grant your request? No. You waited too long to seek court review of the construction of the mall. Although the court had the power to stop construction, doing so was not justified by the facts.[87]

Building and Safety Codes

Most states and cities have adopted building codes that contain detailed requirements on structural safety, fire prevention, the size and number of rooms, exits, lights, heating, ventilation, refrigeration, and sanitary equipment.[88] These include general requirements that apply to all commercial buildings. In addition, foodservice owners and operators must comply with specific health and sanitary rules regarding food handling, equipment, and cooking temperatures, to mention only a few. These rules extend the scope of regulation to operation of the business itself, not merely to the structure in which you operate.

Because you deal with members of the public by inviting them to use your restaurant or hotel, and there is a greater likelihood of harm to more than one person as a result of a failure to comply with local building requirements, your risk of liability may be greater.

When it comes to safety codes, you don't have much room for compromise in most areas. Nor would you want to. The safety of yourself, your patrons, and your employees should outweigh the cost and inconvenience of complying with local codes. A failure to do so not only can cost you your business as a result of a fire or other accident, but can leave you open to lawsuits if it is proved that your negligence was the cause of the incident.

As a manager, you are just as responsible for adherence to safety codes as the owner, if you are on the premises and acting in that capacity.

Whether you are opening a new restaurant, purchasing an existing one, or running the business for an absent owner, one of your first tasks should be to study the local building regulations for businesses. In addition, you may want to institute further safety improvements or procedures. While you will still want to insure the premises, *the best insurance is prevention.*

When new construction of your premises, major repairs, or alterations to an existing building take place, most building regulations require the building to be inspected. Until the inspection is complete and the inspector is satisfied, you may not occupy or use the building. Satisfaction usually means the issuance of a *certificate of occupancy*. If you are going to buy or lease an existing building, don't close the sale or lease until you have that certificate. Without the certificate or other form of approval, you will not be allowed to operate.[89]

Suppose that at the time you purchased or leased your hospitality premises, the building code requirements were relatively light in terms of cost and time allowed for compliance. Later the state or local authorities enacted a much tougher code, which may make it unprofitable for you to stay in business. Are you protected from the new code rules because they were not in existence when you bought or leased your premises? The answer is generally no, unless the new code contains a *grandfather clause* permitting existing businesses to operate under the previous rules.[90] Even in cases where the state or city code contains such a clause, *any* alterations or substantial repairs made usually require the building owner to comply with new code requirements, or to "bring the building up to new code standards." When building codes do not include a grandfather clause, you are compelled to comply with new codes as a condition of staying in business, as long as the rules are reasonable.

When you lease a building, you must insert a clause in the lease protecting you in case the owner fails or refuses to bring the building in which you operate up to new code standards. Additionally, your landlord (owner) may try to shift this burden onto you under a lease. This means that if the owner must comply, he or she may try to obtain reimbursement from you for the costs of compliance.

How can you find out whether the owner or operator of an existing building is in violation of local codes before you buy? Violations are recorded by the department involved in the county or city clerk's office. These records must be checked before you buy or lease the building, since most laws require the seller or landlord to eliminate the violation before a sale. This requirement is extremely important to you. It means that the cost of eliminating the violation is placed on the seller, not on you. However, once you take title to a building with existing violations on record, you become responsible for the costs of removing them. Since you may not wish to delay closing the deal, you may wish to retain a portion of the purchase price as security until the seller removes the violation.[91]

Code requirements are checked periodically by the department (fire, health, electricity) involved. If any violations are found, the owner or operator—the person or party occupying the premises—must bring the premises into compliance by making the changes or repairs required. Failure to do so could result in a fine if the violation is minor, or loss of the certificate of occupancy if it is serious. Failure to comply is also a crime in most states and localities, and a serious, intentional violation can mean a jail sentence.[92]

When a building is structurally unsafe, the state or city may demolish it without being required to compensate the owner. This is so because the owner created a public nuisance, justifying the use of the police power to remove a danger to the public.[93] In some states, the appropriate regulatory body may file a *lien*, or legal claim, on the remaining land to reimburse itself for the cost of demolishing the building. This *demolition lien* often must be paid before other claims against the owner, including mortgages.[94]

Code violations can give rise to lawsuits by patrons or customers injured by a landlord's failure or refusal to comply. Originally such claims were only upheld for violations of state codes, but the present trend is to extend liability to city codes where the city is given *home rule* powers to regulate building construction and maintenance.[95] Home rule is a grant by the state to certain cities to enact their own legislation over certain matters. When the city enacts its own code, it may be tougher than the state code.

You are responsible for structural code violations if you are the owner of a business. As a tenant, you are typically not liable for structural code violations, unless you voluntarily assume such liability by agreeing to keep the building in repair under the lease with your landlord.[96]

PROPERTY MANAGEMENT

Property rights are best enforced in writing. Any purchased items—food, silverware, real estate—should be bought on the basis of a written agreement. The agreement need be no more than a purchase order or it may be as complex as a real estate contract. Not every property deal must be in writing. But if the law is to enforce the buyers' rights, it is easier to have those rights in some tangible form that can be reviewed. Otherwise it is one party's word against the other's in a dispute, and the court can decide in favor of either party.

Before negotiating any property deal, especially in real estate, a buyer must know what is critical. If leasing, is it more important to have the right to remove trade fixtures or to have a lower rent? If buying, the previous owner should settle with his or her creditors instead of leaving them for the new owner. None of this is automatic; it must be negotiated.

Regulatory bodies of all kinds can affect your decision to buy, sell, or lease real estate. Carefully review local zoning, building, and fire regulations to be sure that the kind of building to be built or leased will permit a foodservice business. Also, one may need a license or permit to conduct business, and the cost of the license should be included in financial planning.

Buyers must familiarize themselves with the business tax incentives under the current Internal Revenue Code for buyers of real and personal property. Increased and accelerated depreciation allowances can make it more financially attractive to buy real estate, inventory, equipment, and fixtures. These allowances reduce taxable income, thereby providing additional working capital to finance additions and improvements to your business. By doing so, the buyer is afforded the opportunity to generate higher revenues and greater profits.

SUMMARY

Property is defined as ownership of and legally protected rights in anything of recognized value, and is safeguarded by federal and state laws.

Property includes *real property*, such as land, buildings, and fixtures attached to real property, and *personal property*. Property may be *tangible*, having physical substance, or *intangible*, having no physical substance.

Property may be acquired through possession, purchase, gift, accession, inheritance, confusion, or by finding it.

Protection of ownership rights depends on the type of property and the rights desired. Acquiring trademarks, service marks, patents, and copyrights on formulas, fixtures, or printed material will protect the substance of the property.

To obtain complete rights in property, it must be obtained from the real owner or an authorized agent of the owner. Otherwise the buyer may not have legal rights over the property.

There are various ways to finance property. Usually creditors will want to have a *security interest* in the property or *collateral* to assure that if the loan is not paid or the borrower goes bankrupt, their claims are legally sound.

Land ownership is classified according to type, duration, and the amount of interest held.

Leasing property involves no ownership rights but the right to possess property for a specific period of time. Various types of tenancies, along with varying time periods, are negotiated between the landlord and tenant. Both have rights and responsibilities according to the law. Transfer of ownership to real estate must be in writing in many states, and even if it is not required, it is helpful in preventing disputes. A *deed*, or *title*, is one written form that passes between the buyer and the seller. There are three types of deeds: general warranty, special warranty, and quitclaim.

The *contract of sale* is the other real estate document that is necessary and should be in writing. It is the negotiable part of any real estate transaction.

Whether to purchase or lease property is a business decision that must be considered with factors such as potential expansion, zoning, interest rates, and whether the hospitality business itself is new or well established.

There are several ways to finance the purchase of real estate. Most often the lender will want the option to foreclose or take over the property if the debtor defaults. In some circumstances, a tenant may mortgage a leasehold interest to obtain a loan.

Sale and leaseback is another way of obtaining land. However, the would-be tenant must look carefully at the advantages and disadvantages before entering into this arrangement.

For those seeking a business site, real estate agents may be used to search for and handle real estate negotiations. Buyers, sellers, and agents have various rights and responsibilities.

There are various business tax incentives to buying and using property. The Economic Recovery Tax Act of 1981 provides for recovery deductions for purchases of building and leasing equipment.

Operators may want to purchase business assets in bulk. Such buyers must ensure that any seller creditors do not become the buyer's creditors.

Zoning laws are local laws used to regulate land use. Private controls, such as sales contracts between sellers and buyers, also may control land use. Mostly, zoning is used as a public control of land use by either local, state, or federal governments. Zoning regulation may include eminent domain or use of police power.

In case of new zoning regulations, a foodservice operator may be able to apply for *nonconforming use* designation or for *variance*. If a use is incidental to the designated use, then that use may be legal as an *accessory use*.

Landmark designation is also called preservation zoning, *and has advantages and disadvantages for foodservice operators*. Landowners can appeal this type of zoning before the building is designated.

Incentive zoning is a way of regulating land use by making it attractive to landowners to improve buildings and public areas.

Building codes are used to regulate food-service operations. These local laws may require a certain number of fire exits, regulate the size and capacity of dining rooms, and mandate other safety features.

QUESTIONS

1. What is the difference between *real* and *personal* property? Give an example of each.

2. What is the duty of landlords to patrons of a tenant?

3. Detail a policy, consistent with the law, for yourself and your employees regarding lost property.

4. You have rented a building for your restaurant business for five years and are moving out. Among the items that you've used or have installed are tables, chairs, a built-in bar, paintings, and a heavy crystal chandelier built into the wall. What can you take, and what must you leave? Why?

5. Name some of the advantages and disadvantages of landmark designation.

NOTES

1. See *Sears, Roebuck and Co. v. Seven Palms Motor Inn*, 530 S.W.2d 695 (Mo. 1975); *Roberts v. Yancy*, 165 S.E.2d 399 (Va. 1969) (restaurant lease dispute over what items tenant could remove at end of lease term).

2. Fifth Amendment to the United States Constitution.

3. *Id.*

4. N.Y. Penal Law sec. 165.15 (McKinney Supp. 1978), sec. 165.17 (McKinney 1969), is a representative statute.

5. See *Hebrew Univ. Assn. v. Nye*, 26 Conn. Supp. 342, 223 A.2d 397 (1966).

6. Smith, Robertson, Menn, and Roberts, *Business Law*, Fifth Edition (1982), Chapter 44, p. 901, *Paset v. Old Orchard Bank and Trust*

Co., 62 Ill. App. 3d 534, 378 N.E.2d 1264 (1978) (safety deposit examining booth).

7. Sherry, *The Laws of Innkeepers*, 3d. ed., Chapter 16, see 16:22. *Jackson v. Steinberg*, 186 Or. 129, 200 P.2d 376 (1948), *reh'g denied* 186 Or. 140, 205 P.2d 562 (1949) (hotel employee).

8. Smith, Roberson, *et al.*, *supra*, note 6.

9. Sherry, *supra*, note 7, see 16:23. *Erickson v. Sinkin*, 223 Minn. 232, 26 N.W.2d 172 (1947), citing Minn. Stat. Ann. 622.11 (1945) (hotel guest room).

10. Sherry, *supra*, note 7.

11. *Id.*

12. Aigler, *Rights of Finders*, 21 Mich. L. Rev. 664, 681 (1923) (footnote), cited with approval in *Jackson v. Steinberg, supra*, note 7.

13. 15 U.S.C. secs. 1050–1127 (1946).

14. *Steak and Brew Inc. v. Beef and Brew Restaurant, Inc.*, 370 F. Supp. 1030 (S.D. Ill. 1974); *Howard Johnson Co. v. Henry Johnson's Restaurant*, Civ. Case No. 1258 (D.N.C. 1964). Also see *Tisch Hotels, Inc. v. Americana Inn, Inc.*, 350 F.2d 609 (7th Cir. 1965); *Holiday Inns, Inc. v. Holiday Inn*, 364 F. Supp. 775 (D.S.C. 1973).

15. Lusk, Hewitt, Donnell, Barnes, *Business Law and the Regulatory Environment*, Fifth Edition (1982), at 990.

16. Lusk *et al.*, *supra*, note 15, at 987–988.

17. *Iowa State University Research Foundation, Inc. v. American Broadcasting System*, 621 F.2d 57 (2d Cir. 1980).

18. Superseded by statute, as stated in *Crabshaw Music v. K-Bob's of El Paso, Inc.*, 744 F. Supp. 763 (W.D. Tex. 1990); *Broadcast Music, Inc. v. Claire Boutiques, Inc.*, 949 JF.2d 1482 (7th Cir. 1991).

19. U.C.C. sec. 2–403.

20. *American Restaurant Supply Co. v. Wilson*, 371 So.2d 489, 25 U.C.C. Rep. Serv. 1159 (Fla. App. 1979).

21. Lusk *et al.*, *supra*, note 15, at 610.

22. *Id.*

23. Lusk *et al.*, *supra*, note 15, at 614–16.

24. *Id.*

25. Lusk *et al., supra*, note 15, at 641.

26. *Tortwick v. Lisle*, 268 Minn. 197, 128 N.W.2d 330 (1964).

27. Lusk *et al., supra*, note 15, at 649.

28. Kratovil and Werner, *Real Estate Law*, Seventh Edition (1979), secs. 1054–55.

29. Kratovil and Werner, *supra*, note 28, at sec. 1091.

30. *T. Weaver v. American Oil Co.*, 276 N.E.2d 144 (Ind. 1971).

31. See *Shell Oil Co. v. Marinello*, 307 A.2d 598 (N.J. 1973), *cert. denied*, 415 U.S. 920 (1974).

32. The landlord may insert a clause to the contrary, and then a court must decide whether such a clause is fair. If the clause is upheld, it is binding on the tenant. You must read a proposed lease carefully and negotiate to have such a clause removed. See Kratovil and Werner, *supra*, note 28, at sec. 1085.

33. *Id.* at sec. 1084.

34. *Id.* at sec. 1088.

35. Kratovil and Werner, *supra*, note 28, at secs. 1075–76, 1083.

36. *Id.* Chapter 3, secs. 29–36, 1083.

37. Kratovil and Werner, *supra*, note 28, at sec. 1003 reviews the authorities pro and con.

38. See *Tortwick v. Lisle, supra*, note 26.

39. Kratovil and Werner, *supra*, note 28, at sec. 1081. Also see *Horn & Hardhart Co. v. Junior Building, Inc.*, 40 N.Y.2d 927, 358 N.E.2d 514 (1976); *Reargument denied*, 41 N.Y.2d 901, 362 N.E.2d 640 (1977).

40. *People's Trust Co. v. Schultz Novelty and Sporting Goods Co., Inc.*, 244 N.Y. 14, 154 N.E. 649; *Davis v. Wickline*, 135 S.E.2d 812 (Va. 1926).

41. See Bergfield, *Principles of Real Estate Law* (1979), at pp. 371–72. Sherry, *The Laws of Innkeepers*, (rev. ed. 1981), sec. 12:19 (New York law).

42. Lusk *et al., supra*, note 15, at 649.

43. See *Sommer v. Kreidel*, 378 A.2d 767 (N.J. 1977); *Markoe v. Naiditch and Sons*, 226 N.W.2d 289 (Minn. 1975).

44. Lusk *et al., supra*, note 15, at pp. 625–30.

45. See *Gene Hancock Construction Co. v. Kempton & Sneligar Dairy*, 510 P.2d 752 (Ariz. App. 1973). Disavowed by *Gibson v. Parker Trust*, 22 Ariz. App. 342, 527 P.2d 301 (Ariz. App. 1974). A contract to give a mortgage for the sale of real estate must also be in writing. *Fremming Construction Co. v. Security Savings & Loan*, 566 P.2d 315 (Ariz. App. 1977).

46. See *Old Town Development Co. v. Langford*, 349 N.E.2d 744 (Ind. App. 1976 opinion superseded by 369 N.E.2d 404 (Ind. 1977).); *Pines v. Perssion*, 111 N.W.2d 409 (Wis. 1961).

47. *Service Oil Co., Inc. v. White*, 542 P.2d 652 (Kan 1975); *Van Ness Indust., Inc. v. Claremont Paint Co.*, 324 A.2d 102 (N.J. Sup. Ct. Ch. 1974); *Yuan Kane Inc. v. Levy*, 26 Ill. App. 3d 889, 326 N.E.2d 51 (1975).

48. *United States v. Causby*, 328 U.S. 256 (1946).

49. Lusk *et al., supra*, note 15, at pp. 918–21.

50. *Id.* at p. 919.

51. Kratovil and Werner, *supra*, note 28, at sec. 633 *et seq.*

52. See *FNMA v. Ricks*, 372 N.Y.S.2d 485 (N.Y. Sup. Ct. 1975).

53. See *Morris v. Weigle*, 383 N.E.2d 341 (Ind. 1978).

54. Kratovil and Werner, *supra*, note 28, at sec. 1090.

55. *Id.* sec. 572.

56. See 9 A.L.R. 2d 747 for review of legal authorities.

57. *Bonanza Real Estate, Inc. v. Crouch*, 517 P.2d 1371 (Wash. 1974); *Russell v. Ramm*, 200 Cal. 348, 254 P. 532 (1927); *Hecht v. Mellor*, 23 N.Y.2d 301, 244 N.E.2d 77 (1968); and also 74 A.L.R. 2d 437 for review of legal authorities.

58. *Ellsworth Dobbs Inc. v. Johnson*, 50 N.J. 528, 236 A.2d 843 (1967), cited and followed in *Tristram's Landing Inc. v. Wait*, 327 N.E.2d 727 (Mass. 1975), and in *Shumaker v. Lear*, 345 A.2d 249 (Pa. 1975). *But see Specialty Restaurants, Corp. v. Adolph K. Feinberg Real Estate Co., Inc.*, 770 S.W.2d 324 (Mo. App. 1989).

59. See generally *Pailet v. Guillory*, 315 So. 2d 893 (La. App. 1975) (authority of husband of owner of real estate under lease to cancel lease); *Gunn v. Schaeffer*, 567 S.W.2d 30 (Ct. Civ. App. Tex 1978) (authority of apart-

ment house manager to borrow money from tenants).

60. *Hilgendorf v. Hague*, 293 N.W.2d 272 (Iowa 1980).

61. *Id.* Also see *Montgomery Ward Inc. v. Tackett*, 323 N.E.2d 242 (Ind. App. 1975) (franchise sales agency).

62. *Henderson v. Hassur*, 594 P.2d 650 (Kan. 1979) (Pizza Hut restaurant site location agency); *Taborsky v. Matthews*, 121 So. 2d 61 (Fla. App. 1960); *Sierra Pac. Industries v. Carter*, 104 Cal. App. 3d 579, 163 Cal. Rptr. 764 (1980). *But see Rockwell Engineer Co., Inc. v. Automatic Timing and Controls Co.*, 559 F.2d 460 (7th Cir. 1977).

63. Clarkson, Miller, Blaire, *West's Business Law* (1980), Chapter 53, Zoning, pp. 878–80.

64. Kratovil and Werner, *supra*, note 28, at sec. 689, pp. 305–7.

65. See *Young v. American Mini-Theaters*, 427 U.S. 50, *reh'g denied*, 429 U.S. 873 (1976), where a municipal ordinance confining pornographic movies, bookstores, and nude-dancing establishments to a single "combat zone" was upheld.

66. *United States v. 564.54 Acres of Land*, 506 F.2d 796 (3d Cir. 1974).

67. *Southern Burlington County NAACP v. Township of Mount Laurel*, 336 A.2d 713 (N.J. 1975) (racial discrimination) *cert. denied* 428 U.S. 808 (1975); *Dailey v. City of Lawton*, 296 F. Supp. 266 (W.D. Okla. 1969), *aff'd*, 425 F.2d 1037 (10th Cir. 1970) (same).

68. *Village of Euclid v. Ambler Realty Co.*, 272 U.S. 365 (1926); *Village of Belle Terre v. Boraas*, 416 U.S. 1 (1974).

69. Lusk *et al.*, *supra*, note 1, at Chapter 32, pp. 632–33.

70. However, when the zoning ordinance (law) is found to be totally unreasonable as applied to particular land, the courts will invalidate it as confiscatory (a taking of land without just compensation). *Fred F. French Inv. Co. v. City of N.Y.*, 39 N.Y.2d 587, 350 N.E.2d 381 (N.Y. 1976).

71. Police power in this context means the exercise of the appropriate government's power to regulate land use in the public interest. See Lusk *et al.*, *supra*, note 15, at pp. 631–32.

72. *Urban Renewal Agency v. Gospel Mission Church*, 4 Kan. App. 2d 101, 603 P.2d 209 (1979); also see note 68, *supra*.

73. But you may not substantially alter the structure, meaning a physical change or a conversion of the building into a new or substantially different structure. Such a change will jeopardize the nonconforming use. *Selligman v. Van Allmen Bros., Inc.*, 297 Ky. 121, 179 S.W.2d 207 (1944). An abandonment of such a use also causes the nonconforming use to be lost. *Beyer v. Mayor and Council of Baltimore*, 182 Md. 444, 34 A.2d 765 (1943).

74. See *Newark v. Daly*, 85 N.J. Super. 555, 205 A.2d 459 (1964), *aff'd*, 46 N.J. 48, 214 A.2d 410 (1965).

75. Kratovil and Werner, *supra*, note 28, at sec. 716, pp. 321–23.

76. See Symposium of Historic Preservation, 36 Law and Contemp. Prob. 309–444 (1971); *Rebman v. City of Springfield*, 111 Ill. App. 2d 430, 250 N.E.2d 282 (1969); see also 63 Col. L. Rev. 708, 720.

77. An exception to this general rule has been made in New York, *Lutheran Church in America v. City of New York*, 35 N.Y.2d 121, 316 N.E.2d 305 (1974); and *Penn Central Transp. Co. v. City of New York*, 438 U.S. 104, 98 S. Ct. 2646 (1978), *reh'g denied* 439 U.S. 883, 99 S. Ct. 226.

78. *900 G Street Assocs. v. Dep't of Hous. and Community Dev.*, 430 A.2d 1387 (D.C. App. 1981).

79. Kratovil and Werner, *supra*, note 28, at sec. 710, pp. 318–19.

80. *Temmick v. Baltimore County*, 205 Md. 489, 109 A.2d 85 (1954) (shopping center); *Conner v. Herd*, 452 S.W.2d 272 (Mo. Ct. App. 1970) (firehouse).

81. Kratovil and Werner, *supra*, note 28, at sec. 706, pp. 314–15.

82. *1.77 Acres of Land v. State*, 241 A.2d 513 (Del. 1968); *Blakely v. Gorin*, 313 N.E.2d 903 (Mass. 1974); *Chuba v. Glasgow*, 61 N.M. 302, 299 P.2d 774 (1956); *Schwarzchild v. Welborne*, 186 Va. 1052, 45 S.E.2d 152 (1947).

83. *People ex rel. J.C.Penney v. Village of Oak Lawn*, 349 N.E.2d 637 (Ill. App. 1976).

84. *In re Off-Shore Restaurant Corp. v. Linden*, 331 N.Y.S.2d 397 (N.Y. 1972).

85. *Vasilopoulos v. Zoning Bd. of Appeals*, 340 N.E.2d 19 (Ill. App. 1975).

86. See *Newark v. Daly, supra*, note 57.

87. *Dalsis v. Hills*, 424 F. Supp. 784 (W.D.N.Y. 1976).

88. Kratovil and Werner, *supra*, note 28, at sec. 739, pp. 335–37.

89. *Id. supra*, note 23, at Chapter 28, sec. 739, p. 336; Chapter 34, sec. 1008, pp. 437–38.

90. *City of Chicago v. National Management*, 22 Ill. App. 2d 445, 161 N.E.2d 358 (1959) (transient hotel automatic sprinkler ordinance upheld).

91. If you have paid the full purchase price before discovering the violations, you as buyer may sue the seller for damages. *Gutowski v. Crystal Homes Inc.*, 26 Ill. App. 2d 269, 167 N.E.2d 422 (1960); *Brunke v. Pharo*, 3 Wis. 2d 628, 89 N.W. 221 (1958); *Schiro v. W. E. Gould & Co.*, 18 Ill. 2d 538, 165 N.E.2d 286 (1960).

92. See note 28, *supra*, at Chapter 34, sec. 1008.

93. *Spur Industries, Inc. v. Del E. Webb. Development Co.*, 494 P.2d 700 (Ariz. 1972). Where the nuisance is found to exist, the municipality may demolish the structure without paying compensation to the owner. *City of Honolulu v. Cavness*, 45 Haw. 232, 364 P.2d 646 (1961).

94. Kratovil and Werner, *supra*, note 28, at sec. 739, p. 336.

95. See *Whetzel v. Jess Fisher Management Co.*, 282 F.2d 943 (D.C. Cir. 1960).

96. Liability to your landlord to repair the building must be distinguished from liability to third persons, such as your patrons, who are injured as a result of structural defects. This topic is covered in Chapter 12.

12

Forms of Business Organization

OUTLINE

Not-for-profit and Nonprofit Organizations
Choosing the Right Form of Business Organization
Summary
Questions

OBJECTIVES

After reading this chapter, you should be able to:

1. Describe the basic forms of business organization in the hospitality industry.
2. Compare and contrast the various forms of business organizations within the following framework: 1) supervision and control of business; 2) responsibility for business financing; 3) liability for business obligations; 4) duration of business; and 5) taxation of business income.
3. Outline the legal aspects of purchase, sale, and termination of each form.

Case in Point

Allen and Benny, graduates of a hotel and restaurant school, go into a partnership and open a ski lodge. They build two restaurants: Skis, a quick-service operation, and the Mountain Crest, an evening dining room where wine and mixed drinks are served.

After seven successful years, Mary joins their partnership. Benny and Mary subsequently decide that two restaurants are not adequate for the ski crowd and want to open an after-ski bar that will serve drinks, sandwiches, and appetizers. Allen is opposed, saying the bar will radically change the family atmosphere of the lodge. Benny and Mary outvote Allen and proceed to add the Snow and Slush Tavern.

What remedies are available to Allen? Under partnership law each partner has an equal right to manage the firm business, unless there is a contrary provision in the written agreement. Thus majority rule controls. Allen has the power to dissolve the partnership by withdrawing from it at any time. However, the withdrawal must be made in good faith. Bad faith is illustrated by a withdrawal to take personal advantage of a partnership opportunity, such as the expansion of business voted on by Benny and Mary. In such a case, the two remaining members could sue Allen for any losses they sustained.[1]

This problem could have been prevented in several ways. One would have been to execute articles of partnership or a partnership agreement that required unanimous consent to expand the business. A second method would have been to create a limited partnership agreement, whereby Mary, as a limited partner, would have no voice in management.[2]

SOLE PROPRIETORSHIPS

For those deciding to go into the restaurant business, there are several forms of organizations from which to choose.

Each has advantages and disadvantages as to the degree of responsibility and liability, rights and duties, and taxation. To make the best choice, all the factors must be weighed.

The least complex and, historically, the earliest form of business organization is the *sole*, or *single*, *proprietorship*. This business form is very popular in the foodservice industry. It offers entrepreneurs the independence and challenge of complete management accountability. The full responsibility for success or failure of a foodservice operation depends solely on the skills and people-management qualities of the sole proprietor.

In this form, the owner is also the active operator of the business. The owner voluntarily creates this form by simply starting to do business. No permission to operate as a sole proprietor is required from the federal, state, or local authorities. No fee is required to operate in this form; however, a state license or permit may be required to operate a foodservice business. You may need a health permit to operate a restaurant, regardless of the form of business organization you choose.

Rights, Duties, and Liabilities

As sole proprietor, you remain in exclusive control of the business, although you may hire employees to help you operate it. You are liable for: 1) your own negligent and intentional acts or omissions; 2) all applicable statutory violations (of building, fire, and safety codes); 3) the negligent or willful misconduct of your employees; and 4) all contracts you make with vendors, employees, and other businesses.

It is important to note that *your liabilities are not limited to your business assets, but extend to your personal assets*. You may be personally liable for all debts, losses, and valid claims lodged against your operation. **Exhibit 12.1** shows advantages and disadvantages of this form of business.

Purchase, Sale, and Termination of a Business

As the owner/operator, you determine the kind of operation to run; how, when, and where to run it; and the amount of your investment. You may purchase, sell, give away, or simply stop operating the business at any time. The duration of the business is measured by your own life span. When you die, the business automatically ceases to exist.

Tax Considerations

A single proprietorship is not taxed as a business for income taxes separate from personal taxes. Sales taxes, if applicable, are levied just as with any other business.

Exhibit 12.1   Advantages and Disadvantages of Sole Proprietorship

Advantages

Method of creation   It is often easier and less costly to start a sole proprietorship restaurant than to start any other kind of business. Legal formalities are held to a minimum, and the agreement of others is avoided.

Benefits   The sole proprietor receives all the profits of the business.

Transferability of interest   The sole proprietor has total transfer rights.

Duration   The duration of the business is discretionary with the sole proprietor, but cannot exceed his or her lifetime.

Management   The sole proprietor has unlimited management authority.

Taxation   The sole proprietor escapes corporate income taxation, paying only personal income taxes on profits. However, these taxes are not necessarily lower than those imposed on corporations.

Organizational fees   No fees to create or maintain the sole proprietorship are required by law.

Disadvantages

Liability   The sole proprietor has unlimited liability for all obligations incurred in doing business, extending to personal assets as well as business assets.

Burdens   Entire burden of business, including sole responsibility for losses and mismanagement, rests on a sole proprietor.

Financing   Financing is limited to personal funds and funds of others willing to loan to the sole proprietor.

If you employ other people, federal and state withholding for income taxes on wages, salaries, and employment taxes must be paid, regardless of the form of business you choose. If alcoholic beverages are to be served on the premises, you must obtain a federal alcohol tax stamp. This stamp is required of all those who sell alcohol.

PARTNERSHIPS[3]

A *partnership* is an agreement between two or more people to conduct a business *for profit*. Each partner is a co-owner of the business. The partners have joint operating control over the business and the right to share in its profits. Profit is what distinguishes this form from other forms. The business *must* be set up to make a profit. This is an ideal form for the operation of a food service or hotel. However, the greatest challenge is finding the right partner to assume his or her share of management and financial responsibility.

The partnership agreement establishes the rights and obligations of the partners to each other and to the partnership. The agreement is critical, since the law will look first to the agreement to define these rights and obligations, and will apply general rules of partnerships law only when the agreement does not provide an answer. Because of the importance the law attaches to the agreement, it should be reviewed by legal council and always be in writing. **Exhibit 12.2** lists the advantages and disadvantages of a partnership.

Types of Partnerships

A *general*, or *full partner*, is a partner who has unlimited liability for partnership debts and obligations, has unlimited management powers, and shares in partnership profits. A *silent partner* lacks a voice, and takes no part in the business.

A *secret partner's* presence in the firm is not disclosed to the public, but he or she may help manage the operation. A *dormant partner* is both a silent partner and a secret partner.

Exhibit 12.2   Advantages and Disadvantages of Partnership

Advantages

Method of creation   Nothing more than agreement of the parties is required to create a partnership. This may be oral, though a written agreement is recommended. The cost of drawing up an agreement usually is minimal.

Benefits   The profits of the partnership are pooled and shared equally, unless otherwise provided by agreement.

Financing   The costs of financing may be shared equally or split up by agreement. This reduces the burden associated with the sole proprietorship, and makes it possible to draw upon the financial resources of all the partners.

Duration   The partnership lasts no longer than the lifespan of any one partner or until one partner decides to sell his or her share.

Taxation   A partnership pays no federal income taxes as a business entity. All profits must be distributed to the partners equally, or as set forth in the partnership agreement. The partners add that income to their personal income. Profits are taxable whether distributed or not.

Organizational fees   No fees to create or maintain the partnership are required by law.

Disadvantages

Liability   Each partner is the agent of all the other partners. As such, each partner may be individually liable for any business debts or any liability caused by the negligence of any other partner or any employee of the firm.

Burdens   The losses of the partnership are the individual as well as collective responsibility of all the partners.

Transferability of interest   All partners must consent to the transfer of the partnership interest of any partner.

Management   Each partner is entitled to an equal voice in the management and control of the partnership, regardless of his or her interest in the partnership. This may be an advantage or disadvantage depending on the differing skills of each partner.

A *nominal,* or *ostensible, partner* is one who has consented to be known as a partner, whether or not he or she is a real partner. In practice, the law applies the principle of *partnership by estoppel* to this relationship: though not a real partner, a nominal partner is liable to those who extend credit to the partnership in the belief that the person is a partner.

A *trading partnership* is engaged in buying and selling for profit; hospitality operators belong in this category. A *nontrading partnership* is engaged in providing a service, such as the practice of law or medicine.

Who May Create a Partnership

Any individual or group may create or enter into a partnership. Any number of people can form a partnership. A minor as well as an adult may become a partner, even though a minor may disaffirm such a contract and withdraw at any time until he or she reaches legal age. This is a possibility in the hospitality industry, where family operations may expand.[4]

A corporation and a limited partnership may join a partnership.

The Law Governing Partnerships

Historically, partnership law developed on a case-by-case basis in the form of common law rules developed by the courts in each state.

The Uniform Partnership Act (UPA) was adopted in 1914 to create uniformity and clarity in partnership law, as well as to put into a workable statutory form the many common law court decisions. Only Louisiana and Georgia have not adopted the UPA.

How the UPA Defines a Partnership

The UPA says that: 1) the receipt by a person of a share of the profits is evidence that he or she is a partner; 2) there must be demonstrated *intent* to form a partnership; 3) a partnership must *carry on business* for a reasonable period of time; and 4) the business must be set up to make a profit.

The purpose, to make a profit, distinguishes a partnership from not-for-profit or nonprofit entities, even though the kind of business carried on may be considered identical. A private membership club is not a partnership, even though the club furnishes meals and beverages comparable to those served in public restaurants catering to the same patrons.

The agreement may divide up assets and management responsibilities. Otherwise, in the absence of an agreement, each person in a partnership shares equally in the profits and losses of the business, according to his or her ownership rights and rights to manage the operation.

The following case illustrates how lack of a written partnership agreement can result in problems for both partners as well as time in court.

Barbet v. Ostovar
Superior Court of Pennsylvania
273 Pa. Super. 256, 417 A.2d 636 (1980)

Facts: Daniel Barbet (plaintiff) brought an action against Kurosh Ostovar and his wife, Marjorie Ostovar (defendants), seeking specific performance of an oral partnership agreement and an accounting.

Kurosh Ostovar was a professor at Pennsylvania State University. Barbet had been trained as a chef and had been engaged in the restaurant business all his life. He met Ostovar when he married Ostovar's cousin. Ostovar wanted to establish a French-type restaurant in State College, Pennsylvania. He asked Barbet to assist him. They entered into an oral agreement under which Barbet would participate in the establishment and operation of the restaurant, and Ostovar would provide the capital.

A restaurant property owned by the Meyers Corporation became available. Ostovar and Barbet entered into an option agreement in 1973 to acquire all of the Meyers Corporation stock in Marjorie Ostovar's name. They did this because they had been advised (incorrectly) that aliens could not legally hold stock in a corporation that held a liquor license and neither Kurosh Ostovar nor Barbet were American citizens. However, they signed as partners a lease of the premises from the corporation. They both participated actively in the business, while Marjorie's activities were limited to helping Kurosh keep the books.

In May 1976, Barbet and Kurosh Ostovar met with the attorney and accountant for the Meyers Corporation for the purpose of completing the purchase of the stock. They were represented as the real parties in interest in the restaurant business and as partners. At that time, they agreed that Barbet owned one half of the stock. Shortly thereafter, however, the relationship between Barbet and Ostovar deteriorated. On June 30, 1976, Ostovar had the locks on the restaurant changed to exclude Barbet, who was then "dismissed from his employment."

Barbet filed suit, claiming that he was a partner and seeking the transfer to him of 50 percent of the stock of the corporation and an accounting. At the trial Barbet testified as follows: "It was a very simple agreement. Dr. Ostovar was supposed to invest in this business, and I was supposed to set up the business for him, to supervise it. For disbursement of the profit, Dr. Ostovar was supposed to get his money out of the business, plus an interest on his money in the same amount that the bank would pay for the same. After that, the profit would be shared on an equal basis."

The decree granting specific performance of oral partnership agreement and an accounting was upheld by an equally divided court.

Reasoning: Pennsylvania law holds that the existence of a partnership depends on the intentions of the parties as to being partners and that no formal or written agreement need be executed in order for a valid partnership to exist. As the Pennsylvania Supreme Court wrote in another opinion:

"There is no requirement that partnership agreements be in writing. They may be oral or may be found to exist by implication from all attending circumstances (i.e., the manner in which the alleged partners actually conducted their business, etc.)."

The court found that the evidence supported the Chancellor's conclusion that Barbet satisfied his burden of proving that in 1972, he and Ostovar formed a partnership with the intention of owning and operating a restaurant in State College, Pennsylvania, and said partnership continued in existence until he (Barbet) was "locked out" in June 1976.

The Ostovars contended that the terms of the 1972 oral agreement between Barbet and Ostovar demonstrated, at most, an agreement between them to become partners in the future, but only upon the happening of certain conditions, those conditions being the repayment to Ostovar of his capital investments, with interest, and Barbet's resolution of his immigration problems.

However, the court's examination of the record led us to the conclusion that Barbet and Ostovar had unequivocally entered into an oral partnership. The court found that profits were not to be distributed until Ostovar had been repaid his original investment, plus interest, and that Barbet would receive a stock certificate representing a 50 percent interest in the corporation once he became a United States citizen. The foregoing were merely conditions governing the operation of the partnership business. They were not conditions that had to be fulfilled before the partnership came into existence.

The repayment of capital investments before distribution of profits is an essential element of every partnership agreement, implied by law.

Conclusion: The Pennsylvania Superior Court applied the common law principle that a written partnership agreement is not a pre-condition to creating a valid partnership. As long as all the credible oral evidence establishes all the necessary elements of a partnership, a partnership exists until one of the partners withdraws or is wrongfully prohibited from participating in the partnership business. The court stressed that there was adequate proof of an intent *to carry on a business for profit as co-owners*, further supported by the oral agreement to repay capital investments (return invested capital) before any distribution of profits.

The evenly divided reviewing court in this case illustrates the risk you take by not drawing up a written partnership agreement. Everything rests on the credibility of the opponents when a dispute arises. Each party is at the mercy of the subjective evaluation by a jury or judge of his or her testimony against that of the opponents.

Rights, Duties, and Liabilities

Most partnership matters, including how management decisions will be made, should be included in the agreement for the partnership. Generally, each partner has the power to make independent decisions in the normal business of the partnership. Usually partners will consult each other before making management decisions involving borrowing money, hiring new employees, altering the premises, and the like.

In the absence of contrary agreement, the vote of a majority of the partners controls such decisions, regardless of the partnership share of each partner. By agreement, the majority may delegate management of the business to one partner, or may delegate management of certain defined activities of the partnership to one partner. For example, if one partner is a financial whiz and another likes to be in the kitchen, an agreement may informally divide up these responsibilities. If partners are on fairly equal footing in terms of skills, a joint decision on each major issue might be best.

Any major change in the nature of the partnership business or in the partnership location, such as a decision to purchase the assets of an existing business or to move the restaurant downtown, would require the unanimous agreement of the partners. The test to determine whether such a vote is required is whether the change would substantially alter the risks or financial liability of the partnership.

Each partner is an agent of the partnership relationship, and the partnership relationship is one of trust and confidence. Partners owe each other the highest degree of loyalty and good faith in all partnership matters.[5] This duty is imposed by law and need not be specifically stated in the partnership agreement. Nor may a partner be relieved of this duty by the partnership agreement or any other contract.

A partner is liable to the other partner(s) if he or she:

- uses partnership property for personal purposes without obtaining the approval of the other partner(s).
- misappropriates partnership funds.[6]
- makes a secret profit out of the transaction of partnership business.[7]
- engages in a competing business without the knowledge and consent of the other partner(s).[8]
- accepts a secret commission on partnership business.[9]
- uses information gained as a partner to harm the partnership.[10]

Every partner owes the partnership a duty to exercise reasonable care in the conduct of partnership business. No partner should exceed the authority granted that person under the agreement. Every partner is liable for losses resulting from his or her negligence in handling partnership business and for losses resulting from unauthorized transactions negotiated in the partnership name.

Each partner is entitled to: 1) access to the partnership books and records; 2) a formal accounting of the business; and 3) reimbursement for expenses from personal funds for proper partnership business. Each partner is also required to account for his or her use of partnership funds and property, as well as for any benefit received by the partner without the consent of the others. A partner may not sue a partner or the partnership for damages or failure to receive profits. A partner's only recourse is to obtain a court-ordered accounting and distribution of profits and monies owed.

Partners are called *tenants in partnership* of all the firm's assets. No single partner has the right to sell, assign (transfer), or in any manner deal with any partnership property as a sole or exclusive owner.

In addition, a partner may not sell or convey partnership property unless he or she does so in the regular course of that firm's business. Thus a partner in a real estate partnership could sell real estate on behalf of the partnership, but a partner in a restaurant business could not because a transfer of the land on which the restaurant operates would affect the ability of the business to continue. In that case, all of the restaurant partners would have to agree to the transfer.

Among other business actions that require unanimous consent of each partner are the sale of goodwill, which is actually the sale of public reputation and the right to use recognized product names and identity; an assignment (transfer) of partnership property for the benefit of creditors; any act which would make it impossible to continue the business in the usual way (for example, the sale of the entire food and beverage inventory); a guarantee of the debt of another party; paying or assuming an individual debt of a partner; and providing free services.

Partners are not normally entitled to compensation (salary and wages) for conducting partnership business. Each partner is entitled to a share of the profits of the firm. Only a surviving partner is entitled to compensation for winding up the affairs of a disbanded firm. *However, the partnership agreement may provide for compensation or wages for one partner to run the business.*

Partnership liability differs from sole proprietorship liability in that any legal judgment against the partnership is first discharged out of the assets of the partnership and not from individual income or assets. The personal assets of all general partners are subject to such claims only if the partnership assets are insufficient to satisfy the claims. Individual partners, like sole proprietors, can be personally liable for partnership debts, *but only if the partnership itself is unable to pay the debts.* The law recognizes this dual responsibility to pay by requiring a creditor of a partnership having a legally enforceable claim to exhaust the right to recover before going after the personal property of the partners.

Partners are *jointly liable* on partnership contracts and *severally liable* (individually liable) for *torts* (legal wrongs) committed in the course of partnership business.[11] The partnership is itself liable for contracts and torts, and most states permit the partnership to be sued in its own name.[12]

The crimes of a partner normally do not impose liability on the partnership or other partners, unless those partners participated in the criminal conduct, and the crime was committed in the course of partnership business.

Modern criminal codes do make the partnership liable as a separate entity for partnership crimes, so that upon conviction the partnership must pay a criminal fine.

Partners, *in their relations with third parties*, act as agents of the partnership in the transaction of normal business. Apparent authority is sufficient to bind the entire partnership in its dealings with third parties, unless the third party had knowledge of the partners' lack of authority. The partnership must state any lack of authority to third persons—for example suppliers—with whom it deals in the normal course of its business.[13]

New Partners   The admission of a new partner or partners causes the original partnership to dissolve according to law, as in the case of a withdrawal of a partner. The partnership agreement usually provides for this contingency by permitting the new or remaining partners to continue the partnership business. Nevertheless, in law a new partnership is created. The debts of the dissolved partnership carry over to the new partnership, and creditors of the original entity are creditors of the new partnership.

After a firm is dissolved, no single partner may transfer his or her interest in the partnership (large share of profits earned and return of capital invested) to another person. The consent of all partners is required to admit a new person.

Termination

A partnership terminates at the end of the agreed-upon term or when the partnership objective has been attained. Without agreement on duration or objective, any partner can, by withdrawing, dissolve a partnership. Dissolution by withdrawal is effective, even though it may violate a partnership agreement. A partner who wrongfully withdraws, causing dissolution, is liable to the remaining partners for damages and may not participate in any profits resulting from termination of the partnership.[14] In addition, such a partner loses any claim to the value of partnership goodwill.

The death or personal bankruptcy of a partner automatically dissolves a partnership.

A court may order dissolution if, for example, the partners cannot agree on the business' direction and the partnership is unable to function.[15]

A partner may be *expelled* for cause. For example, if one partner uses funds from a hotel to pay personal gambling debts, the other partners may be justified in expelling that partner. The agreement should state specific practices that may result in expulsion, since, unless the activity is downright illegal, the law may not be on the side of the remaining partners. Liability could result from a wrongful expulsion.

When a partnership is continued by the surviving partners and/or newly admitted partners, the original partners remain liable for all original partnership debts.[16] A newly admitted partner assumes responsibility for all previous partnership debts.[17] For example, say that in a restaurant's dissolution, the partnership assets equal $50,000. A new partner is admitted. The new partner is liable only for his or her share of $50,000, the value of partnership assets. If there were three partners, the new partner would be liable for one fourth of the $50,000.

If dissolution is for just cause and the business continues, the noncontinuing partner is entitled to first claim to his or her partnership interest due at the time. The agreement may specify that the other partners have the right to buy the other interests of the departing partner so they can continue as a partnership. Upon the retirement or death of such a partner, that partner or his or her legal representative can elect to receive either interest on the value of that partner's interest in the partnership or that partner's share in the profits on business conducted after dissolution and wind-up but before termination.[18]

Tax Considerations

Federal Requirements   Each partner in a partnership is taxed on his or her own share of partnership income, whether it has been distributed or not. Accumulated earnings are taxed under this rule. The partnership files an information return only. Capital gains and losses are taxed proportionately to each partner, as are other forms of partnership income. Partners are not taxed on exempt interest received from the firm, such as interest earned on tax-free municipal bonds.

Partners are not eligible to participate in an exempt pension trust. The firm cannot deduct pension payments for its partners under a *Keogh* (self-employed pension) plan. Partners are not subject to social security taxes, FICA, but must often pay a self-employment tax. The Internal Revenue Code does not provide any tax exemption for the payment of death benefits to partners' beneficiaries, excluding those provided by insurance.

State Requirements   Many states exempt partnerships from the payment of income taxes. However, every partnership must pay applicable state sales taxes and unemployment insurance taxes.

Limited Partnerships

In this type of partnership, a degree of limited liability is exchanged for a lack of management of or control over the partnership business. A limited partner who engages in management or asserts control over the business loses the limited liability status and is treated by law as a general partner, regardless of the language of the limited partnership agreement.[19]

For example, when a celebrity permits the use of his or her name in a restaurant's business, it is held that that person is a general partner, with all the accompanying rights and liabilities. Great care must be exercised by anyone wishing to use a limited partnership as an investment opportunity. Under no circumstances must limited partners attempt to manage or represent themselves as managing the partnership business.

Limited partnership is very popular because, unlike a general partner, a limited partner is not fully liable for partnership obligations in the event the partnership is unable to pay them. Rather, the limited partner is liable only up to the amount of his or her investment in the business.

Unlike partnerships and related forms, the limited partnership is not created by agreement but by permission of the state in which the partnership is to be created. The partnership is required to submit its partnership articles or applications for approval and to disclose the names of all partners.

Every limited partnership must have at least one general partner whom creditors and governmental bodies may hold responsible for debts, payment of taxes, and failures to meet legal requirements. As long as the limited partner does not engage in management, that partner's personal assets or estate are subject to liability beyond the original amount invested.

The Uniform Limited Partnership Act   The Uniform Limited Partnership Act, adopted by the District of Columbia and the Virgin Islands and in all states except Delaware and Louisiana, is the law governing limited partnerships. The limited partnership form is created by legislation and not simply by agreement, as in the case of a general partnership. The Act normally governs partnership activities not covered in the agreement.

Joint Ventures

This specialized form of business organization is an offshoot of the partnership. In this form two or more persons agree to join in a *single* business enterprise or activity, sharing profits and losses according to the value of the money or services contributed by each. A joint venture differs from a partnership, which is formed to carry on a business over a continuous, indefinite time period.[20] The joint venture may be formed to start a restaurant, and then may later dissolve or reform into a partnership.

A party to a joint venture is not an agent of the group and does not have authority to bind the others to individual decisions. However, a court may find that authority to bind them does exist.[21] The management and operation of a joint venture are often placed by agreement in the hands of one member. Some states require that one person have authority and be accountable for actions on behalf of the venture.

Unlike a partnership, the death of one party does not automatically dissolve the joint venture, and one member may sue the joint venture or one or more joint venturers to recover damages in a dispute. In all other ways, a joint venture is treated as a partnership and is governed by the law of partnerships.

Unlike a partnership, the joint venture has no legal existence separate from its members. This means that a creditor cannot sue the joint venture in its own name, but must sue all the individual members. Some states have laws that authorize partnerships to be sued and held liable, and these laws may apply to joint ventures. In these states, the joint venture may sue and be sued, collect debts, file for bankruptcy, and convey property. Likewise, creditors may be allowed to collect judgments against the joint venture, rather than its individual members.[22] Members of a joint venture are taxed the same as those in a partnership.

CORPORATIONS[23]

Although the corporate form of organization is normally associated with a large, complex business, numbers are *not* a legal criterion for incorporation. Even one person may incorporate and receive the benefits of corporate organization. State laws usually say how many people may form a corporation.

Public corporations are owned by the public through the purchase of *shares*. These include utility companies. *Private corporations* are the type usually found in the hospitality industry.

Private corporations are established for the benefit of their owners. They include both *business*, or *stock corporations* and *nonprofit, membership corporations*. A typical hospitality corporation is a stock corporation organized for profit. The business corporation is organized to distribute profits in the form of dividends to its owners, called *shareholders*. Business corporations may be *publicly held* (public issue) corporations, which issue stock to the public, or they may be *closely held*, owned by a family or another company.

The basic idea behind a corporation is to pool the assets of the shareholders into a fund to manage the business, make a profit, and repay shareholders for their investments. This may offer a business a larger pool of capital to operate, thus spreading the liability more thinly than in a partnership. Normally, a board of directors is accountable for wrongdoing, as is a chief operating officer. Further, banks and other lending institutions are more likely to lend money to a corporation than to a sole proprietorship, since the corporate form may have a stronger financial base—one that does not rely on the management skills or the funds of one person.

This form may be attractive to an entrepreneur who may wish to expand an idea or restaurant concept at a later point. **Exhibit 12.3** compares the corporate form with the partnership form of organization.

The main features of a corporation are its perpetual existence and the fact that the liability of its owners, or shareholders is limited to their individual investments, or *stock purchases*. Like limited partners, shareholders or stockholders are generally not responsible for the debts of the corporation beyond their own investments.

A board of directors, elected by the shareholders, actually supervises the managerial policies of the corporation. The board, in turn, employs officers to operate the business on a day-to-day basis.

Neither the directors nor the officers of the corporation need to be shareholders unless the corporate articles or bylaws require stock ownership.

Model Business Corporation Act (MBCA)

The Model Business Corporation Act (MBCA), a model act prepared by a committee of legal experts in 1946 and revised in 1969, has been adopted by the majority of states, either totally or in part.[24] It governs all aspects of corpo-

Exhibit 12.3 Comparison of Partnerships and Corporations

Key Elements	Partnership	Corporation
Creation	By agreement of the parties.	By statutory authorization or state approval of charter.
Entity	Dissolved by bankruptcy or withdrawal of a partner.	May be perpetual.
Liability	Partners are subject to unlimited liability for contracts, debts, and torts (legal wrongs) of other partners.	Shareholders are not liable for contracts, debts, and torts of the corporation in excess of their investments.
Transferability of interest in organization	Interest of a partner in the partnership is not transferable without consent of all partners.	Shares of corporate stock are freely transferable.
Management	Each partner is entitled to an equal voice in management of the partnership.	Management of a corporation is governed by a board of directors elected by shareholders. Directors establish policies and appoint officers.
Taxes	Each partner is liable for a *pro rata* share of income taxes on net profits whether or not distributed.	A two-level tax is imposed: The corporation pays corporate income tax on net profits, including dividends; shareholders pay income tax on dividends received and interest received on corporate bonds.
Cost of organization annual fees	No cost or fees are required. However, a limited partnership must pay an organizational fee.	All costs and fees must be paid.
Multistate transaction of business	Generally none.	Must qualify and obtain certificate of authority to do business in other states.

rate formation; powers, duties, and rights and responsibilities of directors, officers, and shareholders to each other and to the corporation; and corporate dissolution.

Doing business as a corporation is governed by statute and not by agreement among the parties, and all states have some laws on this subject. These statutes are usually comprehensive and govern most corporate activities. Only when the statues do not apply to particular corporate actions do the corporate charter and bylaws and the common law of corporations provide the law for court decisions.

Primary Characteristics of a Corporation

The essential features of the corporate form of business organization, as contrasted with the single proprietorship and partnership, are:

- an independent existence as an entity separate from its directors, officers, and shareholders—that is, it is not affected by their deaths or normal withdrawals.
- the right to sue and be sued in the corporate name.
- the right to acquire, hold, manage, and sell property for purposes in the corporate name.
- the right to make bylaws.

The separate legal existence of a corporation means that contracts made by the corporation's agents in their corporate capacity *do not bind the members personally*.[25] Likewise, acts of members in their individual capacity do not bind the corporation.

Members may work in a representative capacity and still avoid individual liability. The common law of agency provides that they must *disclose* their representative or corporate capacity when signing contracts on behalf of a corporation with a third party, such as a vendor.[26]

To illustrate, say you are a buyer and board member for Alpha Corporation, which operates a popular restaurant. A wholesale food and beverage distributor, Omega Foods, approaches you to arrange for the purchase of next month's meat, poultry, and frozen food staples. Omega's representative, Abbott, draws up and presents you with a written purchase order. Assume that both Abbott and you have authority to conclude the deal. How do you sign the agreement as a representative of Alpha? The safest form of execution is to sign as follows.

> Alpha Corporation
> by: Mary Valdez, Buyer

This discloses corporate representative capacity: that Valdez is signing *for Alpha Corporation*, and not for her.

By contrast, if Mary had merely signed "Mary Valdez," the law presumes that Sue alone is contracting to buy from Omega Foods, unless Abbott knew that she was signing for Alpha Corporation. This fact may be disputed,[27] and Valdez may suffer the consequences. The same analysis applies to Abbott's method of signing the order on behalf of Omega.

Rights, Duties, and Liabilities of Board Members

Under the MBCA, the board of directors of a business corporation is authorized, among other things, to adopt initial bylaws, declare dividends, fill vacancies on the board, elect and remove officers, sell and lease mortgage assets of the corporation in the normal course of business, and propose to the shareholders major changes affecting the corporation that require shareholder approval. The last includes amendments to the articles of incorporation; merger or consolidation; sale, lease, or mortgage of major corporate assets other than in the normal course of business; and voluntary dissolution of the company.

The following are duties of the board of directors: 1) to protect assets and other interests of the owners (shareholders) of the corporation; 2) to ensure continuity of the corporation; 3) to ensure sound corporate management; and 4) to make nondelegable decisions, such as the payment of dividends.

An individual board member has no management duties other than possible appointment as a corporate agent by the board. A director has the right to inspect corporate books in order to carry out responsibilities as a director to the board and the corporation.

Although not agents of the corporation by operation of law, as are partners of a partnership, board members are required to: 1) act within the authority granted by the corporation; 2) act with diligence and reasonable care in conducting corporate affairs; and 3) act in good faith for the benefit of the corporation and maintain loyalty to the corporation.

A director who exceeds the authority granted by the MBCA and the corporate bylaws is liable to the corporation for any resulting harm.[28] A director is also liable for losses incurred by the corporation resulting from his or her negligence or failure to carry out corporate responsibilities.[29]

A director is liable for injury to the corporation arising out of any self-interest activity. A director may not profit at the corporation's expense, or direct corporate opportunities to his or her own use.[30] For example, say that a director, acting on behalf of his corporation, comes upon a prime land deal for a new unit. During negotiations for the company, he decides to take a cut of the deal himself. He is no longer acting on behalf of the corporation but in his own self-interest, and, by profiting at the corporation's expense, may be liable for his cut.

Rights, Duties, and Liabilities of Corporate Officers

Officers are agents of the corporation, and derive their rights from the company bylaws and from the board of directors. Officers have the authority to do those tasks necessary to carry out duties assigned to them. The president of the board does not, simply by virtue of the office, have any power to bind the corporation. However, if the president is appointed *chief executive officer* (CEO) with general supervision and control of the company, then as CEO, he or she is vested with broad apparent authority to make contracts and otherwise act in conducting the ordinary business of the corporation. The CEO may hire hospitality managers, handle the budget, and purchase supplies.

The *corporate secretary* usually keeps the corporate minutes of director and shareholder meetings. The secretary has no other authority.

The *treasurer* has custody over corporate funds and has authority to receive and use the funds for authorized purposes. The treasurer, standing alone, has no authority to borrow money or issue *commercial paper* (promissory notes or warehouse receipts, for example).

Officers are personally liable for any willful or intentional misconduct, either under tort law or under an appropriate criminal code. If the misconduct falls within the doctrine of *respondeat superior*, where the employer must account for the actions of employees, the corporation is also held responsible. For example, if an officer acts as an agent in signing a banquet contract the company cannot fulfill, the corporation may be responsible.

Normally, officers are not personally liable for the unintentional negligence of corporation employees unless they authorized or participated in such negligence.

Criminal liability is imposed on a corporation where the board of directors authorizes the wrongdoing or the officer promoting or committing the crime is a high-level officer.[31]

Rights, Duties, and Liabilities of Shareholders

Common shareholders have an interest in: 1) corporate net earnings; 2) control of the corporation; and 3) corporate net assets. They are not truly owners of corporate assets, since the corporation, as a separate entity, owns its assets.

Net earnings are dividends on shares of common stock held by each shareholder. Payment of dividends is up to the board of directors, and rests upon that body's reasonable business judgment. Only flagrant abuse of the judgment will cause a court to order a dividend payment in the face of board refusal to do so.[32]

Control of the corporation means the right to vote shares of stock. Each shareholder is normally entitled to one vote for each share held. Voting rights are based on the statutes of the state of incorporation and the corporate articles and bylaws.

Usually, the common stock of a corporation carries voting rights, and each shareholder is entitled to one vote per common share owned. Voting rights are extremely important, since they enable the shareholder who has a majority of shares to elect the corporation's board of directors and to change its ownership structures, a not uncommon phenomenon in the hospitality industry.

The following proposals are subject to stockholder approval: 1) to merge the existing corporate business with another with only one of the corporations continuing to exist; and 2) to consolidate the corporate business, where two or more corporations combine, with neither surviving and a new one emerging. The proportion of stockholder approval necessary (two-thirds of all eligible shareholders, three-fourths of all such holders, or a mere majority) is determined by state law.

The quality of management is determined in large measure by the intelligent use of voting rights. The shareholders have the right to vote existing management out of office. A merger or consolidation may drastically affect old management policies, and shareholders are free to pool their voting rights by nominating and electing a particular slate of candidates to sit on the board of directors.

Under most state laws, shareholders have the right to amend the *articles of incorporation*, or *corporate charter*. Such a change might involve expanding the types of business the corporation was originally set up to transact. Shareholders have few responsibilities; they elect and remove the directors and they vote on extraordinary transactions.

Because shareholders have so few powers and duties compared with board members and officers, their potential liabilities are not as great. Their major protection against creditor claims is their limited liability for corporate obligations. This limited liability translates into no liability beyond their total investment (the cost of the shares purchased). Illegal dividend payments received by shareholders are recoverable by the corporation, if the shareholder knew of the illegality. Shareholders may be held liable for corporate debts if the corporation piled up debts before completing legal requirements for incorporation.

Purchasing and Selling Corporate Assets

The use of a corporation as a vehicle to buy or sell a business permits greater flexibility than does the purchase or sale of a partnership. The purchase and sale of a majority of the shares of stock of a corporation are sufficient to obtain control. The sale of a majority of shares of a corporation does not normally affect the existence of the corporation. Only consolidation creates a new corporation.

Such alternatives as merger, consolidation, and purchase or sale of assets are readily available to a corporation.

In general, corporate stock represents ownership rights. On the basis of the number of shares owned, each stockholder may transfer those shares either by gift or sale to another person, partnership, or corporation, as well as to nonprofit organizations. If the number of shares owned by any one person or entity exceeds 50 percent of the total number of shares of stock issued by the corporation, the buyer acquires a controlling interest in the corporation. This enables the buyer to elect a board of directors and thus manage the corporation.

Closely Held Corporations[33]

A *closely held corporation*, also called a *close corporation*, is an exception to the general rule that corporate stock is freely transferable by sale or gift. Such a corporation is one whose shares are held by members of a family or a few individuals. It is particularly useful to the small operator starting up a restaurant, because the stockholders are limited in number and the transfer of shares is subject to some restrictions involving sales to third persons.

In many cases, the corporation may not offer its securities to the public. This means that shareholders wishing to sell must first offer their shares to the corporation, the other shareholders, so that they may buy up the shares and maintain control within the group. If the stockholders fail or refuse to buy out the selling shareholder, the seller may offer the shares to any other person or institutional buyer. The right of first refusal by existing shareholders might be contained in the corporation's bylaws.

There are statutory restrictions on the number of shareholders a close corporation may have, so this type resembles a partnership rather than a corporation. The shareholders usually know each other personally and work closely in managing the corporation. Typically, a family or group of close friends would not wish control to be transferred outside, since this would destroy the rapport and close working relationship. The closely held form is treated as a corporation, with all the rights and responsibilities of that form.

There is one substantial risk involved in the use of the closely held corporate form of organization. When the corporation is legally formed by a single person or two or three family members, *it is essential that the statues of the corporation stay separate from the personal affairs of its individual owners.* Hospitality corporations run by family members can become battlegrounds focused on family problems if caution is not exercised. Typical potential problem areas include combining into one account corporate and personal funds, failing to record in proper form what goes on at board meetings, and shareholders' use of corporate property for personal use.

Any conduct that seems abusive of corporate privileges for personal benefit or the use of the corporation to blend personal and corporate activity may result in a court "piercing the corporate veil." The court may hold the owner personally liable to creditors for corporate obligations not fulfilled as a result of such activity. This legal doctrine permits courts to disregard the corporate form when it is used to commit wrongdoing, shield fraud, or evade legal responsibility.[34] The activities of each stockholder will be studied with more care, since the possibility of wrongdoing exists to a greater degree because of the greater control of the corporation than is true in a publicly owned business.[35]

If you seek a high degree of internal and external control over decision-making and protection against full personal liability for business debts, but with the sharing of financing and other responsibilities, you may prefer the *close corporation*. This form survives you. The restriction on stock sales to the public minimizes a take-over by outsiders through the purchase of shares.

The limited liability of corporate owners/shareholders is preserved, and if the corporation qualifies for Subchapter S treatment, corporate income is treated as partnership income, with no double taxation. The one limitation on the close corporate form is the possible need for outside financing through public stock sales if financing by lending institutions becomes prohibitively expensive. Also, you must take great care in selecting your close corporate shareholders, since you will have to work with them. A high degree of personal ownership interaction will be needed, since very important decisions, such as securing major outside financing to open new units, will require unanimous approval.

The following scenario will help show you how the courts may view an action by a corporate stockholder.

Albert lent money to his close corporation, Eats Inc., to finance the purchase of a quick-service franchise. Albert took as collateral a deed to the land he owned on which the franchise would operate. Later, Eats Inc. became insolvent, and general creditors of the franchise sued to set aside the loan and recover the land collateral. They did this to make the land available as an asset of the corporation against which they could satisfy their claims.

Normally, Albert's loan to the company (his own) would give him the status of a preferred creditor, who would be entitled to satisfy *his* loan out of the sale of the land ahead of and in preference to the general creditors.

The court ruled that the loan was not taken out to benefit Albert at the expense of the corporation. He had demonstrated good faith and fair dealing in this case and was entitled to preferred credit status.[36]

However, if Albert had taken the loan out to finance his racing car, which had nothing to do with Eats Inc. or the franchise, the court might have taken a different view. The following case illustrates that view.

Platt v. Billingsley
California Court of Appeals
234 Cal.2d 577; 44 Cal. Rptr. 476 (1965)

Facts: Billingsley, Grimm, and two other persons bought all of the stock in a restaurant corporation. At the time the corporation had a deficit of $51,972. The stockholders did not contribute new money to the corporation. Later, Billingsley and Grimm each sold half of their stock to Button for $17,925. (They did not have a "consent to transfer stock", as required by the California Commissioner of Corporations.) Platt, the contractor, entered into an agreement with the corporation to remodel the building as a Polynesian restaurant. He finished the job at a cost of $12,000. At first Platt was paid regularly, but then checks started to bounce. Platt was assured he would be paid. When he was not, he filed suit and attached $7,999. He released $4,000 when Button and Billingsley told him they would give him a chattel mortgage on restaurant equipment, which turned out to be worthless because of tax liens against the corporation. Platt initiated action to disregard the corporate entity and hold the three personally liable for corporate debts; he won. Billingsley and Grimm appealed.

Reasoning: The court found that they were personally liable as they had disregarded the corporate entity by the following actions:

First, Billingsley and Button retained personal control over the funds, which were never deposited in the corporation account and never became subject to claims of corporate creditors.

Second, $17,925 of the money came from the sale of part of the appellant's corporate stock. They did not risk their own capital; rather, they sold Button's stock in an undercapitalized venture and risked Button's money in an attempt to bolster a faltering business.

Third, use of money to pay debts of the corporation violated representations appellants made to the Corporation Commissioner in seeking permission to sell half of their stock to Button. Appelants represented to the commissioner that: "None of the proceeds from the sale of said shares will be used directly or indirectly for the benefit of said corporation, but are for the personal, individual benefit of the undersigned."

Other acts of appellants that support the trial court's decision to pierce the corporate veil include: a) Billingsley's promise to protect respondent by chattel mortgage on kitchen equipment, a representation upon which respondent relied in releasing $4,000 of a $7,999 attachment of corporate assets; b) an offer by Billingsley and Grimm to pledge their corporate stock, their personal property, to secure the corporate obligation; and c) operation of the corporation with only two directors, a violation of the bylaws of the corporation and of California Corporations Code section 800, requiring that: "all corporate powers shall be exercised by or under authority of, and the business and affairs of every corporation shall be controlled by a board of not less than three directors."

The court said that the corporate veil may be pierced to *prevent* fraud and injustice, and the action does not depend on the proof of fraud. The court agreed with the trial court that at all relevant times, the corporation was influenced, dominated and controlled by defendants Billingsley, Grimm, and Button, and there existed such a unity of interest and ownership that the individuality and separateness of these defendants and the corporation ceased.

If the acts of the defendants, Billingsley, Grimm, and Button are considered those of the corporation alone, an inequitable and unjust result will follow.

Conclusion: The corporate form is not set up to shield its members against legal action or as a cover for wrongdoing. Members may be held liable for corporate debts if their actions are irresponsible and disregard the corporate entity. Another common example of wrongdoing is for members of a corporation to take huge salaries and not obtain capital for the corporation to meet debts. A corporation is not a fund-raising venture for its members but an entity that requires that some profits be channeled back into the corporation so that it remains solvent.

The close corporation might be the vehicle most suitable to the typical hospitality operator. This form provides limited liability to its owners/shareholders and yet retains more centralized ownership and control. It also permits greater financial incentives to investors than either the general partnership or corporate form.

The major drawback is the difficulty of obtaining needed financing from banks in times of very high interest rates, since the close corporation may be prohibited from selling shares to the public without losing its status. *Initial institutional financing* or possible refinancing later may require reorganization of the hospitality business as a general business corporation.

The limited liability of the close corporation form does not always exist in practice, particularly for a new corporation. A bank may be unwilling to lend funds to a small corporation if it is new and without any established financial track record, unless the stockholders co-sign the loan as individuals. This means that each stockholder agrees to be personally responsible for repayment of the loan. Here the limited liability rule normally available to shield stockholders from personal responsibility for corporate debts does not apply because they have assumed repayment responsibility voluntarily. Business necessity in obtaining bank financing may cause corporate shareholders to trade off their legal right to be shielded from personal liability to obtain the corporate loan.

Tax Considerations

For federal income tax purposes, the corporation is treated differently from a partnership. Because the corporation is a separate legal entity, it is required to pay corporate income taxes.

Corporate income is taxed and payable *by the corporation*, which must file a corporate return. Additionally, stockholders are taxed on distributed dividends and interest on corporate bonds or loans. Earnings due but not distributed are not taxed. However, excessive accumulations may result in the corporation paying a penalty.

Corporate capital *gains and losses* are taxed and payable by the corporation. Sales of corporate securities owned by individual shareholders are taxed to the shareholder, with capital gains and losses available to the individual.

Unlike in a partnership, all corporate employees and officers who are stockholders can be beneficiaries and have tax benefits from a pension trust. The corporation can deduct any payments made into the trust on behalf of both groups. Corporate employees and shareholder beneficiaries can receive tax shelter benefits up to $5,000 per person from the corporation.

Whereas partnerships are not required to pay income taxes in many states, corporations are usually subject to state income tax, with the corresponding right to deduct the payments on federal returns.

Subchapter S Corporations

Ideally, some corporations would like to have the tax advantages of a partnership while retaining the limited liability that the corporation offers. The *Subchapter S corporation* is a tax option authorized by the Internal Revenue Code. It permits those corporations that qualify to be taxed as if they were partnerships. Qualifying corporations file only an information return that spreads income among the shareholders regardless of dividend distributions. All corporate income taxes are avoided; the shareholders are not taxed twice. However, unlike other corporations, the amount of income the Subchapter S corporation can transfer to tax-sheltered pension plans is restricted.

Major Requirements to Qualify under Subchapter S   Eight specific requirements must be met to qualify for Subchapter S.

1. The corporation must be chartered under federal or state laws.
2. The corporation must not be affiliated with any other group of corporations.
3. The shareholders of the corporation must be individuals, estates, or trusts completely owned by the creator of the trust. Corporations and partnerships may not be shareholders.
4. The corporation must have no more than 35 shareholders.
5. The corporation can issue only one class of stock. This means that all shareholders have equal voting rights.
6. The corporation may not receive more than 25 percent of its gross income from investments that the company did not actively generate. This means the company may not receive income from securities from other companies held by it in excess of one-fourth of its gross receipts.
7. The corporation may not receive more than 80 percent of its total income from sources outside the United States.
8. The corporation may not issue shares of its stock to nonresident aliens.

The Subchapter S corporation is tailored to meet the needs of the small business that fits the definition of a closely held or family corporation. In addition to other advantages, the Subchapter S corporation is particularly attractive when the individual shareholders are in a lower income tax bracket than the corporation itself. In this case all shareholder income is taxed in the lower income bracket, whether it is distributed or not. This treatment permits the corporation to accumulate more funds than a non-qualifying corporation.

Also, the Subchapter S corporation may select a taxable year other than the normal dividend distribution year and thereby defer some shareholder taxes. Undistributed earnings are not taxed to the shareholder until after the close of the corporation's taxable year.

To illustrate, Alpha, under Subchapter S, declares but does not distribute a stock dividend from its restaurant profits. Its taxable fiscal year runs from January to January. A shareholder, Barnes, uses the calender year for tax purposes. Barnes does not have to include the dividend until after January of the following calender year. The Subchapter S corporation can offer some tax-free benefits not available to other corporations, such as pension plan contributions and death benefits. The fact that they may be restricted still yields some federal tax savings to shareholders.

NOT-FOR-PROFIT AND NONPROFIT ORGANIZATIONS

Not-for-profit and *nonprofit* organizations are organized as membership corporations or cooperatives, and do not have stock or other certificates of ownership that are designed to profit the members or owners.

These entities often operate for religious, charitable, educational, fraternal, or civic purposes. They are formed under state statutes that usually provide for different formation and operation policies than for stock companies. If stock certificates are issued to members, no dividends may be paid to the member holders.

Profits from operation are permitted, but the profits must be left with the corporation rather than distributed to members.

Normally nonprofit and not-for-profit organizations are exempt from income taxes, and are able to reinvest a greater share of earned income than is the case with business or stock corporations.[37] In other respects, the nature of the business they operate is similar to that of corporate entities set up to make a profit.[38] A private country club might be organized as a nonprofit membership corporation.[39] Trade associations also might qualify for exempt income tax status, so long as they do not engage in excessive lobbying activities.[40]

CHOOSING THE RIGHT FORM OF BUSINESS ORGANIZATION

This chapter has explored the choices of business organization available to you when you decide to establish yourself in business, either alone or with others.

You must weigh the advantages and disadvantages of each form. Keep in mind the following.

- Your personal objectives
- The costs of organizing your business
- The need to share control with others
- The amount of outside financing you may require
- Whether you wish others to continue the business on your retirement or death
- How much personal liability you will assume
- How and to what extent your business income will be taxed

No one form will contain the advantages you wish without corresponding disadvantages. Therefore, you must try to list your priorities in the order of their importance and select the form that comes closest to giving you the most pluses and the least minuses. Your goal should be to exercise your best business judgment as to what business form would satisfy your objectives.

Future economic and financial circumstances may cause you to alter your business organization. For example, if your sole proprietorship operation becomes so successful that you wish to operate more than one unit, you may wish to form a partnership or corporation. Delegation of authority may be more easily implemented by a partnership or corporation where supervision of geographically separated units is involved. The partnership and corporate forms may also make it easier to raise the outside financing needed to expand the business. You have the right to change your form of organization to meet future needs.

Whatever your choice of business organization, the decision to adopt one or another form is yours alone to make. The risk of bad judgment rests with you. The law cannot correct the results of a poor choice of organization. However, the law does permit you to alter your form of organization as your business needs change.

SUMMARY

The *sole proprietorship* is a popular form in the foodservice industry. It offers independence and control to the entrepreneur interested in running the operation. It also offers unlimited liability and risk. The owner may sell or give away the business at any time.

The *partnership* divides up the responsibility and the liability among partners. The personal assets of each partner may be taken to satisfy creditor claims if debts cannot be paid out of partnership funds. Other partnerships may be for partners who want the profits, but do not want to help manage or who want their membership in the operation to be kept from the public. The Uniform Partnership Act is the source of law for this form of business. The partnership agreement is the basis for the partners' rights and obligations.

A *joint venture* is formed for a single business purpose, which usually dissolves when the business does.

In a corporation, the assets of the owners or shareholders are pooled to manage the business. The liability is spread among the shareholders. The business is usually run by a board of officers elected by the shareholders, who in turn appoint an executive to handle the day-to-day operations. Each of these parties has rights and duties to each other and the corporation.

A *closely held corporation* adopts the same general rules as the corporation, except that the members are usually either members of the same family or know each other well. Shareholders in this type of corporation usually want ownership and control to stay within the group, and generally do not offer shares to the public.

Corporations are required to pay corporate income taxes; shareholders' dividends are also taxed. Corporations may also be subject to state income taxes.

The *Subchapter S corporation* is run like a corporation but taxed like a partnership.

QUESTIONS

1. Deanne Sanders wants to start a restaurant. She likes to work alone as far as supervision and management activities go. She does not want to have to deal with separate taxation forms for personal and business taxes. She has no children interested in continuing the business. She wants to be able to sell the business when she's ready. What business form might be best for Deanne? Why?

2. What are the characteristics of a closely held corporation?

3. What is the main difference between a partnership and a joint venture?

4. How is a corporation taxed twice?

NOTES

1. See generally *National Biscuit Co. v. Stroud*, 249 N.C. 467, 106 S.E.2d 692 (1959) (grocery store).

2. See *Vidricksen v. Grover*, 363 F.2d 372 (9th Cir. 1966) (auto agency). Superseded by statute as stated in *Briargate Condominium Ass'n v. Carpenter*, 976 F.2d 868 (4th Cir. 1992).

3. Unless otherwise noted, the materials to follow are derived from the Uniform Partnership Act of 1914 (UPA), as amended.

4. A conflict exists among the courts as to whether a minor may recover the full amount he or she invested or must bear a proportionate share of partnership losses, not to exceed his or her total investment. The courts agree, however, that a minor may not recover contributions of capital until all creditor claims have been satisfied.

5. *Waagen v. Gerde*, 36 Wash. 2d 563, 219 P.2d 595 (1950) (commercial fishing business).

6. *Clement v. Clement*, 436 Pa 466, 260 A.2d 728 (1970).

7. *Starr v. International Realty, Ltd.*, 271 Or. 396, 533 P.2d 165 (1975) (real estate venture).

8. *Woodruff v. Bryant*, 558 S.W.2d 535 (Tex. Civ. App. 1977) (finance company).

9. *Starr v. International Realty, Ltd.*, *supra*, note 7.

10. *Estate of Witlin*, 83 Cal. App. 3d 167, 147 Cal. Rptr. 723 (1978) (health center).

11. *McBriety v. Phillips*, 180 Md. 569, 26 A.2d/ #400 (1942) (tavern).

12. *Phillips v. Cook*, 239 Md. 215, 210 A.2d 743 (1965) (automobile dealership); *Vrabel v. Acri*, 156 Ohio St. 467, 103 N.E.2d 564 (1952) (tavern).

13. *National Biscuit Co. v. Stroud*, *supra*, note 1.

14. Where the partnership itself is illegal, the courts will not assist in its dissolution, and will not enforce the rights of any of its partners. *Williams v. Burrus*, 20 Wash. App. 494, 581 P.2d 164 (1978) (tavern).

15. *First Western Mortgage Co. v. Hotel Gearhart, Inc.*, 260 Or. 196, 488 P.2d 450 (1971) (motel and restaurant).

16. *McClennen v. Commissioner*, 131 F.2d 165 (1st Cir. 1942) (law firm).

17. *Wolfe v. East Texas Seed Co.*, 583 S.W.2d 481 (Tex. Civ. App. 1979); *Ellingson v. Walsh, O'Connor & Barneson*, 15 Cal./#2d 673, 104 P.2d 507 (1940) (law firm).

18. *McClennen v. Commissioner, supra*, note 16.

19. *Vidricksen v. Grover, supra*, note 2. Cf. *Frigidaire Sales Corp. v. Union Properties, Inc.*, 88 Wash. 2d 400, 562 P.2d 244 (1977) (real estate firm).

20. See *Travis v. St. John*, 176 Conn. 69, 404 A.2d 885 (1978) (real estate investment).

21. *Misco-United Supply, Inc. v. Petroleum Corp.*, 462 F.2d 75 (5th Cir. 1972) (oil lease venture).

22. *Id.*

23. Unless otherwise noted, the materials to follow are derived from the Model Business Corporation Act (MBCA).

24. These states are Alabama, Alaska, Arkansas, Colorado, Connecticut, Georgia, Illinois, Iowa, Kentucky, Louisiana, Maine, Maryland, Massachusetts, Michigan, Mississippi, Montana, Nebraska, New Jersey, New Mexico, New York, North Carolina, North Da-

kota, Oregon, Rhode Island, South Carolina, South Dakota, Tennessee, Texas, Utah, Vermont, Virginia, Washington, Wisconsin, and Wyoming; also the District of Columbia.

25. *Harris v. Stephens Wholesale Bldg. Supply Co.*, 54 Ala. App. 405, 309 So.2d 115 (1975).

26. *Id.* See also *Rosen v. Deporter-Butterworth Tours, Inc.*, 62 Ill. App. 3d 762, 379 N.E.2d 407 (1978); *Dinkler Management Corp. v. Stein*, 115 Ga. App. 586, 155 S.E.2d 442 (1967) (hotel management corporation).

27. See *Henderson v. Phillips*, 195 A.2d 400 (D.C. App. 1963) (plumbing contract).

28. *Star Corp. v. General Screw Prods. Corp.*, 501 S.W.2d 374 (Tex. Civ. App. 1973).

29. *Neese v. Brown*, 218 Tenn. 686, 405 S.W.2d 577 (1964) (bank).

30. *Morad v. Coupounas*, 361 So.2d 6 (Ala. 1978) (medical laboratory); *Appeal after remand*, 380 So.2d 800 (Ala. 1980). *Lincoln Stores, Inc. v. Grant*, 309 Mass. 417, 34 N.E.2d 704 (1941); *Patient Care Services, S.C. v. Segal*, 32 Ill. App. 3d 1021, 337 N.E.2d 471 (1975); *Aero Drapery of Ky., Inc. v. Engdahl*, 507 S.W.2d 166 (Ky. App. 1974); *Hartung v. Architects Hartung/Odle/Burke, Inc.*, 157 Ind. App. 546, 301 N.E.2d 240 (1973).

31. *United States v. Park*, 421 U.S. 658 (1975) (criminal liability of CEO of national retail food chain for FDCA violation).

32. *Miller v. Magline, Inc.*, 76 Mich. App. 284, 256 N.W.2d 761 (1977) (defense contractor).

33. Lusk, Hewitt, Donnell, Barnes, *Business Law and the Regulatory Environment*, 503–19 (5th ed. 1982).

34. *Valley Finance, Inc. v. United States*, 629 F.2d 162 (D.C. Cir. 1980), cert. denied, 451 U.S. 1018 (1981) (Tongsun Park, Koreagate scandal). See also *Felsenthal Co. v. Northern Assur. Co. Ltd.*, 284 Ill. 343, 120 N.E.268 (1918) (criminal fraud—arson).

35. *Intertherm, Inc. v. Olympic Homes Systems, Inc.*, 569 S.W.2d 467 (Tenn. App. 1978).

36. *Id.*

37. I.R.C. sec. 501(a) *et seq.* require the organization to qualify for tax-exempt status independently of state incorporation statutes that otherwise confer nonprofit status. Clubs organized and operated substantially for pleasure, recreation, and other nonprofit purposes are eligible organizations I.R.C. sec. 501(c)(7); Treas. Reg. sec. 1.501(c)(7)-N1.

38. Unrelated business income, business income that does not contribute importantly to the accomplishment of the exempt purpose of the organization is subject to taxation.

39. See note 37, *supra.*

40. I.R.S. sec. 501(c)(3) makes such lobbying activities a "prohibited act," which can cause the organization to lose its tax-exempt status.

13 Franchising

OBJECTIVES

After reading this chapter, you should be able to:

1. Define a franchise and explain its legal significance.
2. Examine the rights and duties of franchisors and franchisees to each other and to third parties.
3. Discuss the advantages and disadvantages of foodservice franchising.
4. Examine antitrust laws relevant to franchising.

Case in Point

Mr. and Mrs. Frick invested their life savings in a franchise business with Buffalo Bill, a foodservice franchiser. They drew up a written agreement to sell Buffalo Bill's barbecue burgers and other fast foods, using Buffalo Bill's promotional materials, architectural style, food ingredients, and methods of operation in return for a franchise fee. The duration of the agreement was to be five years. Buffalo Bill reserved the right to terminate the agreement prior to the end of that period for specific reasons, such as the franchisees' failure to make fee payments on time or to live up to Buffalo Bill's operating standards. No provision existed giving Buffalo Bill the right to terminate at will; that is, at any time, without reason.

After two years of operation, Buffalo Bill suddenly terminated the franchise. When Mr. and Mrs. Frick refused to give up the franchise, Buffalo Bill sued to force its termination, claiming that as a franchisor it had an implied right to terminate at will independent of any express right contained in the agreement. Who won this legal battle?

The court found in favor of the Fricks. Without any proof of violation of the franchise, the court said that the Fricks could justifiably expect not to lose their substantial investment of time, effort, and money as a result of Buffalo Bill's arbitrary decision to end the franchise. The reasonable expectations of Mr. and Mrs. Frick, as shown by their total commitment to carrying out the franchise, and Buffalo Bill's legal duty to deal with the franchisees in good faith and in a reasonable manner, required a ruling in favor of the Fricks.

FRANCHISE DEFINED

A *franchise* is a license from the owner of a trademark or trade or service name that permits another to sell a product or service under that name or mark. In a franchise agreement the *franchisee*, the party to whom the license is granted, agrees to conduct a business or sell a product or service according to operating procedures specified by the *franchisor*, the party granting the license to sell. The franchisor agrees to assist the franchisee in the management of the product, service, or business through advertising, promotion, and advisory services.

In effect, the franchisor offers such intangibles as goodwill and reputation, as well as tangible products and services, in exchange for franchisee investment and management. The investment is usually a fee, normally a standard amount plus a percentage of the franchisee's annual sales.

The foundation of any franchise is the trademark or trade name of the product or service licensed.[1] The uniformity of the product or service and the control of its quality are incentives for the buying public to patronize a franchised outlet rather than a nonfranchised outlet. The franchisor furnishes a publicly recognizable product or service. A businessperson with limited capital can take advantage of the quality the public identifies with the product.

This public recognition does not, however, guarantee the successful or profitable operation of the franchise. It *may do so*, depending on other factors such as location, adequate market, competition, and quality of management.

A foodservice franchisee is an independent businessperson who furnishes money, time, and management skills and operates his or her own establishment. A franchisee is *not* an employee of the franchisor. The day-to-day operation of the business is the franchisee's sole responsibility. The success or failure of the franchised product or service rests on the business skill and ability of the franchisee. (The franchisee may be a sole proprietor, a partnership, or a corporate form of business organization, although one or two individuals will usually run the operation. Chapter 12 on the forms of business organization covers these types.)

The chain operation is typical of a type of a foodservice industry franchise. This type requires the franchisee to follow the franchisor's prescribed method of operation. The franchisor keeps the right to check up on the franchise operation through frequent inspections to ensure compliance with company procedures. The agreement permits the termination of the franchise for *stated reasons;* requires the adoption of a specific architectural style, layout, and use of signs; and prescribes the purchase, preparation, and service of food.

ADVANTAGES OF FRANCHISING

The basic advantage of the franchise is that for a relatively modest investment, the franchisee obtains the franchisor's trade names, trademarks, goodwill, customer acceptance, and national advertising. Other pluses may be the expertise of the franchisor in operational know-how, training, and bookkeeping methods.

For the franchisor, the major advantage lies in the development of rapid market expansion and exposure with a minimum capital outlay.

DISADVANTAGES OF FRANCHISING

Because most franchisors are large, well-financed, and well-managed concerns in comparison with most potential franchisees, economic leverage in negotiating the terms and conditions of the franchise rests with the franchisor.

Until recently, franchisors were at liberty to disclose very little about financing, method of operation, ownership, and other data to the franchisee. As a result, the franchisee did not have all the facts necessary to make an informed business judgment on how to compare franchisors or whether to accept a franchise at all. Inexperience and lack of equality in bargaining power could result in the franchisee's acceptance of unreasonable terms and conditions. Such problems eventually led to the 1979 FTC Franchise Disclosure Rule.[2]

In some cases, the franchisee must deal exclusively with the franchisor to obtain raw materials and supplies, when the products the franchisor markets is unique or cannot readily be duplicated by others.[3] This can be an advantage, because quality standards are usually consistent and prices are often the same for all franchise buyers. However, when prices are inconsistent with local markets, the franchisee is at a disadvantage compared with competitors, who are free to compare and accept bids from local suppliers. Certain exclusive dealing arrangements (*tying sales*) are illegal under antitrust laws.

THE FTC FRANCHISE DISCLOSURE RULE

Inadequate information furnished by a franchisor to a potential franchisee can spell disaster to an inexperienced person who must finance and operate the franchise independently. The lack of a legal duty of franchisors in the past to disclose pertinent information created a climate that invited fraud and misrepresentation. Further, the common law required that franchisees who felt they had been taken had to prove franchisor fraud. This was extremely difficult, and the common law was inadequate to provide relief to franchisees. Meanwhile, a franchisee might have suffered substantial losses in the initial operating phases of the business.

In 1979 the Federal Trade Commission (FTC), through its authority to regulate business practices found to be unfair or deceptive, issued a trade regulation designed to curb some of the problems encountered by franchisees and franchisors.[4] The regulation requires franchisors to provide the FTC with a disclosure notice. The notice is not verified by the commission; it is merely filed.

Under the FTC regulation, a franchisor is required to give every prospective franchisee a *disclosure statement* 10 days before the franchisee signs a contract or makes any payment for the franchise, whichever occurs first.

Detailed information must be furnished concerning the franchisor's finances, experience, size of operation, and involvement in litigation. Among other required data are total costs of the franchise to the franchisee, including whether any portions are refundable, and the recurring expenses of the franchise. Specific limitations on the franchisee's operations regarding goods and services that may be offered, customers to whom the franchisee may sell, geographic limitations, and territorial protection (if any) from competing company franchises must be explained. Finally, conditions of termination, renewal, and transfer of the franchise, as well as detailed verification of franchisor claims as to sales, income, and gross or net profits, must be provided.

False statements as to sales, income, or profits are prohibited—punishable by a $10,000 fine for each violation. The FTC indicated that franchisees may sue for damages resulting from false statements made in the disclosure statement.

FRANCHISOR RIGHTS AND DUTIES

The franchisor has the right to enforce all terms and conditions included in the agreement with a franchisee, as long as the requirements are not illegal, or illegally obtained through fraud, duress, or undue influence. Every franchise contract is presumed valid and enforceable until declared otherwise, in whole or in part, by a court or government regulatory agency. The burden rests on the party claiming the wrongdoing to prove it in order to be relieved of any duty to perform that contract.[5]

A franchisor has a legal duty to fulfill the terms and conditions agreed to with the franchisee and to do nothing that interferes with the proper performance of the franchisee's duties. If the franchisor agrees to a limit of new franchise operations that might adversely affect the franchisee's business, the franchisor must act in good faith in keeping such promises.[6] The creation of new competition in violation of such a contract would enable the franchisee to recover losses suffered as a result. Further, the franchisor cannot use resulting losses in sales, on which its franchise fee may be calculated, as a means of terminating the franchise. However, the franchisee must prove that the losses were caused by the actions of the franchisor.[7]

Every franchisor has a lawful interest in requiring franchisees to maintain quality standards.[8] This provision is the means by which a franchisor can protect the name, goodwill, and reputation of the company. Any lessening of quality through the fault of the franchisee usually reflects more on the franchisor product than on the franchisee.

However, a franchisor may not, without very good reasons, force a franchise to purchase *all* supplies exclusively from the franchisor or dealers specified by the franchisor. This exclusive dealing arrangement may run afoul of the antitrust laws.[9] As long as franchisees can maintain quality standards while purchasing in the open market, then they must be given this opportunity to reduce costs, and the franchisor must allow other suppliers to sell to its franchisees.

The franchisor has the right to terminate a franchise. However, the law does not favor sudden and arbitrary terminations, unless specifically covered in the franchise agreement.[10] Even so, such clauses are strictly interpreted in favor of the franchisee—in some cases on the grounds that the franchisee lacked equality of bargaining power; and in other cases on the grounds that reasonable notice of the termination was not given.[11] Still other cases examine the length of the franchise and require a reasonable time elapse before termination so that the franchisee may recoup any investments.[12]

However, these cases do not deprive the franchisor of the right to terminate when proper provision for termination exists in the agreement.[13] The mere fact that the termination rights of the franchisor may be hard on the franchisee is not grounds for a court to declare them unenforceable.[14]

FRANCHISEE RIGHTS AND DUTIES

A franchise agreement is usually a standard type of agreement designed to benefit the franchisor. In some cases, there is little the franchisee can do to negotiate a better deal—aside from forgoing the franchise entirely. The FTC disclosure regulation helped to correct one major problem, namely, lack of accurate and complete information on which to base a decision as to whether to invest in a franchise.

The franchisee is under a *continuing* obligation to carry out the agreement with the franchisor. A hard bargain creating a franchise does not, standing alone, prove that the bargain was illegal. Lack of profitability or diminished expectation of profit does not excuse performance by the franchisee. These are business risks voluntarily assumed in order to obtain profits.

The franchisee has the right to expect the franchisor to provide support in advertising, marketing, and management as outlined in the agreement. Each franchisee has the right to the same fair treatment others receive, although fair treatment is hard for the courts to determine. A franchisee should closely monitor the situation to ensure that what he or she is getting from the franchisor is what the agreement promised.

A franchisee has the right to buy some supplies from local dealers, as long as this does not compromise the quality standards of the franchise agreement.

As a foodservice operator, the franchisee is directly responsible to patrons and the public for negligent or insufficient security, and is liable for breaches of implied warranties of food and beverage fitness.[15] The franchisee is responsible for meeting local building, health, and sanitary codes; for correcting faulty or defective premises; and for enforcing alcoholic beverage control laws and other legal requirements. The franchise agreement *does not* insulate the franchisee from liability to the buyer of food and beverages, or from government regulations. Nor does the franchisor assume such responsibilities unless specifically stated in the agreement.

FRANCHISOR RESPONSIBILITY FOR FRANCHISEE'S MISCONDUCT

Standard franchise contracts may contain a clause specifying that the franchisor will not be responsible for any liability arising out of the operation of the business by the franchisee. Additional language may require the franchisee to *indemnify*, or hold the franchisor harmless against any losses suffered by the franchisor in connection with the management of the franchise by the franchisee. Often the contract uses the term independent contractor to define the status of the franchisee.[16] Such clauses are aimed at, among other things, preventing the franchisor from being held responsible for injuries to others due to franchisee negligence or intentional misconduct.

However, two legal theories have been applied to impose liability on franchisors in spite of such language. One theory rests on the exercise of an ample degree of operational control by the franchisor over the activities of the franchisee.[17]

Control by the franchisor usually exists when the agreement: 1) requires the franchisee to build the facility as specified by the franchisor; 2) imposes strict adherence to specific operating methods and gives the franchisor the right to enforce compliance with frequent inspections; and 3) permits the franchisor to cancel the agreement for any substantial violation of the contract. Collectively, such agreement rights can establish enough franchisor control possibly to justify franchisor liability for the misconduct of the franchisee.

The second theory of franchisor liability is that of the degree of authority or control *represented* to patrons through advertising or other representations.[18] The use of the trademark or name, and the fact that the franchisee is part of a national system of franchises with elements of identical interior and exterior decor, are examples of such control. Once these representations, sometimes called *manifestations of authority*, are communicated to third persons, the franchisor is unable to deny control or authority over the franchisee.

However, the law recognizes that the franchisor is not liable for the acts of an independent contractor (in this case, the franchisee) that cause harm to a third party (patron) arising out of the negligence or intentional misconduct of an *employee* of the franchisee. Only the independent contractor is liable under the theory of *respondeat superior*, which means let the employer respond for the legal wrongs of his or her employees, even though the employer may be blameless. However, to allow the franchisor to escape liability, the courts must decide whether the franchisee was indeed an independent contractor.

In the following case, the court of appeals reviewed the factors necessary to hold a franchisor liable for the actions of its franchisee's employee in wrongfully revoking a guest's credit.

Wood v. Holiday Inns, Inc.
United States Court of Appeals, Fifth Circuit, Alabama
508 F.2d 167 (1975)

Facts: Upon checking in at a franchised Phenix City, Alabama Holiday Inn, Wood submitted his Gulf Oil card as proof of credit worthiness. The card was accepted without verification by the front desk and Wood occupied his assigned room without incident. Very early the next morning, the night auditor ran a check on Wood's card and found that it had been revoked. The night auditor phoned Wood at 3:00 AM demanding that Wood leave and using harsh and offensive language. Wood tried to suggest that he would settle his account in the morning, but he was refused.

It was later found that the night auditor's error in punching in the card number had caused the revoked status, and that Wood's card had been valid all along. Wood sued both the franchisee and franchisor alleging slander of credit and intentional infliction of mental distress. The court found the franchisee liable, a necessary element in holding the franchisor liable under the doctrine of *respondeat superior*. Because there was no *express* agency relationship between the Alabama Holiday Inn and Holiday Inns, Inc., the court found sufficient evidence of apparent agency to implicate the franchisor.

Reasoning: A franchisor may be held responsible for a franchisee's legal wrongs that damage a third-party guest only if the franchisor exercised *sufficient operational control* over the conduct of its franchisee, regardless of provisions to the contrary in the franchise agreement. Such factors include construction and maintenance of the franchised premises as specified by the franchisor; strict adherence to the rules of operation, including granting the franchisor the right to inspect the unit to maintain compliance; and, most important, granting the franchisor the right to cancel the franchise for substantial violation of its terms. There was sufficient evidence of the above factors in this agreement to support a judgment for Wood against Holiday Inns, Inc.

The court also ruled that Wood would be entitled to recover under a *representation theory* of apparent agency, under which the franchise agreement required the Phenix City Holiday Inn to be of such an appearance that travelers would believe it was owned by Holiday Inns. Moreover, there was no way that guests entitled to courteous treatment under their contract with Phenix City would know that the employees of the Alabama Inn were employed by Phenix City and not Holiday Inns, Inc.

Conclusion: The issue of *apparent agency*, either under a control or representation theory, rests on the sufficiency of the evidence to support it. There is no unanimity of decision in this area. However, the theories are accepted by most courts and the trend is to allow the jury to decide whether or not to impose liability. Therefore, franchisors may insert bold harmless or indemnity clauses in their franchise agreements to compel the franchise or to cover the cost of any liability judgments, in whole or in part, rendered against the franchisor. This question merits attention and resolution before the franchise agreement is executed.

ANTITRUST RULES AND FRANCHISES

Encouraging competition is an overall objective of economic policy in the United States. Prior to the development of antitrust laws, the business community was guided by the doctrine of *laissez faire*, meaning that survival in the marketplace was achieved by the accumulation of market power rather than by government regulation, subject to the *rule of reason* which regulates monopolies or threats to monopolize markets. Market power was acquired at the expense of competition, through a creation of a monopoly over production or service, by 1) establishing *pooling arrangements* to divide up the market for a product; 2) price discrimination in favor of designated customers; 3) customer discrimination as to quality, quantity, packaging and so forth; and 4) price fixing with other producers.[19]

The first legislation aimed at attacking anticompetitive marketing activities was the *Sherman Act* of 1890. This federal law prohibits contracts, combinations, or conspiracies in restraint of trade or commerce among the several states or with foreign nations, "monopolies or attempts to monopolize."

Later the *Clayton Act* of 1914 and the *Robinson-Patman Act* of 1936 were adopted to curb price discrimination, exclusive agreements, and mergers, but only where the effect of these practices may be to lessen competition substantially or to create a monopoly in any line of commerce.

The most recent federal enactment is the *Hart-Scott-Rodino Antitrust Improvement Act* of 1976, which, among other things, empowers a state's attorney general to sue, on behalf of citizens of that state, businesses that violate the Sherman Act.[20]

REGULATED PRACTICES

Several business practices, interpreted by the courts, are violations of these antitrust acts. Such governmental actions include price fixing and allocations of markets.

Per Se Violations

The first is grouped under the heading *per se violations*, or violations that cannot be justified by good motive, lack of intent to injure competitors, or economic necessity.

1. *Price fixing* is the most serious *per se* violation.[21] A franchisor may not fix the price at which the product or service must be sold by the franchisee. The franchisor may *suggest* retail prices but cannot make the franchisee sell at those prices. Likewise, a group of franchisees within the same franchise may not agree orally or in writing to fix the prices of products sold to the public.

2. A second *per se* violation involves *division or allocation of a given market* by groups of competitors. Competing franchisees cannot agree to divide up the market geographically or territorially among themselves. Only the franchisor may establish and enforce geographic or territorial restrictions with relation to each of its franchisees.[22]

   For example, for the life of a franchisee agreement, a franchisor could agree not to allow establishment of another franchise outlet within a designated area. However, the franchisee could not agree with other franchisees linked with the same company in the area to deny the entry of a new franchise operator.

3. A *group boycott* by competing franchisees is another *per se* violation.[23] Franchisees may not agree to compel a supplier of goods or services to fund a local promotion of the franchised product. Compulsion means *blacklisting* the supplier from further dealings with the group members for refusing to pay the required promotion fee.

The following case illustrates the treatment of various marketing devices, both individually and cumulatively, involving a national hotel franchisor and its franchisees by a federal appeals court.

American Motor Inns, Inc. v. Holiday Inns, Inc.
United States Court of Appeals, Third Circuit, New Jersey
521 F.2d 1230 (1975)

Facts: Holiday Inns, Inc., enforced a policy of inviting its franchisees to determine whether to grant a new franchise to American Motor Inns, Inc., within their respective territories. When the franchisees, in response to this "radius letter procedure," disapproved the application of American Motor Inns, it sued, claiming that this was a "concerted refusal to deal in violation of the Sherman Act."

Another Holiday Inns policy, called the "non-Holiday Inn clause," prohibited franchisees from owning or operating motels other than Holiday Inns. Once again, this clause was applied to American Motor Inns, causing it to sue, alleging an unreasonable restraint of trade.

A final Holiday Inn policy, called the "company town policy," under which potential franchisees could not operate in any area in which a company-owned Holiday Inn was established, was likewise imposed on American Motor Inns. This was also challenged as violation of the federal antitrust laws.

The court of appeals ruled that the radius letter procedure which gave the affected franchisees the right to veto the entry of a new franchise was a *per se* violation of the Sherman Act, since it was a concerted refusal to deal prohibited by law, regardless of whether it was economically reasonable or necessary.

The court next ruled that unlike the radius letter procedure, the non-Holiday Inn clause was not unreasonable as a matter of law, and must be judged by the rule of reason requiring a remand to the trial court to explore the impact on competition of the restraint within the relevant market. In this case, the "relevant market" would depend on: 1) whether Holiday Inns are reasonably interchangeable with other hotels and motels, in the eyes of the traveling public, and 2) whether Holiday Inns' franchises are reasonably interchangeable with other hotels and motels as potential franchisees for other hotel-motel chains.

The court concluded that the three Holiday Inns marketing devices, viewed together, created a horizontal (between franchisees similarly situated) allocation of geographic territories which is *per se* unlawful.

Reasoning: The court of appeals reviewed three Holiday Inn requirements imposed upon its franchisees and upon a franchisee seeking to enter a market occupied either by company town units operated by the franchisor or by existing franchisees who would be affected by the entry into their market of new competition.

The court reasoned that permitting existing franchisees to veto new competition was itself unlawful, because it directly violates the Sherman Act. That Act prohibits groups of motel franchisees from exercising together control over potential competitors. (As an aside, the court noted that Holiday Inns itself had the legal right to veto the new entrant, regardless of whether the competing Holiday Inns affected were company-owned.)

> The court further reasoned that the non-Holiday Inn clause was not itself unlawful, but depended on its reasonableness measured by its impact on competition within the relevant market, noted above.
>
> The court's final ruling, which found unlawful all three franchisor policies taken collectively, is of great significance. This combination of rules tilts the scale in favor of free competition, whereas the two rules (non-Holiday Inn clause and company-town policy) standing alone were deemed lawful activities.
>
> **Conclusion:** The court is attempting to balance the rights of business persons to protect their economic interests against unfair competition with their duty to permit lawful competition. The law is particularly quick to condemn concerted efforts to restrict competition. Where the restraint is not concerted, the role of reason is applied to determine its validity, that is, the court requires *review* of the impact of the restraint on competition within the relevant market. Lastly, while franchisor policies, standing alone may be reasonable, they may when viewed together violate antitrust laws, regardless of proof of economic necessity.

Rule-of-Reason Violations

The second category of antitrust violations is called the *rule of reason*. Rule-of-reason violations include marketing activities that create monopolies or threats to monopolize specific markets. Since in most cases these activities will involve franchisors rather than franchisees, it is enough to point out that such conduct may be justified by proof of economic necessity. Under the rule of reason, a franchisor may not prohibit a franchisee from acquiring a different franchise in the same market, unless the franchisee can prove that such a prohibition is justified on reasonable economic grounds.[24] Likewise, a *combination* of clauses imposed by a franchisor, such as a clause prohibiting a franchisee from obtaining a franchise in any area where a company-owned outlet was established, together with one giving the franchisee veto power over the entry of new franchisees, would be a *per se* violation—without economic justification—of the antitrust laws.[25]

Tying contracts to the exclusive purchase of a franchisor's products in exchange for obtaining the franchise is a rule-of-reason violation.[26] In other-words, the franchisor may not tie the franchise license to a franchise commitment to buy the franchisor's products exclusively, unless the franchisor can justify such arrangements.

Such contracts are not automatically illegal, but become so when the following conditions exist.

1. There are separate products, one the *tying product* and the other the *tied product*.

2. There is enough economic power for the tying market (trademark or formula) to force the purchase of the tied product.
3. There is plenty of competition in that market.
4. The arrangement would affect competitors in the tied market.[27]

In the event of a court action, the franchisor may try to prove that any or all of the conditions do not exist, and thus be excused from liability. The franchisor may also try to prove that the arrangement was reasonable on economic grounds. It is then left to a jury to decide these issues.

The following case gives an example of a tying arrangement that the court found to be both unjustified and in violation of the Sherman Act.

Siegel v. Chicken Delight, Inc.
448 F.2d 43 (9th Cir. 1971)
cert. denied 405 U.S. 955 (1972)

Facts: Siegel, a Chicken Delight franchisee, brought a *class action* suit (an action on behalf of himself and other Chicken Delight franchisees) to have a federal court declare illegal certain requirements contained in the franchise agreement.

The requirements were that the franchisees buy all of their cooking equipment, dry mix food items, and other products exclusively from Chicken Delight as a condition of obtaining a Chicken Delight trademark franchise. Siegel argued that Chicken Delight violated federal antitrust laws by tying the purchase of essential operating items to obtaining a franchise. Chicken Delight argued that the tying arrangements were economically justified and, therefore, legal.

The federal district court found that the arrangements were tying contracts; that the Chicken Delight trademarks used to tie these purchases by Siegel and other franchisees were monopolistic; and that Siegel had been injured. A jury found separately that there was no economic justification for these tying arrangements. The district court's findings of tying arrangements and of Chicken Delight's economic clout to create a monopoly were upheld. The jury's findings of no economic justification for Chicken Delight's contractual requirements with Siegel were also upheld. The damage judgment in favor of Siegel was *reversed* and sent back for a new trial on that question only.

Reasoning: Chicken Delight maintained that the requirement that franchisees buy their products exclusively from Chicken Delight was economically justified and reasonable because it helped to measure and collect revenue and preserved the distinct flavor, uniformity, and quality of the food product. The court ruled that the tie-in was not justified because it simply exerted the power and leverage of franchisor over franchisee.

> **Conclusion:**   In this case, no justification for a tying arrangement could be established by Chicken Delight. The issues relevant to such tying arrangements are whether they can be justified by the franchisor or whether they are an unreasonable restraint of free trade and violate antitrust laws.
>
>   A franchisor may require a seller of a product available to franchisees to meet the franchisor's required quality standard, or require the franchisee to request approval of another product.

State Antitrust Laws

Some states, such as California, New York, Texas, and Wisconsin, have enacted their own antitrust statutes to supplement and reinforce the federal laws. This is important because these states are able to deal with activities within their own borders, whereas much of the federal efforts are limited to activities *among* the states.

   Foodservice operators who do business within a single state and serve mostly intrastate patrons must comply with the state antitrust regulations. However, if a significant number of patrons are from out of state, or a significant portion of basic foodservice ingredients is purchased from out-of-state sources, then both federal and state antitrust laws apply.

TRADEMARK INFRINGEMENT

Franchisors are interested in protecting their trademark and products from *infringement* by those who would illegally cash in on a recognized name. Infringement may take the form of using recognized marketing strategies, or capitalizing on the franchised name, if only by changing it slightly or copying other aspects of the franchised company's products and services.

   The following case is one illustration of how the courts may view claims of trademark infringement by taking into account the widespread use of the name as well as the geographic location.

> *T.G.I. Friday's, Inc. v. International Restaurant Group, Inc.*
> *International Restaurant Group, Inc. v. T.G.I. Friday's, Inc.*
> *United States Court of Appeals, Fifth Circuit, Mississippi*
> *569 F.2d 895 (1978)*
>
> **Facts:**   T.G.I. Friday's, a franchisor whose restaurant motif included Tiffany lamps, among other turn-of-the-century fixtures, granted a franchise license to the Tiffany company to operate a unit in Jackson, Mississippi. A year later, Tiffany requested an agreement to operate another unit in Baton Rouge, Louisiana. Friday's denied their

request. In response, the two soul shareholders of Tiffany formed another corporation (International Restaurant Group) and opened an Ever Lovin' (E.L.) Saturday's Restaurant, whose decor closely resembled Friday's in Baton Rouge.

Friday's sought an injunction to stop operation of Saturday's seeking damages for breach of contract, service mark infringement, and unfair competition. The franchisees claimed no violation on their part.

The district court ruled that there was no violation by the franchisees. On Friday's appeal, the court upheld that ruling, saying that Friday's had not expressly forbidden use of the word "Saturday" in its franchise agreement and that the new restaurant did not infringe on Friday's trademark.

Reasoning: The appeals court addressed the issues of breach of contract, trademark infringement, and unfair competition.

As for breach of contract, it stated that Friday's franchise agreement, which forbade use of " . . . the names of the days of the week singly or in combination with other words in connection with the operation of a business" and Tuesday, Wednesday, Thursday, and Sunday in particular, was ambiguous and did not prohibit use of the word "Saturday's" by anyone who was bound by the agreement.

As to trademark infringement, the court ruled that the use of the name "E.L. Saturday's," though similar to "T.G.I. Friday's" did not constitute trademark infringement, since the two were dissimilar visually and phonetically, and, since the Friday's chain had no reputation in the Baton Rouge area, the physical similarities between the restaurants were irrelevant to the issue of infringement.

As for unfair competition, the court reasoned that "Friday's" did not compete in the Baton Rouge area and that "Friday's" did not own exclusive rights to a "turn-of-the-century" theme.

Conclusion: Franchisors tend to be very protective of their trademarks since they are the center of their licensing agreements and the basis of their profitability. Operators must be aware of the difference between trademark infringement, which is very serious, and using knowledge of common marketing strategies and services to appeal to customers. Geographic location must also be taken into account. Had the franchisees established "Ever Lovin' Saturday's" in areas where other Friday's franchises were located, the court might have taken a different view.

FRANCHISE GUIDELINES

A franchise gives an operator the right to sell recognized products and services, giving a franchisee an advantage over new operators just starting out. At the same time, the foodservice franchisee remains an independent operator of the business. The franchisee is responsible to state and local authorities for complying with all building, health, and sanitary codes; to patrons for personal injuries and property damage or losses caused by negligence or intentional misconduct by employees; and to federal and state regulatory agencies that enforce antitrust activities.

Most franchise agreements strongly favor the franchisor, making the franchisee responsible for any losses suffered by the franchisor due to failure to carry out legal responsibilities.

Any potential franchisee must weigh the pros and cons of franchise against the lease or purchase of foodservice premises. There is protection under the FTC Franchise Disclosure Rule, which requires the franchisor to provide information openly. Nonetheless, disclosure does not guarantee success. The FTC will not make good losses in the event the franchise is not profitable.

Court judgments notwithstanding, it is up to the franchisee to read the franchise agreement carefully to see whether it contains tying arrangements, and to decide whether this will place the franchisee in a less competitive position because of them.

The only sure guide to the rights and responsibilities of both franchisor and franchisee is the agreement. Independent of antitrust laws, courts are not likely to impose any more responsibilities on the franchisor than are contained in this contract. Although most franchise agreements are standard, it would still be a good idea to have a lawyer go over the agreement. In the final analysis, only an individual can decide whether a franchise agreement meets his or her management standards, is flexible enough, and offers the best return on investment.

SUMMARY

A franchise is a license from the owner, the *franchisor*, of a trademark or service mark that permits another, a *franchisee*, to use the name and conduct the business according to specific procedures.

A franchise usually offers standardized products and goodwill as well as management help to its licensees. Still, the day-to-day management is up to the franchisee. The success of the operation rests as much on the skills of the franchisee as on the name and product identity of the franchisor.

There are advantages and disadvantages to franchising. Often contracts may be standardized in favor of the franchisor. It is up to the franchisee to negotiate the best deal. The FTC Franchise Disclosure Rule offers some protection in that it requires franchisors to disclose information on company finances, size, and involvement in litigation.

Franchisors and franchisees have rights and duties, usually spelled out in the contract. Franchisors may not be liable to the franchisees' patrons unless the law provides for liability. Usually the franchisee is directly responsible to patrons for injuries or illness caused on the premises.

A number of federal antitrust laws affect both franchisors and franchisees and are designed to prevent monopolies and to protect free trade and competition. Violations of these laws can result in heavy fines, or even jail terms, for violators. In addition, several states have antitrust laws.

QUESTIONS

1. What tangible and intangible advantages can a franchisor offer a potential franchisee?

2. An agreement for a hamburger franchise requires you, as a franchisee, to purchase all supplies from the franchisor. Should you sign? Why or why not?

3. What type of information should a potential franchisee expect to obtain from the franchisor? How will this information help the would-be franchisee decide?

4. Give an example of a franchisee violation of an antitrust law. Categorize the type of violation.

5. How does the law make a franchisor liable for franchisee misconduct to guests or patrons? How is the franchisee liable?

NOTES

1. H. Brown, *Franchising, Realties and Remedies* 1–15 (rev. ed. 1981). But see *McDonald's Corp. v. Markim, Inc.*, 209 Neb. 49, 306 N.W.2d 158 (1981), holding that a franchisee's contractual right to be given first consideration for a renewal of the franchise did not require the franchisor to give its franchisees thoughtful or sympathetic regard in preference to everyone else.

2. Disclosure Requirements and Prohibitions Concerning Franchising and Business Opportunities, 16 C.F.R. sec. 436 (1982).

3. But such exclusive dealing arrangements have been held to violate the Sherman Antitrust Act. See *Siegel v. Chicken Delight, Inc.*, 448 F.2d 43 (9th Cir. 1971), *cert. denied*, 405 U.S. 955 (1972). Contra see *Principe v. McDonald's Corp.*, 631 F.2d 303 (4th Cir. 1980), *cert. denied*, 451 U.S. 970 (1981).

4. The FTC was established to provide expert and ongoing enforcement of antitrust policies lacking under the Sherman Act. The commission is an administrative regulatory agency, not a court of law. Its purpose is to investigate unfair methods of competition, which include any of the anticompetitive activities found to violate the Sherman Act, to issue, cease, and desist orders in appropriate cases, and to issue trade regulation rules. Its order and trade regulation rules are subject to judicial review. Congress recently set limits upon the FTC's authority to issue trade regulation rules.

Unlike a private cause of action or lawsuit for an antitrust violation, which enables the person or party injured to recover damages, the FTC does not create any damage remedy. The FTC is established to protect the public interest; the injured person or party must find his or her damage remedy within the Sherman Act. A court of law, not the FTC, decides this and related issues.

5. Brown, *supra*, note 1, at 8–23.

6. But no such rights exist in the absence of a contract clause precluding the franchisor from operating a company-owned unit nearby. *Snyder v. Howard Johnson's Motor Lodges, Inc.*, 412 F. Supp. 724 (S.D. Ill. 1976). See *T.G.I. Friday's, Inc. v. International Restaurant Group, Inc.*, 569 F.2d 895 (5th Cir. 1978). See also *Druker v. Roland Wm. Jutras Assoc., Inc.*, 370 Mass. 383, 348 N.E.2d 763 (1976), where the Supreme Judicial Court of Massachusetts adopted the following language: "[I]n every contract there is an implied covenant [promise] that neither party shall do anything which will have the effect of destroying or injuring the right of the other party to receive the fruits of the contract which means that in every contract there exists an implied covenant of good faith and fair dealing."

   Uproar Co. v. National Broadcasting Co., 81 F.2d 373, 377 (1st Cir.), *cert. denied*, 298 U.S. 670 (1936), quoting from *Kirk LaShille Co. v. Paul Armstrong Co.*, 263 N.Y. 79, 87, N.E. 163 (1933).

7. *Id.* at 385, 348 N.E.2d at 765. Where intentional fraud is established as the cause of the breach of contract, both compensatory and punitive or exemplary damages may be recovered. See *Slater v. KFC Corp.*, 621 F.2d 932 (8th Cir. 1980) (case set back for new trial to determine whether jury verdict based on actual fraud or only on concealment) (seafood franchise); *Clinco v. Carvel Corp.* (N.Y. Supp. 1977) found in 7 IFA Franchising World (April 1977) (ice cream franchise).

8. Brown, *supra*, note 1, at 3–21, –22, –23, *Kentucky Fried Chicken Corp. v. Diversified Packing Corp.*, 549 F.2d 368 (5th Cir. 1977).

9. See note 3, *supra*.

10. Decisions to terminate or not to renew a franchise, provided for by the franchise agreement, must be made in entirely good faith and not capriciously or arbitrarily. *McDonald's Corp. v. Markim, supra*, note 1.

11. See generally *Arnott v. American Oil Co.*, 609 F.2d 873 (8th Cir. 1979), *cert. denied*, 446 U.S. 918 (1980). Contra *Zapatha v. Dairy Mart, Inc.*, 1980 Mass Adv. Sh. 1837, 408 N.E.2d 1370 (1980). Also see *Milsen Co. v. Southland Corp.*, 454 F.2d 363 (7th Cir. 1971).

12. See Brown, *supra*, note 1, at 1–15.

13. *McDonald's Corp. v. Markim, Inc., supra*, note 1.

14. *Ungar v. Dunkin' Donuts of America, Inc.*, 68 F.R.D. 65 (E.D. Pa. 1975), rev'd, 531 F.2d 1211 (3d Cir.), *cert. denied*, 429 U.S. 823 (1976).

15. *Eastep v. Jack-in-the-Box, Inc.*, 546 S.W.2d 116 (Tex. Civ. App. 1977); *Zabner v. Howard Johnson's Inc.*, 201 So. 2d 824 (Fla. Dist. Ct. App. 1967).

16. A typical indemnity clause would read as follows: "Under no circumstances shall **FRANCHISOR** be liable for any act, omission, debt, or any other obligation of **FRANCHISEE. FRANCHISEE** shall indemnify and save **FRANCHISOR** harmless against any such claim and the cost against defending against such claims arising directly or indirectly from, or as a result of, or in connection with, **FRANCHISEE'S** operation of the franchised business." Brown, *supra*, note 1, at A-26.

   An *independent contractor* is a party over whose activities the other party has no control or supervision concerning the performance of the contract existing between them. A typical franchise contract provides that: "This Agreement does not constitute **FRANCHISEE** as an agent, legal representative, joint venturer, partner, employee, or servant of **FRANCHISOR** for any purpose whatsoever; and it is understood between the parties hereto that **FRANCHISEE** is an independent contractor and is in no way authorized to make any contract, agreement, warranty or representation on behalf of **FRANCHISOR,** or to create any obligation, express or implied, on behalf of **FRANCHISOR** . . ." *Id.* at A-25.

The fact that the contract establishes an independent contractor status between franchisor and franchisee is not binding on a third party patron who wishes to pursue legal remedies against both parties for injuries suffered as a result of an act or omission on the part of the franchisee itself or its employees. The question as to whether the franchisee is or is not an independent contractor is for the court and jury to decide. The contract language is evidence that the parties intended to establish that relationship, but the patron was not a party to that contract and therefore did not agree to be bound by the contractual status of independent contractor. See *Peters v. Sheraton Hotel and Inns*, N.Y.C.J., July 6, 1979, at 7 (N.Y. Civ. Ct. 1979).

17. *Billops v. Magness Constr. Co.*, 391 A.2d 196 (Del. 1978) (motel franchisor for intentional torts of its franchisee's banquet director in disrupting a party). See also *Wood v. Holiday Inns, Inc.*, 508 F.2d 167, 175–77 (5th Cir. 1975) (franchisor liable for its franchisee's night clerk who negligently revoked motel guest's credit card and breached motel franchisee's common law duty to provide its guests with courteous and considerate treatment). *Peters v. Sheraton Hotel and Inns, supra*, note 16 (franchisor must defend negligence action based on defective bed involving infant guest of franchisee).

18. *Sapp v. City of Tallahassee*, 348 So.2d 363 (Fla. Dist. Ct. App. 1977), citing *Wood v. Holiday Inns, Inc., supra*, note 17, on the law of apparent authority conveyed by representations the franchisor makes to the public. (Either theory states good cause of action against franchisor where motel guest alleges inadequate security precautions.)

19. A pooling arrangement is an unofficial collective agreement among suppliers of facilities or services to combine there facilities, goods, or services so as to exclude nonmember competitors from participation in the subject matter of the pool. See *Associated Press v. United States*, 326 U.S. 1 (1945). To illustrate, the pooling of a unique computer hotel reservation service, not readily duplicated, by a group of hotels for the purpose of shutting out competing hotels would constitute a pooling arrangement. As such, every pooling arrangement constitutes a per se violation of the Sherman Act, even though the total activities of pool members do not create or threaten to create a monopoly.

20. 15 U.S.C. Sec. 1–7, 12–17, 13–13a, 16 (1976 & Supp. v. 1981).

21. *United States v. Socony-Vacuum Oil Co.*, 310 U.S. 150 (1940).

22. *American Motor Inns, Inc. v. Holiday Inns, Inc.*, 521 F.2d 1230 (3d Cir. 1975). The franchisor may act individually to protect his or her legitimate business interests free of restraint under section 1 of the Sherman Act. The franchisor commits no *per se* violations. The rule of reason must be applied to determine whether the franchisor exerts or threatens to exert monopoly power in the relevant market, under section 2 of the Sherman Act. Economic considerations may properly be introduced by the franchisor as a defense to any such claim of monopoly.

By contrast, the franchisees in combination violate section 1 of the Sherman Act, because they compete with each other and act to exclude the entry of new competition within their market area. It is their combined restraint of trade that the court in *American Motor Inns v. Holiday Inns, Inc.*, found an illegal restraint of trade, without justification on any grounds.

23. *Id.*

24. *Northern Pacific Ry. Co. v. United States*, 356 U.S. 1 (1958).

25. *American Motor Inns, Inc. v. Holiday Inns, Inc., supra*, note 22.

26. *United States v. Hilton Hotels Corp.*, 467 F.2d 1000 (9th Cir. 1972), *cert. denied*, 409 U.S. 1125 (1973).

27. See *Siegel v. Chicken Delight, Inc., supra*, note 3; *Kentucky Fried Chicken Corp. v. Diversified Packaging Corp., supra*, note 8.

Bankruptcy and Reorganization

OBJECTIVES

After reading this chapter, you should be able to:

1. Explain the purposes of bankruptcy.
2. Compare and contrast the different types of bankruptcy.
3. Explain who may file for bankruptcy, rights and responsibilities of the parties involved, procedures, tests used to determine bankruptcy status, and what debts are most affected by bankruptcy discharge.

Case in Point

Quik-Stop, operator of a self-service vending machine food and beverage outlet, failed to make the required payments on its vending equipment lease with Leaseafoods, and Leaseafoods obtained a court judgment for the total amount due. Afterward Quik-Stop filed a petition to be declared bankrupt, but failed to inform the court of its debt to Leaseafoods or Leaseafoods' identity. The court declared Quik-Stop bankrupt. When Leaseafoods tried to collect payment, Quik-Stop argued that the equipment debt was wiped out when the court declared the operation bankrupt. Leaseafoods responded that its debt was not wiped out because it was never notified of the bankruptcy court proceedings, and the debt was not given to the court. Which has bankruptcy law on its side?

The reviewing court ruled in favor of Leaseafoods. Every creditor is entitled to challenge a discharge of bankruptcy, and must be given notice of a bankruptcy hearing in advance to allow time to prepare for any challenge.[1]

PURPOSES OF BANKRUPTCY

Contrary to popular belief, bankruptcy is not intended to give debtors an open-ended, revolving escape hatch through which to thwart their creditors. If that were truly the case, all credit transactions would virtually end, and our credit-based economy would suffer permanent damage.

Bankruptcy laws serve to protect creditors by permitting an orderly, equal distribution of a debtor's estate. Bankruptcy laws also protect creditors from a debtor's temptation to conceal its finances or to favor one creditor at the expense of others. Last, bankruptcy proceedings give debtors a new financial lease on life, freeing them from an otherwise endless obligation to pay debts beyond their ability to do so.[2]

In addition to absolute bankruptcy, the law also allows some debtors additional time to pay outstanding debts free of the pressures of creditors.

Viewed realistically, bankruptcy is a rescue device established to protect creditors and debtors from the consequences of circumstances beyond their control, such as accidents, natural disasters, illness, divorce, severe economic problems, and reduction or end of government benefits. Bankruptcy may relieve the severe financial hardship to which these unexpected events can lead.[3]

Both businesses and individuals can file for bankruptcy. Hospitality operators should be aware of all their options of staying in business.

FORMS OF BANKRUPTCY

The Bankruptcy Reform Act of 1978 includes three forms of bankruptcy available to businesses. The most common is straight bankruptcy, or chapter 7 bankruptcy, with any assets distributed to creditors by a bankruptcy trustee.

Chapter 11 bankruptcy allows the businessperson to reorganize under a repayment plan, assembled by a trustee and approved by creditors.

Chapter 13 allows for a voluntary repayment plan by the debtor.

We will examine each in detail and present the advantages and disadvantages of each.

WHO PETITIONS FOR BANKRUPTCY?

All bankruptcy proceedings are started by the filing of a *petition*. Filing for bankruptcy can be done in one of two ways: 1) a voluntary petition is filed by a debtor; or 2) an involuntary petition is filed by a creditor or creditors of the debtor.

Individuals, corporations, and partnerships may file a *voluntary* petition. However, a partnership petition does not protect the individual partners. To obtain individual relief in bankruptcy, each partner must file an individual petition. Otherwise each partner will be held fully responsible for partnership obligations not satisfied by partnership assets. Debtors wishing to file a voluntary petition do not have to prove *insolvency*, meaning that their debts exceed their assets. All that is required is proof that the debtor has debts.

Exhibit 14.1 shows a voluntary petition for bankruptcy.

An *involuntary* petition may be filed by any single creditor having a valid claim exceeding $5,000, which does not include security interests or collateral held by the creditor. Three creditors must sign an involuntary petition when the debtor has 12 or more creditors. In an involuntary proceeding, the debtor must be unable to pay debts as they become due, or must have had a court-appointed custodian take charge of his or her property within the previous four months.

In an involuntary proceeding, the court may permit foodservice debtors to continue to operate if they can establish that they do not intend to dismantle the business or sell the assets at less than fair market value. An involuntary petition may not be filed against nonprofit organizations.

This is usually a last resort for creditors when large sums of money are owed, and creditors are encouraged to try to settle out of court if possible.

WHERE TO FILE A PETITION

Every debtor may file a petition in the federal judicial district[4] in which, for the longest portion of a 180-day period before filing, the debtor had either: 1) a residence; 2) a *domicile* (a person's true, fixed home, to which he or she always intends to return); 3) principle assets; or 4) a principle place of business.[5] These are called *venue*; that is, where a petition is filed.

For individuals and corporations, venue is the place where the individual or corporation resides. Corporations may file in any federal district in the corporation's state of incorporation or the district where the company's principle assets or principle place of business is located.[6] An independent operator must file where the restaurant business and assets are located.

Partnerships and *unincorporated associations* must file in the district where their principle place of business or their principle assets are located.[7] Once a partner has filed an individual petition in one district, all other partners may file their petitions on behalf of the partnership in that district. Likewise, a filing in one district on behalf of a partnership permits all general partners to file individual petitions in that district.[8]

If a petition is filed in the wrong federal district, the court has the power to transfer the petition to the correct district or to hear and determine the petition in spite of the wrong filing, called *improper venue*. The court choice is made in the interests of justice and the convenience of the parties.[9]

Exhibit 14.1   Voluntary Petition for Bankruptcy

A Voluntary Petition for Bankruptcy

[CLERK'S STAMP]

Name _____

Address _____

_____

Telephone _____

Attorney or Assistant or Representative for Debtor(s)

United States Bankruptcy court for the

_____ District of _____

In re

Debtor(s)

Social Security No. _____	[Set forth here all names, including trade names, used by debtor(s) within last 6 years]
Social Security No. _____	
Debtor's Employer's Tax Identification No. _____	

Case No. _____

→ VOLUNTARY PETITION UNDER CHAPTER _____

(If this form is used for joint petitioners wherever the word "petitioner" or words referring to petitioners are used they shall be read as if in the plural.)

1. Petitioner's mailing address, including county, is _____

_____

2. Petitioner has ❏ resided (or has ❏ had his domicile or has ❏ had his principal assets) within this district for the preceding 180 days (or for a longer portion of the preceding 180 days than in any other district).

3. Petitioner is qualified to file this petition and is entitled to the benefits of title 11, United States Code as a voluntary debtor.

4. (If appropriate) ❏ A copy of petitioner's proposed plan, dated _____, is attached (or ❏ Petitioner intends to file a plan pursuant to Chapter 11 of title 11, United States Code).

5. (If petitioner is corporation) Exhibit "A" is attached to and made part of this petition.

6. (If petitioner is an individual whose debts are primarily consumer debts) Petitioner is aware that he (or she) may proceed under Chapter 7, 11, 12, or 13 of title 11, United States Code, understands the relief available under each such Chapter, and chooses to proceed under Chapter 7 or Chapter 11 or Chapter 12 of such title.

7. (If petitioner is an individual whose debts are primarily consumer debts and such petitioner is represented by an attorney) A declaration or an affidavit in the form of Exhibit "B" is attached to and made a part of this petition.

WHEREFORE, petitioner prays for relief in accordance with Chapter 7 or Chapter 11 or Chapter 12 of title 11, United States Code.

Signed: _____   Petitioner signs if not represented by attorney

Attorney for Petitioner

_____

Petitioner

Address: _____

_____   _____

Petitioner

I, (We), _____ and _____, the petitioner(s) named (or the president or other officer or authorized agent of the corporation named as petitioner, or a member or an authorized agent of the partnership named as petitioner) in the foregoing petition, do hereby declare, under penalty of perjury, that the statements contained therein are true and correct, (and that the filing of this Petition on behalf of the corporation [or partnership] has been authorized).

Executed on _____, 19 _____   _____

Petitioner

_____

Petitioner

STRAIGHT BANKRUPTCY

For foodservice businesses that are seriously in debt, *straight* or *chapter 7 bankruptcy* may be the only route. The end result of straight bankruptcy, however, is that the business usually terminates, and any remaining assets are distributed to creditors by a court-appointed *trustee*. Straight bankruptcy should be the last resort for businesspersons.

Straight bankruptcy may be voluntary, if filed for by the debtor, or involuntary, if filed by the creditor(s). Either way, if a straight bankruptcy petition is granted, the operator may be effectively out of business. Even if the debtor could start up again, the inability to obtain credit after a straight bankruptcy, with the resulting need to pay cash up front, would present a serious cash flow problem.

The main advantage of straight bankruptcy is that it results in a "clean slate" for the debtor. The debtor is freed of the requirements to pay the debts erased under the bankruptcy petition, and the creditors are paid, although bankruptcy laws allow only so many cents to be paid on each dollar owed.

When a bankruptcy petition is filed, the bankruptcy court must decide whether to grant or deny relief. Relief is automatic when a voluntary petition is filed or the debtor does not contest the filing of an involuntary petition. Should the debtor argue against the granting of an involuntary petition, the court will hold a trial to decide whether a declaration of bankruptcy is appropriate. Relief of this kind will be granted only if the debtor is unable to pay his or her debts as they become due, or in the event a trustee was appointed within four months of the filing.[10]

The court will require the bankrupt to file a list of assets, liabilities, and creditors, and a statement of his or her current financial condition. The debtor must list *all* of his or her creditors on the schedule required by law, or face the likelihood that an unlisted creditor will successfully move to revoke the discharge. The court will then call a creditors' meeting. At that meeting the creditors may elect a committee or appoint a trustee to manage the estate. The court must approve this appointment. The debtor must appear at the meeting and answer questions about his or her financial circumstances.[11] One of the most frequently asked questions is whether the debtor has concealed or wrongfully disposed of assets.

The court-approved trustee takes charge of the debtor's property and has it appraised for value. The trustee sets aside those items of the debtor's property that are exempt from distribution under state or federal bankruptcy laws. In some cases the trustee has temporary authority to operate the debtor's business.[12]

The trustee evaluates creditor claims and objects to those he or she thinks improper. All secured and exempt property is segregated from unsecured property. Unsecured property is sold and the proceeds held for the benefit of creditors in the name of the bankrupt's estate. All the money received by the trustee must be deposited in that account, which the trustee must keep and report to the creditors. The trustee must make a final accounting report to the creditors at their last meeting.[13]

At this point, unless the bankrupt party files a waiver of the right to a discharge, he or she is entitled to a discharge in bankruptcy. The debtor is released from obligation for the debts involved. An individual must wait six years from the last discharge he or she received to be eligible for another discharge.

Corporations are not eligible for a discharge in bankruptcy.[14] They are eligible for reorganization under bankruptcy laws, whereby they remain in business and pay off their debts gradually.

Criteria for Grant or Denial of Bankruptcy Status

Eligibility for discharge of a bankrupt is not the same thing as *granting a discharge* in bankruptcy. All creditors will allow the trustee or a United States attorney to file objections to any pending discharge. A final hearing may be held to determine whether or not a discharge should be granted. A discharge will be denied if the debtor fails to appear at the hearing, or refuses to submit to questions of creditors at a creditors' meeting called by the court.[15] Additionally, the court must decide whether the bankrupt has committed any act that is a legal obstacle to discharge, or creates a bar to discharge.

A *bar to a discharge* is intended to prevent dishonest debtors from using bankruptcy as a means of benefiting from their own wrongdoing. Acts that require the court to deny bankruptcy are: 1) intentionally falsifying, concealing, or destroying debt records; 2) falsifying statements of financial condition to obtain credit or extensions of credit; 3) making fraudulent transfers, removals, or concealment of property to hinder, delay, or defraud creditors; 4) failure to account satisfactorily for assets; and 5) failure to obey court orders or to answer questions the court approves and directs the debtor to answer.[16]

It is possible to repay debts after they have been discharged in bankruptcy. Before a discharge is granted, the creditor needs to negotiate an agreement with the debtor and get the court to approve the agreement. The reaffirmation agreement is allowed as long as the debtor does not cancel within 30 days of the date of court approval.[17]

Debtor Rights and Responsibilities

Every debtor has rights and responsibilities in bankruptcy proceedings.

Exemptions   Certain items of property are not considered debtor assets and are exempt from court disposition, which means they may be kept by the debtor free and clear of creditor claims, unless the debtor conceals or fraudulently transfers such property in violation of law. Federal and state laws govern such exemptions.[18]

Federal exemptions include: 1) real or personal property used as a residence not to exceed $7,500 in value; 2) one motor vehicle not to exceed $1,200 in value; 3) household goods, furnishings, apparel, appliances, books, animals, crops, or musical instruments primarily for personal, household, or family use of the debtor or dependents, not to exceed $200 in any one item; 4) personal or family jewelry not to exceed $500 in total value, and used by the debtor or dependents; 5) implements, professional books, or tools of trade not to exceed $750 in total value; 6) life insurance contracts; 7) medically prescribed health aids; 8) any other property of the debtor's choosing not exceeding $400 in value; and 9) social security, disability, alimony, and other benefits reasonably necessary for the support of the debtor or the debtor's dependents.

Liens[19]   *Liens* are legally created rights that certain creditors may use to seize and sell a debtor's assets. Debtors may avoid certain liens that would otherwise apply to defeat or impair their legal exemptions, such as life insurance contracts. Such *voidable liens* are court-ordered liens.[20]

Surviving Obligations

The Federal Bankruptcy Act makes certain obligations and debts nondischargable under chapter 7, meaning that the debtor remains liable to pay them even though the bankruptcy court has discharged all other allowable, provable claims. These surviving obligations follow.[21]

1. Federal, state, and local fines or taxes
2. Any obligations arising out of obtaining money by false statements, pretenses, or representations
3. Obligations arising out of willful or malicious injury to a person or that person's property
4. Alimony and child-support obligations
5. Obligations created by the debtor's criminal misconduct, such as *larceny* (theft), *embezzlement* (wrongful taking of property of another entrusted to one's care), or *fraud* committed by the debtor in a fiduciary or trust capacity, such as wrongful use of money entrusted to a debtor as trustee for another person or party
6. Certain types of educational loans that come due within five years prior to the filing of the bankruptcy petition
7. Obligations not included in the bankruptcy petition in time, because the creditors involved did not have enough notification of the proceedings even though the debtor new he or she owed them money

Prohibited Practices

Every debtor is prohibited from the following practices.

Preferential Payments[22]   A *preferential payment* is a voluntary payment made by a debtor to one creditor over other creditors. This preference means that one creditor receives more than his or her fair share of the debtor's asset. A preferential payment is defined as one made within 90 days of the filing for bankruptcy, and enables the favored creditor to obtain a greater percentage of a debt than is obtained by the other creditors. The fact that the creditor was unaware of the debtor's bankruptcy petition makes no difference.

A one-year period rather than 90 days applies to a payment to an "insider," a relative of an individual debtor, or an officer, director, or advisor of a company debtor. Such payments can be recovered from the favored creditor by the bankruptcy trustee for redistribution to other debtors.

Preferential Liens   The same definition and recovery procedures apply to *preferential liens*, since they, too, seek to create preferential legal rights for one creditor and are unfair to all other similar creditors.[23] A preferential lien is a preferential transfer by the debtor of assets, either in cash or property, by prior agreement with a creditor. The lien may be in the form of a prior security interest in the property, such as collateral for a bank loan. In other words, a debtor may give a bank the right to sell a pizza delivery truck if he or she fails to repay the loan. However, if the debtor goes bankrupt, the trustee will seize the truck to help repay all the creditors. The lien may take the form of a *judicial lien*, meaning the right of a *judgment creditor* (one who has recovered a court judgment against the debtor) to be paid out of the judgment before other creditors are paid.

Fraudulent Transfers   Under the federal Bankruptcy Act, a fraudulent transfer is any gift, sale for less than fair market value, or other disposition of a debtor's property with the intent to hinder, delay, or defraud creditors. Such a transfer is voidable, meaning that the bankruptcy trustee may take the transferred property for distribution to all creditors. For example, if the owner of Quik-Stop, the debtor in the Case in Point, had made a transfer of vending machine foods and beverages to a spouse for $50 when the fair market value of the inventory was $500, in order to remove the inventory from the reach of Quik-Stop creditors, that sale could be voided and the spouse forced to return the inventory to the bankrupt's estate.

Bankruptcy Crimes

The Bankruptcy Code defines the following bankruptcy crimes, which are felonies punishable by a fine of not less than $5,000 or imprisonment of not more than five years, or both.

1. Knowingly and fraudulently making a false statement under oath
2. Knowingly and fraudulently presenting or using a false claim
3. Knowingly and fraudulently giving or receiving a bribe
4. Knowingly and fraudulently withholding records

These criminal acts are in addition to the court's blocking or refusing to discharge debts involving any debtor's own bankruptcy case.

These provisions of the Bankruptcy Act do not prohibit a debtor from continuing a hospitality business. Once the bankruptcy court authorizes the debtor to use current funds to buy materials to carry on the operation, that person is in business; the total assets are not being reduced. The debtor is merely trading current cash for current inventory and supplies, and not building up debts he or she cannot pay.

Creditor Rights and Responsibilities

Creditors also have certain rights and responsibilities under the federal Bankruptcy Act.

1. Every creditor whose debt is not secured by collateral or property of the debtor has the right to file a *proof of claim*, usually within six months after the first meeting of creditors. When the collateral or security is insufficient to satisfy the entire debt, the secured creditor must file a proof of claim for any unsecured portion of that debt.[24]
2. Merely filing a claim does not automatically entitle a creditor to share in the distribution of the debtor's assets. The claim must be valid, meaning that the trustee must allow the claim. Any legal defenses that the debtor could have made against the creditor may be used by the bankruptcy trustee to reduce or eliminate the claim.[25]

   For example, if Quik-Stop could argue that Leaseafood breached a warranty of fitness in the sale to Quik-Stop of its vending machines and inventory, that breach of warranty would be available to the trustee, and Leaseafoods might be denied a portion or all of its claim. In this sense, the bankruptcy trustee steps into the shoes of the debtor. Otherwise, creditors could obtain a remedy to which they are not legally entitled, simply by filing a petition for involuntary bankruptcy.
3. A secured creditor has a preference to receive payment, but only if the creditor's claim is legally enforceable.[26] Every creditor must prove a claim before any payment is made.
4. Certain creditor claims are given priority whether or not they are secured by collateral. These claims include: a) expenses and fees required to administer the bankrupt's estate; b) $2,000 worth of wages for each wage earner for wages earned 90 days before the filing of the bankruptcy petition; c) contributions to employee benefit plans arising out of work or services performed within 180 days of the filing of the petition; d) $900 worth of claims for each individual debtor for deposits made on goods or services for personal use not delivered or provided; and e) taxes.[27]
5. The creditor is entitled to sufficient notice of a debtor's bankruptcy claim. This is because the creditor must be allowed time to challenge the discharge or have equal access with other creditors to debtor assets. A creditor who fails to protect himself or herself after such notice cannot share in the bankrupt's estate.

6. The creditor may not continue to harass or use other methods to obtain payment once a debtor has been granted bankruptcy.[28]

The order of payment gives all allowable secured claims priority, but only to the extent of the security or collateral, over the priorities mentioned above. Afterward, all unsecured creditors are paid. Unsecured creditors receive very little, if anything.

Any loan made to someone else should be secured by property or other debtor assets that can be used to reduce or wipe out that debt if the debtor is unable to pay. Otherwise the lender stands to lose any real chance of repayment.

What advantage, if any, does a creditor whose obligation is nondischargeable have over other creditors? A creditor with nondischargeable debts has the right to recover the balance of the debts in full, as well as to share with other creditors in the distribution of the debtor's estate.

CORPORATE REORGANIZTION UNDER CHAPTER 11[29]

Individual, partnership, and corporate debtor's are eligible for reorganization of their business by filing voluntarily for such relief or by having a creditor petition the debtor into involuntary bankruptcy under chapter 11.

Bankruptcy courts will order relief for a properly filed petition. Once this is accomplished, the court will normally appoint a committee of unsecured creditors, a trustee, and a committee of corporate shareholders, who would also be involved in the case of a corporate bankruptcy.

The trustee may run the debtor's business as well as develop a plan for disposing the claims of creditors and shareholders. The plan may include refinancing the business with existing treasury shares (shares held by the corporate debtor), or exchanging shares for debts held by creditors. All reorganization plans require that creditors be divided into classes, explain how each class of creditor claims will be handled, say which claims may receive less than full satisfaction, and provide equal treatment for all creditors in each class, unless those creditors agree otherwise. The plan is then submitted to all creditors for approval.

Approval usually requires agreement of creditors holding two-thirds in amount and one-half in number of each class of claims receiving less than full satisfaction (full dollar value of claims submitted). The court must then confirm the plan as approved. The debtor is then charged with responsibility for carrying out the reorganization plan.[30] For example, the total amount of claims is $1,200, the agreement is needed for $800 of the total debt and one-half of the number of creditors. If there were 100 creditors, 50 of them would have to vote yes on the plan. The following case is a chapter 11 bankruptcy and illustrates how the court dealt with one creditor's objection to the plan.

Nite Lite Inns and Grosvenor Square Restaurant
U.S. Bankruptcy Court, S.D. California
17 Bankr. 357 (1982)

Facts: In the chapter 11 reorganization proceeding involving three hotels and a restaurant, a creditors committee (the Burke Investors) strongly objected to a plan of reorganization submitted to the bankruptcy court by the debtor for confirmation. The bankruptcy court conducted a confirmation hearing to accept or reject the plan of reorganization. The plan of reorganization—which included the three hotels, the restaurant, and individual stockholders of the corporate debtors who had guaranteed payment of corporate obligations—was accepted.

Reasoning: At the hearing on confirmation, Burke Investors objected to confirmation of the plan on the following grounds: 1) that it is not feasible; 2) that it was not fair and equitable; 3) that it was not proposed in good faith; 4) that it provided for an improper classification; 5) that it did not provide this creditor with property of a value that was not less than the amount such holder would receive if the debtors were liquidated; 6) that consolidation of the cases was not in the best interest of creditors. Burke Investors also moved to block the use of $300,000 on deposit pending a final determination as to whether Burke Investors was entitled to such sums. This opinion was filed to deal with each of Burke Investors' objections and to determine whether the debtors' amended plan should be confirmed.

In answer to Burke's claims that the debt would not be satisfied in the event of liquidation, the court said the sale of the properties would satisfy the debts. Therefore, the plan was feasible.

The court said that the *good faith* requirement was met by the reorganization plan. The court found the plan fair and equitable.

The approval of the plan was in effect a "*cram down*." In other words, since most of the creditors agreed to the plan and the plan met all the legal requirements, the plan would be approved over Burke's objections. The court also said that without the cram down it was questionable whether many plans would be confirmed where there were disputes between creditors and the debtor. The presence of the cram down provision actually facilitated settlement in many cases, while at the same time it allowed for confirmation where a settlement could not be reached. Having already determined that the plan met the applicable requirements of 11 U.S.C. sec. 1129(a) and that the plan did not discriminate unfairly, the court needed only to determine whether the plan was fair and equitable.

Conclusion: Chapter 11 permits a debtor to continue in business, either with or without the supervision of a court-appointed trustee. Chapter 11 differs from straight bankruptcy in that the debtor is 1) not forced to terminate operations, 2) not required to liquidate remaining assets, and 3) not required to distribute the proceeds to existing creditors.

A plan of reorganization will not be confirmed if it is likely that the debtor cannot make payments to creditors under the plan without going into straight bankruptcy (liquidation). In this case, the San Diego hotel would be sold if the debtor failed to pay creditors. This sale was determined to be more than enough to cover all outstanding debts owed all creditors.

A reorganization plan submitted by a debtor in good faith may be confirmed. *Good faith* means that the plan is prepared to achieve a result consistent with the purposes of the federal bankruptcy law. Here the debtors were making an honest effort to keep the San Diego hotel operating to benefit all the creditors as well as themselves. This was sufficient proof of good faith to reject the creditors' objection.

In an interesting moment of reflection, the court noted a basic fact of bankruptcy: "In any bankruptcy case there are seldom any winners, just survivors." In addition, keep in mind that a major objective of a bankruptcy proceeding should be to remain in business and attempt to reorganize. The business may very well, then, be a survivor.

DEBT ADJUSTMENTS UNDER CHAPTER 13[31]

Chapter 13 permits individual debtors to pay their debts in installments with court protection against *attachments*—seizure and sale—of property by the creditors. The debt ceilings fixed by the federal Act are individual liquidated (acknowledged), unsecured debts of less than $100,000, and secured debts of less than $350,000 for each individual and his or her spouse.[32]

Sole proprietorships of business and individuals may apply for chapter 13 protection.

Unlike either chapter 7 or chapter 11 proceedings, *chapter 13 permits voluntary debtor petitions only.* No creditor may petition a debtor into involuntary debt adjustment. Debtors may petition for either a composition of debts or an extension of debt payments, or both, out of future earnings. A composition is a reduction of total indebtedness. An extension gives the debtor a longer period of time in which to pay debts in full. Usually, in filing for such relief, the debtor provides a list of creditors as well as assets, liabilities, and contracts not yet performed by the debtor.[33]

Once a petition is filed, the court will hold a meeting of creditors, and then review proofs of claim filed by each. The court questions the debtor, who then submits a payment plan. The plan is submitted to the secured creditors for their approval. If they accept—and after finding that the plan was made in good faith, meets legal requirements, and adequately protects all creditors—the court approves the plan.

Next, the court appoints a trustee to administer the plan. All payment plans require repayment in three years or less, unless the court approves a longer period of time. In no case may the court approve a plan scheduling repayment beyond five years from the date of its approval.[34]

If the plan is approved, the creditors may not later force the debtor involuntarily into straight bankruptcy. Under chapter 13, the court may order a discharge of the bankrupt even though the bankrupt has failed to meet all requirements within three years, if the court finds that the debtor's failure to do so was due to circumstances beyond that person's control. However, if the debtor fails to file an acceptable plan, or fails to pay for unjustified reasons, the court may dismiss the chapter 13 proceedings.[35] Then creditors may petition for involuntary bankruptcy and require a total liquidation of the debtor's assets.

The following foodservice bankruptcy example will illustrate the procedures and criteria for obtaining a petition of bankruptcy. Chapter 13, the Debt Adjustment for Individuals with Regular Income, is chosen because it is the most favorable form of bankruptcy discharge for the independent foodservice operator.

Chapter 13 bankruptcy is only available to a single proprietor, not corporations or partnerships. However, this does not disqualify a husband and wife team, if they are not operating a foodservice partnership or corporation. Our operators, John and Lois Hilleck, run the Devon Park Restaurant.

The Hillecks file a voluntary petition together. Only they can file for bankruptcy; no creditors can petition for their involuntary bankruptcy under chapter 13.

The Hillecks submit a list of all creditors, including their names and addresses, and schedules an A and B listing all of the restaurant's assets and liabilities. There is a filing fee that must be paid. An original plus three copies of the petition are filed.

Schedule A contains a list of creditors, including priority secured creditors and unsecured creditors nonpriority. Priority creditors include those providing administrative services, and claims for wages, contributions to employee benefit plans, consumer deposits, and taxes. Secured creditors are those holding security interests in collateral owed by the single proprietor, either John or Lois, or creditors with judicial liens against the restaurant. Nonpriority, unsecured creditors are the remaining creditors.

Schedule B contains separate lists of the Hillecks' real property (home or other real estate), personal property, and property claimed by them as exempt.

It is very important for the Hillecks to file the petition for bankruptcy as soon as possible, even without Schedules A and B. These schedules can be filed within 10 days after the petition is filed, if a list of the names and addresses of all creditors has been furnished. An additional 10 days or more may be granted by the court for good cause.

Providing accurate schedules is of vital importance. Carelessness or negligence on the Hilleck's part may be viewed as an attempt to falsify information, with very severe criminal penalties, not to mention that the claim of the creditor will be omitted from any discharge granted.

The automatic stay or hold on debt collection and lien enforcement by creditors protects the Hillecks. However, the stay only suspends collection temporarily. It does not permanently excuse them from payment of those debts. Moreover, the stay applies only to debts that John and Lois plan to pay off *in full* at a later date, and not to debts they want to pay only in part. For example, John and Lois owe Joe, the vegetable vendor, $1,000. If they submit a debt-adjustment plan to pay off only $700 in full later, Joe can take immediate steps to recover the $300 difference. Joe's request to the bankruptcy court to do so will probably be granted.

One special advantage under chapter 13 is that the Hillecks can continue to operate the restaurant themselves. Usually no trustee will be appointed to run the business.

Trustees are given certain duties in chapter 13 proceedings, such as investigation of financial condition and the operation of any debtor's business. The court always has the right to order that the restaurant be operated by a trustee, when the trustee's investigation reveals mismanagement, incompetence, dishonesty, or other business irregularities.

The heart of every chapter 13 case is the repayment plan. Only the Hillecks may file a repayment plan. A plan may not be forced on them by their creditors, as is true under chapter 11 reorganization or under chapter 7 liquidations. The creditors may challenge the plan, however. Failure to file on time gives any interested party the right to ask the court to dismiss the petition or to convert the case to a liquidation under chapter 7.

The role of the trustee is only to advise the Hillecks to help them keep the restaurant open without breaking themselves financially.

Three provisions must be contained in the Hillecks' repayment plan.

1. Provision for submission of all or a portion of future earnings or income to the trustee to execute the plan. All other income remains with the Hillecks to support themselves and their dependents.
2. Provision for payment in full of all priority claims listed on Schedule A.
3. Provision for equal treatment of all creditors if the plan classifies creditors.

A confirmation hearing on the repayment plan will be held only if a creditor or other interested party objects to the plan. Only the trustee may appear at the hearing.

Confirmation under chapter 13 does not give unsecured creditors the right to vote to accept or reject the Hillecks' plan. Nor does chapter 13 confirmation require a finding that either John or Lois is not guilty of any bankruptcy act that would otherwise cause an objection to their discharge.

The Bankruptcy Code under chapter 13 requires the Hillecks' plan to meet six requirements as conditions of confirmation.

1. Compliance with the Bankruptcy Code
2. Allowance for payment of fees and charges
3. A good faith proposal
4. Provision for some distribution to all unsecured creditors
5. Equal treatment of secured creditors
6. Feasibility

The effect of confirmation of the repayment plan makes it binding on the Hillecks and all creditors, whether or not the creditor claim is provided for by the plan and whether or not the creditor has objected to, accepted, or rejected the plan. However, *under chapter 13, the Hillecks are not discharged until the plan is fully carried out.* Under chapter 13 reorganization, confirmation results in an immediate discharge of debts and their replacement with new debts as spelled out in the plan.

USING BANKRUPTCY

One stark reality confronts those who wish to use bankruptcy protection. Bankruptcy is not a painless legal escape from one's debts. Unfortunately, bankruptcy stigmatizes those who use it by requiring anyone discharged in bankruptcy to state that fact on any future application for credit, and thus face the likelihood of having the credit application legitimately rejected. The federal Act does not require such disclosure, but any potential creditor is permitted to ask.[36] However, in most states a creditor need only call the local credit bureau to get a credit record, which includes information on bankruptcy.

A discharge in bankruptcy necessarily makes the debtor a less than worthy credit risk. Therefore, a straight bankruptcy should be used only as a last resort, when no other viable means of settling debts is available.

Although a reorganization under chapter 11 or consumer debt adjustment under chapter 13 does not carry the same stigma as straight bankruptcy, it still may cause a creditor to place a credit application in a less desirable category. Naturally this result can worsen business as well as personal financial prospects.

Most creditors and suppliers want to avoid the prospect of a debtor's straight bankruptcy or some other plan as much as the debtor. This is true because creditors usually receive so much less in straight bankruptcy than in other forms, or through their own payment plan. To avoid this, creditors will usually accept a voluntary repayment plan without court supervision. A reasonable, mutually agreeable repayment plan has two advantages: 1) it avoids the stigma associated with bankruptcy; and 2) it gives the creditor more assurance of total repayment than would be likely under federal and state bankruptcy laws.

A hospitality operator might be eligible to reorganize a business under chapter 11 or chapter 13 of the federal Bankruptcy Act. These procedures will allow the operation to stay in business rather than liquidate assets to satisfy creditor claims. These less drastic procedures should be used whenever possible, since a reorganized business is better than no business at all. **Exhibit 14.2** lists advantages and disadvantages of the three types of bankruptcy.

Bankruptcy is not a debtor's haven; it should be used only as a last resort when all other avenues of voluntary settlement of debts are off limits.

Creditors who seek to compel bankruptcy of a debtor must be aware that if debts are not secured by other assets, they stand little chance of recovering anything. Hospitality managers should carefully check the credit standing of any potential debtor *and* obtain security from the debtor as a hedge against unforeseen economic circumstances.

Exhibit 14.2 Advantages and Disadvantages of the Three Forms of Bankruptcy

---

ADVANTAGES AND DISADVANTAGES OF TYPES OF FEDERAL BANKRUPTCIES TO DEBTORS

ADVANTAGES	DISADVANTAGES
Voluntary Liquidation under Chapter 7 (Straight Bankruptcy)	
May be sought voluntarily by debtor.	Discharge (forgiveness of debts) available only to individual debtors, not to partnerships, corporations, or unincorporated associations.
Provides most complete relief to widest types of debt.	
Six-year discharge rule does not prevent filing of bankruptcy petition, proceeding as debtor, and obtaining orderly liquidation of assets.	Discharge may be opposed and denied—discharge not automatic.
Debtor receives automatic stay (hold) on debt collection and enforcement of liens by creditors.	Assets of debtor must be sold and distributed to creditors.
	Usually results in very poor credit rating.
Reorganization under Chapter 11	
Primarily available to all businesses.	Creditors may petition for involuntary reorganization of debtor, using same procedures established under chapter 7.
Permits businesses to continue operating, thus minimizing unemployment and waste of business assets.	Debtor must perform functions and duties of trustee.
Makes better use of business assets: It enables use of assets for intended purpose, rather than forcing sale of assets at depressed prices.	When liquidation is granted by chapter 11, confirmation of the plan will not discharge the debtor. This means that corporations and partnerships otherwise entitled to liquidate will not be discharged.
Debtor normally remains in control of business (as a debtor in possession), with same powers as that of court-appointed trustee.	
Debtor receives automatic stay (hold) on debt collection and enforcement of liens by creditors.	
Debtor can rehabilitate without terminating the business.	
Authorizes confirmation of reorganization plan by court, preventing creditors from blocking plan by failure to approve plan unanimously.	
Unlike chapter 7, corporations and partnerships are entitled to discharge in bankruptcy.	
Discharge is not refused by reason of fraud, or willful misconduct of debtors, making such debts nondischargeable under chapter 7.	
Tax claims of corporations and partnerships are discharged.	
Liquidation is authorized.	
Debt Adjustment for Individuals* with Regard Income Under Chapter 13 **Applies to sole proprietors and employed persons.*	
Debtor is free from many debts not exempt under chapter 7.	Only available to persons with regular income.
Debtor may not need to make periodic payments on old debts, as required under chapter 7.	The pay-off of old debts may be required, which would be avoided under chapter 7.
Debts not otherwise dischargeable (willful injury, fraud, income taxes) can be reduced and then paid off in installments (this procedure is not available under chapter 7).	Debt ceiling limits: Less than $100,000 of noncontingent, liquidated *unsecured* debts.
The six-year discharge rule does not apply as it does under chapter 7.	Less than $350,000 of noncontingent, liquidated *secured* debts.
The stigma attached is not as severe as under chapter 7, since no liquidation and complete discharge is provided.	
Encourages debtor to pay as a moral duty.	
For small business persons, creditor approval is not required.	
Simpler, more flexible procedure than under chapter 7.	
The individual debtor cannot be compelled by creditors.	

SUMMARY

There are three forms of bankruptcy foodservice operators may use. *Chapter 7* bankruptcy is *straight bankruptcy*, with any assets distributed to creditors by the trustee. *Chapter 11* is designed for a businessperson who wants to stay in business. It allows for reorganization of the business, which is supervised by a trustee and must be approved by a majority of creditors. *Chapter 13* bankruptcy allows for a voluntary repayment plan by the debtor.

Petitions for bankruptcy may be filed for *voluntarily* by the debtor, or *involuntarily* by creditors for chapter 7 and chapter 11.

Jurisdiction for bankruptcy proceedings rests in the federal courts. The debtor must file in the district in which the business is located.

Bankruptcy is not granted automatically. The person must first be eligible for discharge. Then creditors have time to object to the discharge. Dishonest acts, such as fraudulent transfers or false information, will usually bar a discharge.

Debtors have rights as well as surviving obligations. Debts such as alimony and taxes survive a discharge of bankruptcy.

Creditors also have rights to protect claims, and courts will usually decide which claims have precedence over others. It is not up to debtors to decide how many assets will be distributed.

QUESTIONS

1. Norma and John Sweeny have owned and operated the Courtyard Inn as a corporation for 10 years. Some major equipment purchases made at the same time that the interior of the restaurant was remodeled have put them in severe debt and they cannot continue to operate until their debts are erased. They had planned to run the restaurant for five more years before turning it over to their children. Assuming they would like to stay in business, which form of bankruptcy would you recommend? Explain your answer.

2. Miriam Duffy is going through bankruptcy proceedings. She files on September 3. On November 1 she pays the full amount to one of her creditors, Buck Wholesalers, whose debt is discharged along with those of the other creditors. Has she done anything wrong? What could be the result of this? Explain.

3. Name several obligations that cannot be discharged in bankruptcy.

4. List three criteria or tests according to which a bankruptcy court may deny a discharge in bankruptcy to an otherwise eligible debtor.

5. How does chapter 13 bankruptcy of the federal Bankruptcy Act differ from a straight bankruptcy proceeding? Who may file?

NOTES

1. *Moureau v. Leaseamatic, Inc.*, 542 F.2d 251 (5th Cir. 1976).

2. Article 1, Section 8, Clause 4 of the United States Constitution grants Congress the power to establish uniform bankruptcy laws throughout the United States. The Bankruptcy Reform Act of 1978, Pub. L. No. 95–598, 92 Stat. 2549 (1978), created a new Bankruptcy Code, giving bankruptcy courts broad powers to hear and determine all controversies affecting debtors or their estates.

    The Code is found in a new Title 11 of the U.S./Code, 11 U.S.C. sec. 1 *et seq.* (1976 and Supp. IV 1980 and Supp. V 1981). The states are permitted to enact bankruptcy laws as long as they are not in conflict with federal law. This chapter will deal exclusively with federal bankruptcy law.

3. Lusk, Hewitt, Donnel, Barnes, *Business Law and the Regulatory Environment* 932–33 (5th ed. 1982).

4. Bankruptcy Reform Act of 1978, 11 sec. 201(a), amending 28 U.S.C. See 28 U.S.C. secs. 151, 1471 (Supp. IV 1980). In *Marathon Pipeline Co. v. Northern Pipeline Constr. Co.*, 12 Bankr. 946 (D. Minn. 1981), the bankruptcy court held section 1471 unconstitutional because of the extensive jurisdiction over regular civil actions granted to judges

who lack lifetime appointments to the federal courts. The U.S. Supreme Court affirmed that ruling on appeal. 102 S. Ct. 2858 (1982), stayed 103 S. Ct. 199 and 200.

5. 28 U.S.C. sec. 1472(1) (Supp. IV 1980).

6. *Id.*

7. *Denver & R.G.W.R.R. v. Brotherhood of R.R. Trainmen*, 387 U.S. 556 (1967), interpreting prior federal bankruptcy law (Chandler Act of 1938). No case has yet interpreted the new code on this question.

8. 28 U.S.C. sec. 1472(2) (Supp. IV 1980).

9. *Id.* sec. 1477(a). See also Bankruptcy Rule 116(b)(2).

10. 11 U.S.C. sec. 301, 303(h) (Supp. V 1981).

11. *Id.* Sec. 343, 521(1) (Supp. V 1981); Bankruptcy Rules 204(a)(2), 205(b).

12. 11 U.S.C. sec. 303(g) (Supp. V 1981) governs the appointment of an interim trustee by the bankruptcy court until the creditors elect a permanent trustee, subject to court approval. The powers of the interim and permanent trustee to take charge of and appraise the debtor's property are the same. Normally, however, this is done by the interim trustee prior to the confirmation of a permanent trustee by the court. *Id.* sec. 544(a)(1), (2), 721 (Supp. V 1981).

13. *Id.* sec. 363(a), (c)(4), 544(b), 1106(a).

14. *Id.* sec. 727(a)(1), (8).

15. Unlike prior law, the Bankruptcy Code permits a debtor to refuse to testify or answer questions on grounds of the constitutional privilege against self-incrimination. *Id.* sec. 727(a)(6)(B).

16. *Id.* sec. 727(a)(2), (4)(A), (4)(D), (5), (6).

17. *Id.* sec. 524(c).

18. The states are permitted to establish bankruptcy laws not in conflict with the federal Bankruptcy Reform Act of 1978, the latest major revision of the Chandler Act.

19. Liens given this protection are defined as *statutory liens* under 11 U.S.C. sec. 101(38) (Supp. V 1981), and must be distinguished from *security interests* created by the parties under sec. 101(27). *Id.* sec. 101(27).

A typical security interest is a bank's interest in restaurant tables and chairs purchased by means of a bank loan. The interest is protected by giving the bank a lien or right to seize and sell the tables and chairs in the event of failure or refusal to pay back the loan in the manner agreed upon with the bank.

An example of a *statutory lien* is a *tax lien* to ensure the payment of federal or state taxes due, or a *mechanic's lien* created by law to protect a contractor by giving the contractor the right to sell a business if the operator does not pay for work performed by a contractor, such as installing a new kitchen floor.

A *judicial lien* is created by a court order giving the court a right to sell property to pay a money judgment rendered by the court.

20. Federal law permits a bankruptcy trustee to avoid statutory liens, otherwise recognized in three situations: 1) Liens that first become effective upon financial disaster. 11 U.S.C. sec. 545(1) (Supp. V 1981). 2) Liens that do not meet a hypothetical *bona fide* purchaser test. The test requires the lien to be perfected or enforceable against a *bona fide* purchaser who purchases the property at the time the bankruptcy case commences; that is, when the bankruptcy petition is filed. The purchaser need not actually exist; only the lien need by perfected. If this test is not met, the lien is not valid in bankruptcy. *Id.* sec. 545(2). 3) Landlord's liens to secure the payment of rent. *Id.* sec. 545(3), 545(4).

21. *Id.* sec. 523(a) *et seq.*

22. *Id.* sec. 547.

23. For example, a preference may take place where a creditor (bank) obtains a judicial lien on restaurant tables and chairs by getting a court order of execution and *levy* (the right to seize and sell) upon those assets after the operation has filed for bankruptcy. Giving the bank a mortgage on the restaurant, land, and building would also create a preferential transfer as a debtor in bankruptcy. See generally *Glessner v. Massey-Ferguson, Inc.*, 353 F.2d 986 (9th Cir. 1965), *cert. denied*, 384 U.S. 970 (1966).

24. 11 U.S.C. sec. 501(a) (Supp. V 1981). Under the Bankruptcy Code, filing a claim is not required. However, in order to share in the liquidation of a debtor's estate or in a chap-

ter 13 debt adjustment, a creditor must file a proof of claim with the bankruptcy court. *Id.* sec. 502.

25. A presumption exists that every claim that is filed is valid, placing the burden on the bankruptcy trustee to object. *Id.* sec. 502(a), (b)(1); Bankruptcy Rule 306(b).

26. 11 U.S.C. sec. 506(a) (Supp. V 1981).

27. *Id.* sec. 507(a)(1), (3)–(6).

28. *Moureau v. Leaseamatic, supra*, note 1.

29. 11 U.S.C. sec. 1101 *et seq.* (Supp. V 1981).

30. *Id.* secs. 1102(a)(1), 1104. The appointment of a trustee is not mandatory. In appropriate cases the bankrupt may continue in business, as a "debtor in possession," during reorganization, *id.* secs. 1101(1), 1102(a)(2), 1106(a)(5) (plan), 1108 (operate business), 1122(a), 1126(c), 1128(a). The statutory requirements for confirmation are set forth in sections 1129(a), 1129(d), and 1141(b).

31. *Id.* secs. 1301–1330.

32. *Id.* sec. 109(e).

33. *Id.* secs. 303(a), 103(a).

34. *Id.* sec. 1322(c).

35. *Id.* sec. 1328(b). Section 524(d) requires the bankruptcy court to hold a discharge hearing at which the debtor must appear in person. Absent the granting of a hardship discharge under section 1328(b), a debtor's failure to complete payments will cause the courts to dismiss the application for discharge.

36. Chapter 13 adjustment cases do not require an individual debtor to make an adequate disclosure statement. Chapter 11 reorganization cases do require such statements. See *Id.* sec. 1125.

15 Choosing and Managing Your Attorney

OUTLINE

OBJECTIVES

After reading this chapter, you should be able to:

1. Use advertising, referral lists, personal references, appointments, and interviews to choose the most appropriate lawyer to fit your needs.
2. Discuss the common ways lawyers are paid by clients: by the hour, by the day (per diem), by contingent fee or by retainer agreement.
3. Describe some of the considerations important in developing an effective working relationship with a lawyer.
4. Discuss ethical issues, such as conflict of interest and attorney-client privilege, involved in working with lawyers.

Case in Point

Tom and Sue Doyle hired Darren Varney to represent them in a case in which they were being sued for liability for injuries caused by a patron's fall. The patron maintained that the Doyles failed to provide proper railings on the staircase leading to the rest rooms and that he had fallen as a result.

The Doyles failed to tell their attorney that they had been warned a week before the fall occured by the building inspector that the railing on the right side might come loose because of the way it was bolted to the floor. They were given 10 days to have it fixed. They had already contracted and scheduled a carpenter to repair the railing. Varney had asked the Doyles at their first interview if they had ever been warned or cited for any violation, and they both said no. During the trial Varney was "surprised" by the warning, courtesy of the opposing attorney. Varney had prepared no defense for this factor, although the 10-day rule might have been sufficient.

The Doyles lost the case.

Can the Doyles sue Varney because he had no defense for the warning issue? Hardly. Their own failure to provide this information when asked was just that—their failure. Allowing the attorney to be surprised in court by something that could have been clarified, and possibly defended, was their fault, not Varney's.

CHOOSING A LAWYER

You are very likely to require the assistance of a lawyer at some time during the course of your hospitality career. No one is immune from the possibility of a lawsuit brought by a patron, competitor, or the government. You also will need legal guidance in the organization and operation of your foodservice business, and in those situations where you may wish to bring a lawsuit against a patron, a competitor, or the government. Both large and small operations need competent, unbiased legal advice at a reasonable cost.

Finding a Lawyer: The Advertising Factor

Unlike other businesses, the legal profession has traditionally frowned on public advertising of legal services as being unbecoming to the profession. Advertising in the past was limited to professional announcements for lawyers or lists maintained primarily to benefit other lawyers. Such advertising was virtually unavailable to the public. Lawyer referral lists maintained by local *bar associations* (associations of lawyers in a particular geographical area who are supervised by the highest court in that area) were seldom publicized, with the result that potential clients had to rely on word-of-mouth recommendations.

The absence of public advertising often deprived the potential client of adequate legal services simply because the recommended lawyer might not practice the branch of law the client required. Another referral to the right lawyer was possible, but meant loss of time, effort, and money.

The traditional policy of forbidding public advertising of legal services was eventually overturned by the United States Supreme Court in 1977.[1] However, in denying state bar associations the right to prohibit any advertising of legal services directly to the public, the Supreme Court did not outlaw state bar regulation of such advertising. Reasonable regulation was approved.

While advertising is now allowed, neither the larger, well established firms nor storefront law offices have made much use of major advertising. It is still very low-key, and you will probably have to do your own research.

Generalists versus Specialists

Another factor that affects public knowledge of the availability of legal services is the great degree of specialization that has developed in recent years. Jack-of-all-trades lawyers are being replaced by lawyers who specialize in narrow legal areas. These specialists tend to be in lucrative areas such as real estate, tax, or contract law, or in business fields where the type of business generates more legal work at higher fees.[2]

Lawyers who specialize exclusively in hospitality law are still a minority, mainly because hospitality law does not command the higher fees in terms of legal effort that other businesses do. Lawyers who do practice in this field usually represent big business interests, quick-service franchisers, or other corporate clients who are better able to pay hefty legal fees.

The types of problems that a small business generates are better suited to the general practitioner, because the operator's needs are likely to be more varied and numerous than those served by the specialist.

One thing you should keep in mind when searching for and working with your lawyer is that law schools may not teach either general business law or hospitality law. This means that any lawyer willing to help, but lacking practical exposure to hospitality management, may learn the fine points at your expense.

In spite of these problems, you owe it to yourself to get the best legal advice you can afford. There are plenty of lawyers available to help you, and media advertising makes finding them somewhat easier than before.

Practical Guidelines to Choosing a Lawyer

Some guidelines to follow when choosing a lawyer include the following.

1. Seek the recommendations of trusted business and personal friends who have or had problems similar to your own, but be sure they are recommending someone who has had experience in handling hospitality business problems.
2. Contact your local bar association, usually located in your county seat or the city where your trial courts are. Ask for the client referral office. Explain your needs as specifically as you can.
3. Go to your public library and verify the list of attorneys provided by the bar association office by checking the *Martindale-Hubbell Legal Directory*. This directory comes in a series of volumes, broken down by states. Look in the front of the volume for your state to see if any of the lawyers from your bar referral list are listed there. If so, check to see whether the lawyer or law firm with which he or she is associated has a separate listing in the back of the volume. These listings include the type of law practiced by all partners, and biographical information about all partners and senior associates.

    Martindale-Hubbell carefully screens all attorneys listed in each volume. *However, Martindale-Hubbell does not guarantee the competence of the lawyers who are listed. They merely provide a list of lawyers who meet their high screening standards.* A Martindale-Hubbell lawyer listing is a fairly reliable recommendation for that lawyer, however.
4. Make appointments to see those lawyers, three if possible, who pass muster on the criteria mentioned above.
5. At each interview, briefly outline your legal problem and let the lawyer ask you questions. Ask yourself: Do I feel comfortable with this lawyer? Can I divulge my needs fully and frankly, without hesitation? What is my gut reaction to the way the lawyer conducts himself or herself? Am I getting an honest, objective initial opinion?

Lawyers are generally cautious and conservative in their professional relations with clients. Do not be put off by this. Lawyer flattery and over-confidence in the initial interview ought to be viewed skeptically. Remember that you are purchasing honest, objective, and competent legal advice. A lawyer who appears more anxious to please you than to perform usually is trying to cover a lack of experience, competence, or both. Such a lawyer may not serve your best interests.

6. Make no final decision on your choice of a lawyer until you have had an opportunity to review your reactions to all three interviews. Watch out for a legal fee that seems too good to be true, or promises of results that sound like pie in the sky. *Do not allow an attorney to force you into signing a retainer* (a contract of services) *agreement at the first interview*. Give the matter careful thought to ensure a good choice.

Advertising by lawyers does not guarantee the competence of legal services. It is still necessary to evaluate the legal services of any lawyer you wish to employ. State and local bar association referral lists are also no guarantee of competence, since some states require all licensed attorneys to join. However, limited competence of licensed attorneys (those who passed the bar) can be assumed because of the rigors of legal education and the passing of a bar examination as required by virtually all states. Some bar associations are better than others for referrals. Bar associations are not required by law to establish and maintain referral lists. Even so, you are advised to contact your local bar association for assistance. Most bar associations do list lawyers who will volunteer to meet with you at a reasonable initial fee. These lists are usually broken down into legal specialties to help you locate a lawyer competent to serve your particular needs.

LAWYERS' FEES

The practice of law is a business as well as a licensed profession. The object of every businessperson in our private enterprise system is to make a reasonable living; every lawyer's professional fees reflect this fact of economic life. However, a lawyer has an ethical obligation to put your interests above his or her own interests in setting the price of legal services. The legal profession is a branch of the administration of justice and not a mere money-making trade.

Fee Schedules

Legal fees can and do take various forms. There are voluntary fees and scales for certain types of services, such as writing a will. Generally you will have to negotiate with your lawyer. There are four general methods attorneys use to fix fees.

Lawyers generally work by the hour. This means that your lawyer's fee may be made up of an hourly rate plus expenses for writing a hospitality employment contract or handling a real estate contract. This is the normal method of payment for legal work other than courtroom representation.

Lawyers usually charge by the day for courtroom trial and appeal work. Such a fee usually consists of a per-day (*per diem*) amount times the number of trial days. An appeal is usually heard during part of one day, and the fee charged includes preparing and filing the records on appeal and legal briefs. This preparation time is lengthy. Filing appeals is costly, making them available only to those hospitality operators who have the means to afford what is actually a review of the original decision.

Lawyers may work on a *contingent fee* basis, meaning that their fee is a percentage of any favorable judgment or award, with no fee paid if they are unsuccessful. These contingent fees are often controlled by court rules to prevent lawyers from taking the entire award and leaving you empty-handed. Lawyers who handle liability cases are more likely to enter into this type of arrangement.

Lawyers enter into *retainer agreements* for services of a general kind over a longer period of time. Here you pay a fee for having the services of a lawyer. A written retainer agreement is a contract stating your rights and responsibilities to your lawyer and his or her rights and responsibilities to you. It outlines the legal advice or services you may request of that lawyer or law firm. Essentially the retainer agreement includes what is to be done, when it is to be done, and how much the work will cost you. Retainer agreements are used to provide general advice, and the fee reflects this by usually being less for a single case or problem. Very large hospitality operations, or operators getting into franchising or other areas of the hospitality business, may use this arrangement. It is of questionable value to small businesspersons with only sporadic need for lawyers.

WORKING WITH YOUR LAWYER: YOUR RIGHTS AND RESPONSIBILITIES

Once you have selected a lawyer, it benefits both of you to interact in a cooperative, effective, and efficient manner. Some major points require attention.

1. You have the right to receive timely, honest, competent, and confidential advice from the beginning. You have the right to compare the costs of legal services.
2. You have the right to obtain independent legal advice, without conflict of interest or a biased viewpoint that might cause your lawyer to be less than totally committed to serving you. You have the right to receive competent legal advice. Your lawyer cannot deceive you about professional skills and

experience or take on a case he or she is not competent to handle. A lawyer cannot limit responsibility to you for his or her own malpractice, or disclaim that responsibility by prior agreement.

3. Your lawyer has the right to expect truthful answers to questions, your cooperation in meeting court and other deadlines, and the freedom to exercise his or her best professional judgment in providing you with legal advice. Your lawyer has the right to payment within a reasonable time, or according to your agreement.

Both you and your lawyer have the right to terminate the relationship at any time, either as per the agreement or for just cause. If this is done, the lawyer is entitled to receive the reasonable value of services provided to date, and you are entitled to the return of any papers or other documents you gave to the lawyer.

Getting on the Right Track with Your Lawyer

The following practical tips will help create a mutually beneficial working relationship with the lawyer you select.

1. At the first interview or meeting, state clearly, concisely, and honestly your needs, what you expect from the lawyer, and how much you can afford to pay. If you need to employ a lawyer on a long-term basis, to cover all types of problems, a written retainer agreement may be negotiated, to be signed later. A lawyer retained by you should charge you less overall than a lawyer you employ to solve a single problem.

    If you need to employ an attorney for one-time or nonrecurring advice, then obtain an agreement on a fee for that advice and a commitment letter indicating the type and duties of service.

2. Before signing a retainer agreement or a commitment letter, decide how you feel about developing a close working relationship with the lawyer. Do you feel comfortable giving very private information to the lawyer? Remember that your lawyer will be one of your closest advisors.

3. Explain your particular legal needs in the context of your *hospitality operation*, not just in terms of a business. Thoroughly explain not only what happened, or what you want to happen, but *why*. This will enable your lawyer to tailor the right legal remedy to your problem or to prevent a legal snarl.

    Don't be afraid to mention what you know about hospitality law. This is especially useful regarding local regulations that other small businesses do not encounter. The hospitality industry is somewhat unique in the scope and amount of local regulation, and your lawyer may not be aware of this.

4. Inform your lawyer about hospitality experts who can help on specific issues. For example, you may know that your local hospitality association has an expert in security matters, financial management, credit controls, or real estate.

5. Encourage your lawyer to contact specialists who teach at major schools of hotel and restaurant management and have published books or articles in your problem area. They can be invaluable to the general practitioner.

6. When your legal problem involves an insured risk, always provide your lawyer with the name of your insurance carrier. This step should be taken even though the legal defense as well as the claim may be covered by that carrier. Cooperation between your lawyer and your carrier's legal staff is needed to prevent future risks as well as to defend claims against you.

7. Be candid and truthful with your lawyer. Holding back information because it is unpleasant or personally damaging is shortsighted. Doing so will only hamper your lawyer, create surprises at negotiations or at trial, and make both you and your lawyer lose credibility with opposing council, the court, and a jury. Knowing the worst of your problem will enable your lawyer to meet that problem, and not be caught by surprise without adequate time to deal with it.

The following guidelines should help you maintain a good working relationship with your lawyer.

1. Always reach a definite understanding of what you want to discuss before calling or seeing your lawyer. Frequent, unnecessary phone calls or visits are time consuming, expensive, and accomplish little. Set aside enough time and be prepared to discuss important matters. Being unprepared wastes valuable time and money.

2. Set a reasonable time table for results, with appropriate deadlines or *benchmarks*. Monitor the results and deadlines promised by your lawyer.

3. Always meet deadlines proposed by your lawyer; otherwise he or she will be unable to perform any assigned tasks properly.

4. Always document the length of telephone calls you make to your lawyer, so you can verify any fees for the calls.

5. Require your lawyer to give you a detailed statement supporting each bill for professional services. Examine each statement against your original agreement for accuracy. Check to see that all fees are within your fee agreement, and that any overcharges are correct. Do not hesitate to question your lawyer about fee statements. It is your money, and you deserve to pay only for the services you agreed to, at the rate agreed on.

6. When you feel that your lawyer is deliberately trying to take advantage of you, such as billing you for fictitious services, or serving you or another person interested in the outcome without informing you, you should notify your local state or local bar association. Each bar association has the power to investigate violations of its professional codes and to discipline guilty attorneys. You have the right to expect honest as well as competent legal representation.

7. Work together with your lawyer as a team. You are the team captain, responsible for making all business decisions. Your lawyer is your advisor, not your superior. Properly used, your lawyer will hear you out, give you

his or her best legal judgment on the pros and cons of your plans, and offer alternatives to accomplish your goals that can stand muster legally, should that be necessary.

You are free to disregard your attorney's advice and to use your own best business judgment instead. Your lawyer has no power to override your decision, but may terminate his or her association with you if your decision in any way violates the law.

Ultimately you, and you alone, must take responsibility for the consequences of your decisions. Your lawyer is your agent, and is not liable to you or to others for your mistakes or lapses in judgment.

You have the right to expect that your lawyer will follow the proper legal procedures to protect you, or carry out specific promises made to you. You have the right to check periodically to see that your lawyer does so, or gives you a reasonable explanation as to why it was not done.

Good management of your lawyer means reasonable and periodic inquiries on the progress of your case. It does not mean badgering your lawyer with unnecessary calls for which you may be charged an additional fee.

Judge your lawyer by results obtained, not by his or her manner. Remember that no lawyer can guarantee a favorable outcome. What a competent, trustworthy lawyer *can* do is to promise to use his or her best efforts to ensure the best possible result, or to tell you honestly that your chances of success are unlikely. Be wary of the lawyer who is too quick to promise you perfect results. Such a lawyer is either ill-informed, dishonest, or both.

ETHICAL ISSUES

A final word of warning is needed in choosing and working with any lawyer. First, you must make certain that the lawyer you select is free from any *conflict of interest*. For example, you retain lawyer Jones to represent you in the purchase of land for your new business. Jones operates as a real estate agency (which Jones may do as an attorney in some states), and owns a parcel of land and a building that fit your needs. Jones is under an ethical and legal duty to tell you that, or disqualify himself from representing you in the purchase of his own property. However, you should also make independent inquiry of Jones as to any interest in the property you're thinking of buying, so that any concealment or false statement Jones may make can be used to deny Jones any legal fee, and also to bring Jones up on any charges of professional misconduct.

Second, all lawyers are bound under the rules of professional conduct not to disclose any communications you make to them in their role as lawyers, without your approval. This duty is called the *attorney-client privilege*.

For example, you hire Jones to represent you in a liquor license revocation hearing, where you are accused of selling liquor to minors. You tell Jones that you were found guilty of selling to minors in another state some years ago. Jones casually mentions this information to a friend at dinner. The friend calls the local newspaper, and the following day the newspaper contains a lead article titled, "Convicted Restaurateur Martin Again up on Charges of Selling Liquor to Minors." That unauthorized disclosure by your attorney of damaging information is a violation of the attorney-client privilege, whether the information is true or not. Jones could be professionally disciplined for the disclosure. You in turn can sue him for malpractice for any damages you suffer as a result of his disclosure.

Most problems businesspeople have with their attorney can be traced to breakdowns in communication. If you follow the guidelines in this chapter for establishing and maintaining good communication with your attorney, you should have few problems. You must be able to differentiate between a problem caused by your own failure to provide critical information, or to notify your lawyer of necessary deadlines, and a lawyer's negligence, willful misconduct, or incompetence in handling your case.

The American Bar Association's (ABA) Code of Professional Responsibility and Canons of Legal Ethics establish rules of conduct for all lawyers, which each state bar association or disciplinary board is urged to follow when reviewing individual cases of misconduct. (See **Exhibit 15.1**.) The overall aim of the code is to establish and maintain honesty and fairness among lawyers in their dealings with the government, other lawyers, and their clients. The ultimate penalty for misconduct is *disbarment*, or revocation of the guilty person's license to practice law.

What happens if a lawyer injures you by acting negligently, failing to act properly, or intentionally misrepresenting your interests? Is your only recourse to complain to the local bar association? No. You may sue the lawyer to recover compensation for any losses the misrepresentation or negligence caused you.

You would sue for *malpractice*, or for the negligence or misconduct. For example, Ann Conley employs a lawyer, Sam Smith, to appeal a ruling of the

Exhibit 15.1   American Bar Association Canons of Legal Ethics

1. A lawyer should assist in maintaining the integrity and competence of the legal profession.
2. A lawyer should assist the legal profession in fulfilling its duty to make legal council available.
3. A lawyer should assist in preventing the unauthorized practice of law.
4. A lawyer should preserve the confidences and secrets of a client.
5. A lawyer should exercise independent professional judgment on behalf of a client.
6. A lawyer should represent a client competently.
7. A lawyer should represent a client zealously within the bounds of the law.
8. A lawyer should assist in improving the legal system.
9. A lawyer should avoid even the appearance of professional impropriety.

Excerpted from *Model Code of Professional Responsibility and Code of Judicial Conduct* with permission. Copyright © 1982 by the National Center for Professional Responsibility, American Bar Association.

Zoning Board. The Board turned down her application for a variance she needs to build a restaurant on land she's contracted to buy. Conley's purchase contract requires she obtain a variance within six months or forfeit her down payment. Smith, who knows of this requirement, negligently fails to file the appeal within the required time limits. Unless Smith can prove that his failure to do so was beyond his control or was otherwise justified, he may be liable for Ann's loss of the down payment. As in other negligence actions, Conley would be required to prove the existence of a legal duty, a breach or violation of that duty, and proximate cause and damages.

To protect the public further when a lawyer cannot make good a client loss, some states have created state-administered client compensation funds whereby lawyer licensing and registration fees may be used to purchase insurance against such risks. Lawyers often insure themselves by purchasing professional liability insurance, but it is not required.

Your lawyer is only responsible for injuries inflicted on you by his or her own willful misconduct or negligence in the handling of your legal affairs. The law differentiates between injuries caused by your lawyer's failure to act and injuries caused by your lawyer's exercise of legal judgment, a discretionary act. Failure to act, such as not filing a legal complaint in time to satisfy the statute of limitations, may constitute malpractice. Exercises of professional judgment, however poor they may seem afterward, do not constitute malpractice, unless the advice given is known to be false or is made intentionally to injure you.

The Code of Professional Responsibility requires every lawyer to deal fairly and truthfully with his or her clients. An ABA Disciplinary Rule prohibits lawyers from intentionally making false statements as to their skills or experience. Every lawyer is a *fiduciary*, meaning in a position of special trust, in dealing with clients. A violation of that trust makes a lawyer subject to *censure* (reprimand), or even disbarment in very serious cases.

The following case is unique in that the client encountered problems with two attorneys. The malpractice suit against the second attorney for negligence is the subject here.

Basic Food Industries, Inc. v. Grant
Court of Appeals of Michigan
107 Mich. App. 685, 310 N.W.2d 26 (1981)

Facts: This is a legal malpractice lawsuit for negligence brought by Basic Food, a corporate foods producer, against its lawyer. Basic Food accused the attorney of negligence in defending the company against a lawsuit for legal fees by another lawyer. That previous lawsuit resulted in a jury verdict against Basic Food in the amount of $25,000. The judgment was affirmed by the Court of Appeals.

Basic Food then started this lawsuit for malpractice, arguing that Grant's failure to obtain the sworn statement of the chairman of the board of Basic Food for use at the trial despite repeated written promises to do so, to investigate or ascertain the facts so as to establish a defense to the attorney fee lawsuit, and to engage in any pre-trial discovery whatsoever resulted in a judgment against Basic Food in an amount greater than that for which Basic Food properly should have been held liable.

The judgment of the trial court based on a $5,000 jury verdict for Basic Food was upheld.

Reasoning: The court evaluated the following criteria in coming to its decision: In an action against an attorney for negligence or breach of implied contract, the plaintiff has the burden of proving: a) the existence of the attorney-client relationship; b) the acts which are alleged to have constituted the negligence; c) that the negligence was the proximate cause of the injury; and d) the fact and extent of the injury alleged. . . .

The court held as follows.

1. The defendant's contention that Basic Food had to prove it would have won the case had it not been for his negligence, was not required in this case. The court said, "Rather, the attorney's liability, as in other negligence cases, is for all damages directly and proximately caused by the attorney's negligence." That the attorney's negligent conduct maximized the chances of Basic Food's prior opponent led to a higher damage award than would otherwise have been returned by the jury.
2. The attorney's defense that he had made errors of *professional judgment*, which are not grounds for a malpractice suit, was held to be insufficient in this case. The court found the errors too serious to be classified as errors of professional judgment.
3. The court found that the evidence which Basic Food presented at the lower court trial again was sufficient to justify that court's decision for the plaintiff.
4. The court also pointed out that a challenge by the defendant that he was entitled to a new trial because of the previous trial judge's failure to charge the jury properly could not be substantiated, and that it was up to the trial court to determine whether the defendant was entitled to a new trial.

Conclusion: This case illustrates a number of legal rules regarding the suing of a lawyer for malpractice or negligence resulting in injury to a business.

1. A lawyer is bound to use reasonable care, skill, and judgment in the conduct and management of legal affairs entrusted to him or her.
2. There are two types of cases relevant to this malpractice suit, and Basic Food either could have sued only on the basis that it was prevented from filing on time *or* on the basis that the filing caused Basic Food to be liable for a larger settlement than otherwise might have been decided. A lawyer's negligence is a *proximate cause* of your injuries if the value of the judgment rendered against

> you may not have been set if the lawyer had performed properly, not whether you lost because you were prevented from suing or filing an appeal within the required time. In the second type of case, you must prove that you would have won your case, or won your appeal. In the first case, illustrated here, Grant was found to have caused Basic Food to suffer a higher jury award than would have been decided had Grant done what he was supposed to do in his representation of Basic Food.
>
> 3. A lawyer's defense that he or she merely made errors of judgment, which would otherwise exclude liability, does not apply to *very serious* errors. Grant's negligence in not even offering any defense at the trial was very serious.

MANAGING YOUR LAWYER

Manage your lawyer's services with the same vigilance as you would manage any other professional person you employ, such as a certified public accountant, architect, or engineer. This means that you are entitled to have all services agreed to performed on schedule, or with an explanation and reasonable alternatives provided if the services are delayed or must be changed. You are *not* entitled to supervise the methods your lawyer employs in his or her professional capacity, but merely the accomplishments of the promised result.

You are entitled to your lawyer's best professional effort on your behalf. You are not entitled to a guarantee of success unless the lawyer makes one. The honest attorney knows better than to guarantee services against all risks.

Successful management of your lawyer requires maintaining good communication throughout your association. Good communication means knowing when to contact your lawyer and when to leave your lawyer alone, and rests on a foundation of mutual trust and respect.

Legal success or justice is not assured. Your ultimate decision as to whether to retain the services of a lawyer should rest on the quality *and* honesty of his or her performance.

SUMMARY

You should know how to choose an attorney. This means more than just finding an attorney. It means finding an attorney able, willing, and experienced enough to provide you with the best service per dollar you can afford to spend.

Increasingly, lawyers specialize in areas such as tax law or real estate. Since hospitality operators may encounter a variety of legal problems, a general practitioner might be the best choice.

Although advertising is allowed for lawyers, it is still uncommon, but there are more varied sources of information about legal services currently available than in past years. First, there is the traditional word-of-mouth referral by a family member or friend. Second, the referral services of your local or state bar association should be explored. Third, mass media advertisements are now available. Fourth, the local chapter of your state hospitality association may be able to refer you to local or national hospitality law specialists. In each of these cases, you must remember that the referral is not a guarantee of successful performance, but only notice that the lawyer is available for consultation.

You must next determine whether the attorney you find through any of those sources is personally compatible, and is willing to meet your specific needs and to work for you at a fee you can afford to pay. It is critical that you know what your needs are, that you communicate your needs to your attorney, and that both of you agree on all essential terms and conditions. To avoid misunderstandings later, it is to your advantage to draw up a letter of commitment.

There are four methods lawyers use to fix fees. They may charge by the hour, by the day (per diem) for courtroom trials or appeals, by contingent fee, or a by retainer agreement.

Both attorney and client have rights under the law. The attorney has the right to expect honest answers and cooperation. The client has the right to expect competent service, free of any conflict of interest.

There are a number of guidelines in this chapter for working with lawyers to obtain the best results. The objective is not to interfere with or direct an attorney's work, but to monitor the work, document expenses, and cooperate.

It is important to ensure that an attorney does not have any conflicting interest in the outcome of your case and will not violate the attorney-client privilege.

If a lawyer acts unprofessionally or incompetently, or violates a client's trust, the client may have the option both to sue the attorney and to complain to the local bar association. Generally the misdeed must be very serious before a lawyer is disbarred or loses a malpractice suit.

QUESTIONS

1. List three sources of information available to you in selecting a lawyer for the first time.

2. What type of lawyer would you want to help you with the legal areas of running your operation—a general practitioner or a specialist? Explain your answer.

3. What steps should you take to manage your lawyer's services after hiring one?

4. The Donaldsons are buying land on which to build a restaurant. They have decided on some land owned by the local school board and they hire attorney Rhonda Killian to help them with the contract. She is also one of five attorneys representing the school board. Is there a conflict of interest? If so, what should the Donaldsons do?

5. When should you consider complaining to your state or local bar association about your lawyer's misconduct?

NOTES

1. *Bates v. State Bar of Arizona*, 433 U.S. 350 (1977). *Reh'g denied*, 434 U.S. 881.

2. Additionally, some states are planning to certify legal specialists, meaning that these states will license lawyers as specialists. As yet, most states have not adopted this certification procedure. Moreover, the special practice areas are limited in those states that have done so or are moving in this direction.

16

The Court System and Out-of-Court Settlement

OBJECTIVES

After reading this chapter, you should be able to:

1. Explain the use of judicial review.
2. Compare the two major U.S. court systems.
3. Explain how jurisdiction imposes limits on cases and courts.
4. Evaluate going to court, and explain the use of small claims court.
5. Analyze how out-of-court settlement occurs, and compare three settlement methods: compromise, arbitration, and mediation.

Case in Point

Bob Baker is the sole owner of Bob's Beefsteak and Brew, a high-volume, quick-service operation situated on private property adjoining Powtochie Park, which is operated by the United States Park Service. Baker is a resident of Wyoming, where his restaurant is situated. One day Baker is served with a United States government notice that the government is going to acquire a portion of his private parking lot to widen an adjoining federal highway. The government offers Baker an amount he considers too little to compensate him for the purchase of his land. Moreover, this widening will cut off existing access to his operation from the highway, and the government's offer does not include adequate provision to build an alternate access road. Negotiations with the federal Highway Administrator do not resolve the dispute.

Baker decides to sue. Which court, federal or state, will have jurisdiction over this case?

As a citizen of Wyoming, Baker can sue the federal government in the federal district court for the District of Wyoming. Baker satisfies the requirements of federal jurisdiction.

COURT TRIALS

Most legal disputes are settled out of court by negotiation between the parties. A court trial is usually a last resort, when all other efforts to resolve a dispute have been unsuccessful. Going to court is expensive and time consuming, and the outcome is uncertain. You may have a good case based on the law, but if the jury or judge does not believe your version of the facts, you may not recover if you sue, or you may be found liable to pay damages if you are sued.

Our trial system is an *adversary* system. The object of a trial is to present the case in such a way as to persuade the trier of the facts (the jury or judge) that one side is right and the other is wrong.[1]

The judge hearing a case to be decided by a jury plays an important role as mediator. The court makes rulings on the admission of evidence and instructs the jury on the law.[2] The court may also dismiss a case if it does not justify a legal remedy, or if there is insufficient evidence to prove it.[3] The court may even render a judgment in the person's favor in spite of a contrary jury verdict if the judge finds the evidence insufficient to support the jury's verdict.[4]

The court functions as an umpire, permitting each side to plead, prove, and persuade the jury (or court sitting without a jury) of the truth of the facts, or a reviewing court of the sufficiency of its legal argument.

Court Systems

There are two major court systems in the United States: federal and state. Each state has its own court system. The federal government has a court system as well, with different types of jurisdiction.

State Courts In a typical state court system, courts are divided into a triangular group. (See **Exhibit 16.1**.) At the bottom of the triangle, and supporting the rest of the court structure, are *trial courts* of *general* or *limited jurisdiction*, the latter also called *special inferior courts*. Trial courts are the courts in which all controversies are first heard. Most hospitality lawsuits originate in a trial court.

Exhibit 16.1 State Court System

State Supreme Court	Probate
State Appellate Court	Criminal
State Trial Court of General Jurisdiction	Family
Special Inferior Courts	County

General jurisdiction is the power to hear all cases under state law, such as corporate law, contract law, agency law, and tort law, often without regard as to the money value of the case being heard. Courts of general jurisdiction may exist within each county of the state, and in large cities. A county that is too small to qualify for a trial court of general jurisdiction will be served by a court from a neighboring county, provided by state law. A state judge may be assigned to hear cases in each county at specified times of the year.

Limited jurisdiction means that a court has power to decide only certain types of cases. Jurisdiction can be limited to specific subjects, such as family matters or landlord-tenant cases. These courts are also called *state inferior courts*.

Jurisdiction can also be limited by the amount of money the party suing seeks to recover. *Small claims courts* typically have a monetary ceiling for jurisdiction of between $250 and $1,000, which is set by state law. This means that if the amount the claimant seeks to recover exceeds the state ruling, the small claims court cannot hear the case.

The next level of courts is called *intermediate appellate* or reviewing courts; their jurisdiction lies between that of trial courts and the highest state court. Each state has at least one reviewing court. Most heavily populated states also have intermediate appellate courts, as well as a supreme court, a "court of last resort." Less populated states may have only trial courts and one court of appeals or supreme court.

The reviewing courts sit to hear appeals—they never function as trial courts. The courts primarily decide only questions of law. Questions of law refer to errors of the trial court claimed by the party that appeals. A typical error is an improper or erroneous charge to the jury on the law made by a trial judge. If the error is serious, the reviewing court has the authority to send the case back for a new trial. If the error is harmless, the court has the authority to uphold the judgment given by the trial court.

Federal Courts   Like the state courts, federal courts are divided into trial courts, intermediate appellate courts, and a supreme court (See **Exhibit 16.2**.) There is more specialization in federal courts, including administrative tribunals for various federal regulatory agencies.

Exhibit 16.2   Federal Court System

Supreme Court of the United States	Court of Claims
United States Court of Military Appeals	Court of Customs and Patent
United States Court of Appeals	Appeals
Special Administrative Tribunals	Tax Court
NLRB, FTC, etc.	Customs Court and other
United States District Court	Administrative Tribunals

Federal trial courts are called *district courts*. District courts of limited jurisdiction include bankruptcy courts, tax courts, customs courts, courts of claims, and a court of customs and patent appeals. All of these trial courts are located in federal judicial districts established by Congress, and vary in size and number from state to state.

The federal system also has *intermediate courts of appeals*. This is a change from the state court system, where such courts are optional. This is because there is only one Supreme Court of the United States, and that court could not function if it were required to hear all appeals from every state. The federal courts of appeals, found in 11 judicial circuits, each with a fixed number of states, operate in much the same way as state reviewing courts. These courts review decisions of the federal district courts within their geographic circuits, and decisions of federal administrative agencies, such as the Food and Drug Administration.

The United States Supreme Court sits at the top of our federal court system. Because it is the only court of last resort within the federal system, appeals to the Supreme Court are heard and decided with permission of the court by a procedure called *petition for certiorari*. Only the most important matters are accepted and heard by the Supreme Court.

Why do we have a system of state courts and federal courts? Would one court system be sufficient to handle all legal controversies? The answer is this: Each state is sovereign, an independent political body, over all of its citizens in their activities and conduct within its borders. Each state court normally lacks the power to settle controversies that affect citizens of other states that arise in another state.

The federal court system was created to fill this void in the state court system. The federal courts under the federal Constitution have two sources of judicial power or jurisdiction: 1) the power to settle federal questions, meaning controversies totally or partially involving the United States Constitution, a treaty, or a federal law; and 2) the power to decide controversies involving *diversity of citizenship*. Diversity of citizenship means a case involving: 1) citizens of different states; 2) a foreign country suing citizens of a state, or different states; and 3) citizens of a state and citizens or subjects of a foreign country. *Citizens* has been interpreted to mean corporations as well as persons.

In *diversity jurisdiction* cases—cases in which a citizen of one state sues the citizen of another—the amount in controversy must exceed $75,000. Only when the defendant is the federal government or a federal employee is the jurisdictional amount eliminated. Diversity jurisdiction cases can be heard in the federal courts, although the law applied is that of the state in which the federal court is located (unless a federal law is involved).

In the Case in Point, Bob's legal right to sue is based on the federal constitutional requirement that the federal government adequately compensate U.S. citizens for land acquired for legitimate federal purposes. The dispute does not question the government's *right* to acquire Bob's parking area. It instead centers on the adequacy of the compensation by the government for the exercise of that right.

After trial the district court will issue a money judgment for Bob. If either Bob or the government is dissatisfied with the amount of the award, either party may appeal to the federal Court of Appeals for the Tenth Circuit, which includes Wyoming. That court will review the adequacy of the district court's findings on which the award was based. If the award is clearly erroneous, the Court of Appeals may send the case back for a new trial, or may increase or decrease the award, modifying the judgment of the district court.

At this stage either party may petition for a *writ of certiorari* (permission to hear the case) to the Supreme Court, asking for further review. In all likelihood the petition will be denied because the High Court usually reserves review for important federal questions; to resolve conflicts of law interpretation between federal courts of appeal; and to review cases which can only be resolved by the Supreme Court, such as interpretation of a federal treaty with a foreign government. In such cases the Federal Court of Appeals is really the court of last resort. Bob's controversy does not fit any of the necessary criteria.

Jurisdiction

Jurisdiction is a very important factor in a legal case. Courts are limited in jurisdiction by geographical location, but there are also important limitations related to the case and procedure used by both parties.

Jurisdiction refers to the power of a court to decide a case. If a court lacks jurisdiction over the parties or over the subject of the case, the court cannot make a binding determination of the dispute.

Jurisdiction over the parties means that the parties are all properly before the court. Parties to a lawsuit are properly before the court either because they voluntarily appear or because they have been served with a legal process (legal papers such as a summons, complaint, or petition). The service of process gives the court the power to act in the dispute even if the party served fails or refuses to appear.

Jurisdiction over the subject of the case means that the court has the power to decide the controversy before it. Not all courts have the power to decide all controversies brought to them for determination. For example, a bankruptcy court can only hear and determine bankruptcy cases. It has no power over, say, a liquor liability lawsuit. A family court has no power to resolve a dispute in a management contract case.

Jurisdiction can also involve the money value of the claim or case. Minor courts, such as small claims courts, cannot hear cases involving a sum greater than that fixed by each state, even though they have the power to hear the controversy itself. In the complaint the party suing must state the amount of money he or she wishes to recover from the other party. If the amount claimed in the complaint exceeds its ceiling for jurisdiction, the court must either dismiss the case or transfer it to the proper court.

GOING TO COURT

At some point you may enter into a dispute that can only be settled in court. If you do, remember that there are procedures for court trials that can pave the way for you to win your case, or at least ensure you a fair trial. These steps are usually handled by your lawyer on your behalf, but knowing them will help you monitor his or her progress.

A trial seeks to resolve disputes by finding for or against one of the parties on the *facts* and the *law*. Facts are presented to the jury, which decides to accept or reject them. The law is explained to the jury by the judge at the end of the case, to enable the jury members to apply the facts they found to the law and to enter a *verdict* in conformity with the law. A verdict is a jury's decision.

Both the *plaintiff*, the complaining party in a litigation, and the *defendant*, or answerer to the complaint, have basic rights in United States courts. Each has the right to have an attorney represent him or her in court, to gather evidence, to participate in judge selection, to make opening statements, to call friendly witnesses, to dispute evidence given by the other side, to examine and cross-examine witnesses, and to give closing arguments. The right to a jury trial is limited in some states by the amount of damages in dispute.

A wronged party has the right to sue if another party has committed a legal wrong that affected the plaintiff, and for which damages are due. It is up to the judge to determine whether these factors exist, and if they do not, he or she may dismiss the suit.

Small Claims Courts

Going through a court trial with a lawyer can be complex, time consuming, and costly. To meet the criticism that the traditional court system does not address the needs of the "average person," *small claims courts* have been established in most major cities. These courts, usually part of the city or municipal court system, were created to provide justice quickly, conveniently, and inexpensively in cases not requiring a lawyer.

The following are the major features that distinguish this special court from other courts.

1. The procedures used by small claims courts are very informal. A simple form of *summons* and *complaint* is issued by the clerk of the court, which may be served by any adult person other than the plaintiff. A simple answer by the defendant is then filed, and the case is immediately set for trial. No discovery procedures or interrogatories before trial are used, nor are pre-trial motions permitted. On the trial date the plaintiff tells his or her version of the case, and the defendant gives his or her side. Usually no lawyers take part, although in some cases they may represent clients as in regular court proceedings. No objections to evidence or other trial

maneuvers are allowed because the case is heard by a judge without a jury. At the end of the testimony, the judge makes a decision from the bench.

2. The monetary jurisdiction of small claims courts is limited, and the ceiling amount varies according to state law. The court has no authority to issue *injunctions*—court orders halting a labor dispute, for example—or to order specific performance of a contract, such as the sale of a house. These jurisdictional limits are intended to keep the disputes simple and to provide prompt relief.

3. Normally no appeals from small claims judgments are allowed. This is because no record of testimony or of the proceedings is kept by small claims courts for review by a higher court. Where an appeal is permitted, it takes the form of a new trial.

4. The small claims judge functions more as an arbitrator than a trial judge. In some states the judge is not elected or appointed, but is an attorney assigned to hear the cases and sits part-time. As an arbitrator the judge tries to protect the interests of both sides of the controversy, since attorneys are not usually present to perform this function. Once satisfied that he or she has heard enough testimony to reach a decision, the judge does so.

5. Small claims courts have certain *advantages:* a) They are accessible to more members of the public, since an attorney need not be hired. b) They hear cases sooner, with much less formal pre-trial and trial procedure than is found in most other courts. c) They are cheaper to use since the small claims judge acts as a legal arbitrator for both sides. d) They render immediate decisions, which in most cases are final, because they are nonappealable.

6. Small claims courts have certain *disadvantages:* a) They are set up to handle minor disputes, and can issue only money judgments. b) They only deal with small money disputes, which compels most parties to hire an attorney to take their case to a regular court. The money ceiling on cases heard means that only a small fraction of the total cases filed can be heard by small claims courts. c) They lack the power to provide anything other than money relief. d) They deprive the parties of the opportunity to appeal an unfavorable judgment. The losing party has little choice but to accept the decision given.

Small claims courts have attempted to simplify the process of resolving minor disputes to avoid expensive, formal, and time-consuming court trials. Small claims courts provide "homespun" justice without the legal niceties of procedure normally used before and during a full trial. They are available to hospitality operators as well as patrons and should be considered as part of the legal checklist of useful methods of settling disputes.

Let's look at an example to illustrate the use of small claims courts. Henry LaFrance owns and operates La Bastille, a three-star New York City restaurant. The establishment prides itself on impeccable French food, wine, and service, and tries in all cases to cater to the whims of its upscale clientele. One morning, LaFrance received and confirmed a telephone reservation for a party of four in the name of Claude Renfrew. Renfrew did not specify which table he wanted reserved, but mentioned that the table should be suitable for an important business meal. LaFrance assured Renfrew that he would make every effort to seat his party at a suitable table, but told him that the restaurant would be filled that night and to please understand the circumstances.

When Renfrew and his party arrived at La Bastille, they were seated at a table not readily noticeable to other patrons. Renfrew complained but was not able to be seated at another table. He became outraged and stormed out. Renfrew later tried to sue for $1,000 in damages for humiliation and mental anguish he suffered as a result of the incident. Because of the small amount involved, the New York State small claims court would have jurisdiction.

OUT-OF-COURT SETTLEMENT OF DISPUTES

The process of trial and appeal of court cases is costly and lengthy, and may be disappointing. As a method of settling business disputes, going to court should be viewed with caution and used only as a last resort.

True, we must defend ourselves against lawsuits others bring against us. But even in such cases thought should be given to resolving the dispute out of court, for a few reasons: 1) By doing so you have more control over the outcome of the dispute than by leaving matters in the hands of a trial judge and/or jury. You do not run the risk that the jury will not believe your testimony and will give a verdict in favor of your opponent, or that the judge will not like your case and unconsciously make those feelings known to the jury. 2) It is better to settle for less now than run the risk of paying more later. 3) An out-of-court settlement saves time and energy needed to manage your hospitality business. 4) Settlement may preserve the goodwill of the person or party suing or being sued.

What disadvantages are there to settling a dispute out of court? 1) You may be tempted to settle a case just to get rid of the dispute even when no settlement is warranted. This gives your opponent the opportunity to take advantage of you. Your opponent may be encouraged to do so repeatedly, knowing that you will settle rather than go to court. 2) Settlement will deprive you of the opportunity to test a legal theory or rule that should be, and might be, changed by the courts in your favor. 3) Settlement may work a hardship if the dispute involves costly and permanent economic harm. You may not be able to recover sufficient money in an out-of-court settlement to make up for the long range costs to you. 4) Settlement can deprive you of the opportunity to pursue an unscrupulous person who will only settle under the threat of a pending lawsuit. Settlement presumes that both parties have something to gain by settling out of court. When your opponent has everything to gain and nothing to lose by refusing to settle, then a lawsuit may be needed to persuade that person to change his or her thinking.

There are different types of out-of-court settlements: compromise, arbitration, and negotiation.

Compromise, Arbitration, and Mediation

Compromise is a voluntary settlement of a dispute by the disputing parties, without the aid of third parties. An out-of-court settlement is a compromise of a dispute. For example, in a contract dispute with a vendor, you may want to reach a negotiated settlement. Honest mistakes can be remedied without going to court.

Arbitration differs from a compromise in that an arbitrator is a third party selected by the disputing parties to *decide* the issue(s) of the dispute. An arbitrator's decision *is binding* on the parties and enforceable by the courts. The parties to an agreement may select those issues that will be left to arbitration.

The major advantage of arbitration is that it keeps certain issues out of court for the sake of a timely and less costly settlement. The major disadvantage of arbitration is that once a decision is made by the arbitrator, it will be enforced by the courts without any review of its soundness or sufficiency.

The following case was decided by an arbitrator, and the decision is consistent with legal principles. This is a labor case, and many such employer-union cases are decided by either arbitration or through mediation. Both parties were represented by attorneys here, but this not necessary.

Facts: A San Diego, California, restaurant offering European cuisine required its cocktail waitresses to wear classically styled gowns reminiscent of the Roman or Grecian mode and to wear their hair blond and upswept. Management believed that this requirement was consistent with its image of fine European dining. The appearance rule, in existence for over 18 years, had become a city-wide trademark, according to the restaurant, and had been instrumental in its receiving numerous national awards for fine European cuisine, elegant decor, and superior service.

One waitress, who had dyed her hair blond in order to get the job, and been forced to purchase two blond wigs at a cost of $60 each when she decided to return her hair to its natural color, charged that the restaurant's policy was unreasonable and that, in any event, the restaurant should pay the cost of purchasing and maintaining her wigs.

Award: The restaurant's hair rule was reasonable, according to arbitrator Thomas Christopher, who concluded that it has been applied in a fair and consistent manner and without violating any express or implied term of the collective bargaining contract. Thomas found that the waitresses' appearance was a significant component of management's business concept, since an employee who did not wish to dye her hair could wear a wig and thus have her appearance off the job remain unaffected. Although the arbitrator concluded that he was without authority under the contract to order the restaurant to pay for the purchase and maintenance of wigs, he asked the restaurant to consider proposals outlined by the union for relieving the waitresses of some of the costs of complying with the hair rule (Mister A's Restaurant, 80 LA 1104).

Discussion: Arbitrator Christopher agreed that management had the fundamental right unilaterally to establish reasonable work rules that were not inconsistent with law or the collective bargaining contract (55LA 306, 54LA 942, and 54LA 129). However, he noted that work rules must be reasonably related to a reasonable objective of management (55LA 283). In this situation, the record established that the restaurant had achieved an outstanding reputation for its elegant and distinctive dining atmosphere in part because of the appearance of its waitresses. Although Thomas believed that any rule governing the appearance of employees must consider the impact the rule may have on employees' freedom of expression, he pointed out that where the sales of the business "are highly sensitive to the image portrayed, . . . the balance tends to weigh heavily in the favor of the employer" (56LA 597).

Although the arbitrator was barred by contract from forcing the employer to pay the waitress for the cost and maintenance of the wigs, he noted:

> Nevertheless, this case does raise some equitable considerations and the arbitrator will therefore request that the company consider proposals outlined by the union in its post-hearing brief for relieving the waitresses of some of the costs of complying with the hair rule. The proposals are: 1) The company could purchase the wigs and contract for their upkeep at probably a discount rate. To insure their longer use, the wigs could be stored on the company premises. 2) For waitresses who dye their hair, the company, at no cost to itself, could make an arrangement with a particular salon to provide a discount to the waitresses if a certain amount of them regularly patronize the salon.

The major benefit of arbitration is that the decision is final and binding. There are no appeals. Out-of-court settlements in liability cases may be the result of arbitration. It is generally less costly and time consuming than court trials.

Mediation is the action of a third party, usually agreed to by the parties in a dispute, to encourage opposing sides to come to agreement. The mediator makes no decision for the party, but rather acts in the role of a neutral expert to help the parties continue their own efforts to resolve the problem. The mediator issues recommendations, which are not binding to the parties. *Conciliation* is another term for mediation.

Mediation falls between compromise and arbitration. Mediation is less costly in time and expense, but does not force the parties to settle. The settlement of the dispute rests solely with the parties.

RESOLVING LEGAL DISPUTES

You may find it necessary to resort to litigation to resolve disputes either as a plaintiff or as a defendant. Normally you will settle controversies with your patrons, your suppliers, your competitors, and the government at all levels by negotiation and compromise. Litigation should be used when all other methods of settlement have failed.

Whether you settle out of court or after court trial and appeal, you must *always* make a written record of every major dispute. This is essential for a number of reasons: 1) It will enable those responsible for a settlement (lawyer, judge, or jury) to know the nature of the dispute, how it occurred, why it occurred, and how frequently it has occurred. In this way you may not only settle the current dispute, but, wherever possible, be able to plan ahead to prevent or minimize future disputes. 2) You will be able to locate and summon people involved if litigation should prove unavoidable. 3) You will be able to use your insurance coverage better as a hedge against insurable risks and thereby effectively control your insurance costs. 4) You will be maintaining all records required by law to establish your rights with greater certainty in the event of a government agency investigation and possible court action. Disorganized records can destroy an otherwise winnable court case. 5) It will demonstrate that you are professional in your business transactions and merit the confidence and respect of all who do business with you. This objective alone may result in fewer lawsuits and a better business climate.

If you decide to bring a lawsuit against someone, you must be prepared to state and prove your case. This burden is normally on the plaintiff. You must not only produce concrete facts, rather than hypothetical theories, but you must consult your attorney to determine whether your facts are legally sufficient to justify court action. If they aren't, then you know you must settle the case out of court.

Your physical time, effort, and financial resources are not inexhaustible, and you must, as in your other business dealings, weigh the cost of litigation against the benefits that you may receive. In court cases where a jury is involved (that is, cases not decided on the law alone), you cannot predict the outcome. Even if successful in court, you may find that the costs of litigation are too high to justify a favorable judgment. You can only make such a critical business judgment by understanding the basic judicial process.

SUMMARY

Our trial system is based on an adversary justice system. Each side endeavors to prove that that side is right and the other is wrong. To accomplish this usually requires, in addition to a competent lawyer, witnesses and evidence. Obtaining these can be costly.

The states and the federal government have a trial system that functions according to which courts have jurisdiction geographically, or over the subject, or as to the amount of damages (money) involved.

Courts of general jurisdiction hear all types of cases within a geographic jurisdiction. Courts of limited jurisdiction deal only with certain subject matter such as bankruptcy or tax law.

Small claims courts exist in major cities to settle disputes involving damages set by each state. These courts do not require lawyers and involve fewer procedures than regular courts.

Out-of-court settlements are always preferable to a trial. Relevant methods include compromise, mediation, and arbitration.

QUESTIONS

1. Mort Quentin of Mort's Delicatessen chain is declaring bankruptcy in Arizona. Will the federal or state court have jurisdiction over this case? Why?

2. What is the difference between a trial court and an appellate court?

3. Who usually has the burden of proof in a trial, the plaintiff or the defendant?

4. Karla Kemp ordered $300 worth of potatoes for her first-class steak restaurant and specified Idaho baking potatoes. When her shipment arrived, it consisted of Wyoming potatoes suitable for frying. Now Karla has potatoes she cannot use and is out $300. What, if any, legal steps should Karla take to get her money back from the supplier? Explain your answer.

5. Compare and contrast compromise, mediation, and arbitration from the standpoint of advantages, disadvantages, and the degree to which each is binding.

NOTES

1. Lusk, Hewitt, Donnell, Barnes, *Business Law and the Regulatory Environment* 21–24 (5th ed. 1982).

2. Throughout the trial of a case, the trial judge rules upon or determines questions of law. The trial judge is not bound by legal arguments made by opposing lawyers. The trial judge is free to disregard those arguments and interpret the law independently, subject to review if one or the other side makes an appeal, claiming an error of law committed by the trial judge.

3. *Wallace v. Shoreham Hotel Corp.*, 49 A.2d 81 (D.C. 1946) (complaint by patron, alleging hurt feelings inflicted by defendant's dining room waiter, insufficient; only hotel guest entitled to courteous and considerate treatment); *Buck v. Del City Apartments, Inc.*, 431 P.2d 360 (Oakla. 1967) (hotel guest fell on icy walkway; risk of fall within knowledge of guest; demurrer to guest's evidence affirmed).

4. *Apper v. Eastgate Assoc.*, 28 Md. App. 581, 347 A.2d 389 (1975) (*res ipsa loquitur* doctrine applied to overturn directed verdict in favor of motel keeper in case involving defective bathroom shower handle). See also *Haft v. Lone Palm Hotel*, 3 Cal.3d 756, 478 P.2d 465, 91 Cal. Rptr. 745 (1970) (reversal of trial court verdict, notwithstanding the jury verdict for the motel keeper, in a motel pool drowning death).

Federal Acts Affecting Foodservice Operators

Age Discrimination in Employment Act (1967)
Prohibits discrimination against job applicants and employees between the ages of 40 and 70.

Americans with Disabilities Act (1990)
Requires all public accommodations and employers to allow access to customers and employees through "reasonable accommodations."

Bankruptcy Reform Act (1978)
Resulted in clarification of alternatives to straight bankruptcy for businesses and individuals, simplified procedures, and changed the previous court and trustee system.

Civil Rights Act (1964)
Prohibits discrimination in employment and public accommodations on the basis of race, color, religion, or national origin. Sex and pregnancy are covered in the employment section.

Clayton Act (1914)
Antitrust law which prohibits exclusive-dealing contracts, trying arrangements, and stock acquisitions designed to lessen competition.

Economic Recovery Tax Act (1981)
Provides accelerated recovery deductions for depreciation of buildings and equipment.

Equal Employment Opportunity Act (1972)
Prohibits discrimination based on race, color, religion, sex, or national origin. Amended Civil Rights Act of 1964.

Equal Pay Act (1963)
Requires employers to provide employees of both sexes equal pay for equal work.

Fair Labor Standards Act (1938)

Establishes requirements for minimum wage, work time, overtime pay, equal pay, and child labor, and enforcement of such requirements.

Federal Insurance Contributions Act

Source of federal patrol tax law, especially regarding Social Security.

Federal Unemployment Tax Act

Source of tax law for unemployment compensation.

Hart-Scott-Rodino Antitrust Improvement Act (1976)

Empowers states' attorneys general to sue businesses that violate the Sherman Act.

Internal Revenue Code

Major source of tax laws. Regularly amended.

Labor Management Relations Act, a.k.a. Taft-Hartley Act (1947)

Regulates employee and employer actions regarding unionization.

Model Business Corporation Act

Regulates all aspects of corporation formation and dissolution, and rights of directors, shareholders, and officers.

Nutrition Labeling and Education Act (1990)

Requires packaged foods to carry complete nutrition labeling, and nutrition claims (including those on foodservice menus) to be substantiated and conform to clearly defined use of nutritional claim terms.

Robinson-Patman Act (1936)

Prohibits price discrimination designed to lessen or injure competition, or to create a monopoly.

Sherman Act (1890)

Prohibits contracts, combinations, or conspiracies to restrain trade among the several states or with foreign nations.

Truth-in-Lending Act, a.k.a. Consumer Credit Protection Act (1969)

Regulates extension of credit in the absence of similar state legislation.

Uniform Commercial Code

Sweeping law covering all commercial transactions, particularly aimed at sales of consumer products.

Uniform Limited Partnership Act (adopted 1961; revised 1976)

Source of law for setting up and dissolving limited partnerships. It is the law only in states that allow limited partnerships.

Uniform Partnership Act (1914)

The law regulating partnerships, adopted in most states.

Vocational Rehabilitation Act (1973)

Prohibits discrimination against job applicants or employees because of mental or physical disabilities. Applies only to those who contract or subcontract with the government.

B Accuracy in Menus

REPRESENTATION OF QUANTITY

Proper operational procedures should make misinformation on quantities virtually nonexistent. Steaks are often listed on menus by weight, and the generally accepted practice of declared quantity is that prior to cooking.

The more obvious claims are understood by most people. For instance, a double martini should be twice the size of a regular one. A term such as "jumbo eggs" is a recognized egg size. "Petite" and "supercolossal" are among the official size descriptions for olives.

There is no question about the meaning of a "three-egg omelet" or "all you can eat." But the use of terms such as "extra large salad" or "extra tall drink" may invite problems if they are not qualified. Also remember the implied meaning of words—a bowl of soup should contain more than a cup of soup.

REPRESENTATION OF QUALITY

Federal and state standards of quality grades exist for many restaurant products including meat, poultry, eggs, dairy products, fruits, and vegetables. Terminology used to describe grades include Prime, Grade A, Good, No. 1, Choice, Fancy, Grade AA, and Extra Standard.

Care must be exercised in preparing menu descriptions when these words are used. In certain uses, they imply certain quality. An item appearing as "choice sirloin of beef" connotes the use of USDA Choice Grade Sirloin of Beef. One recognized exception is the term "prime rib." Prime rib is a long-established, well-understood, and accepted description for a certain cut of beef (the "primal" ribs, or 6th to 12th ribs) and does not represent the grade quality, unless USDA is used in conjunction.

Ground beef should contain no extra fat (no more than 30 percent), water, extenders, or binders. Seasonings may be added as long as they are identified. These requirements identify only meat ground and packaged in federal- or state-inspected plants.

REPRESENTATION OF PRICE

If an operation's pricing structure includes a cover charge, service charge, or gratuity, these must be appropriately brought to customers' attention. If extra charges are made for requests such as "all white meat" or "no ice drinks," this should be stated at the time of ordering.

Any restrictions when using a coupon or premium promotion must be clearly defined. If a price promotion involves a multi-unit company, clearly indicate which units are participating.

REPRESENTATION OF BRAND NAMES

Any product brand that is advertised must be the one served. A registered or copywrited trademark or brand name must not be used generically to refer to a product.

A "house" brand of a product may be so labeled even when prepared by an outside source, if its manufacturing was to the specific operation's specifications. Containers of branded condiments and sauces placed on a table should contain only the product appearing on the container label.

REPRESENTATION OF PRODUCT IDENTIFICATION

Because of the similarity of many food products, substitutions are often made. These substitutions may be due to their not being delivered on time, their availability, merchandising considerations, or price. When such substitutions are made, they should be reflected on the menu. Common substitutions include the following.

Common Menu Item	Common Substitution
Maple syrup	Maple flavored syrup
Boiled ham	Baked ham
Veal cutlet	Chopped and shaped veal pattie
Ice cream	Ice milk
Fresh eggs	Powdered eggs
Ham	Picnic style pork shoulder
Milk	Skim milk
Pure jam	Pectin jam
Whipped cream	Whipped topping
Turkey	Chicken
Black Angus beef	Hereford beef

Common Menu Item	Common Substitution
Ground sirloin	Ground beef
Chicken	Capon
French style ice cream	Standard ice cream
Haddock	Cod
Egg noodles	Plain noodles
White meat tuna	Light meat tuna
Haddock	Pollack
Sole	Flounder
Cheese	Cheese food or processed cheese
Cream sauce	Nondairy cream sauce
Tuna fish	Bonito
Roquefort cheese	Blue cheese
Peanut oil	Corn oil
Calf's liver	Beef liver
Cream	Half & Half, nondairy creamers
Butter	Margarine
Tenderloin tips	Diced beef
Mayonnaise	Salad dressing

REPRESENTATION OF POINTS OF ORIGIN

A potential area of error is in describing the point of origin of a menu offering. Claims may be substantiated by the product, packaging labels, invoices, or other documentation provided by suppliers. Sources and availability of products change often, and it is the responsibility of the operation manager to inform customers. The following are common assertions of points of origin that must be used accurately.

Lake Superior whitefish	Gulf shrimp
Idaho potatoes	Florida orange juice
Maine lobster	Smithfield ham
Imported Swiss cheese	Wisconsin cheese
Puget Sound Sockeye salmon	Danish bleu cheese
Alaskan king crab	Louisiana frog legs
Imported ham	Colorado brook trout
Colorado beef	Florida stone crabs
Long Island duckling	Chesapeake Bay oysters
Bay scallops	

There is widespread use of geographic names used in a generic sense to describe methods of preparation or service. Since these terms are readily understood and accepted by customers, their use should in no way be restricted. Examples include the following.

Russian dressing	French toast
New England clam chowder	Country-fried steak
Irish stew	Denver omelet
Country ham	French dip sandwich
French fries	Swiss steak
Danish pastries	German potato salad
Russian service	French service
English muffins	Manhattan clam chowder
Swiss cheese	

REPRESENTATION OF MERCHANDISING TERMS

Merchandising terms are a difficult area to define clearly. "We serve the best gumbo in town" is understood by the dining public for what it is—boasting for advertising's sake. However, to use the term "we use the finest beef" implies that USDA Prime Beef is used, as a standard exists for this product.

Advertising exaggerations are tolerated if they do not mislead. When ordering a "mile-high pie," a customer would expect simply a pie heaped with topping, but an advertised "foot-long hot dog" should be one foot long.

Mistakes are possible in properly identifying steak cuts. Use industry standards such as provided in the National Association of Meat Purveyors Meat Buyer's Guide.

Terms such as "homestyle," "homemade style," and "our own" are preferred over "homemade" in describing menu offerings prepared according to a home recipe since most foodservice sanitation ordinances prohibit the preparation of foods in home facilities.

All of the following terms should be used accurately.

Fresh daily	Milk-fed chicken
Fresh roasted	Corn-fed pork
Flown in daily	Finest quality
Kosher	Center-cut ham
Black Angus beef	Own special sauce
Aged steaks	Low-calorie

REPRESENTATION OF MEANS OF PRESERVATION

The accepted means of preserving foods are numerous, including canned, chilled, bottled, frozen, and dehydrated. If these terms are used, they must be accurate. Frozen orange juice is not fresh, canned peas are not frozen, and bottled apple sauce is not canned.

REPRESENTATION OF FOOD PREPARATION

Preparation is often the determining factor in a customer's selection of a menu item. Absolute accuracy is crucial. Readily understood terms include the following.

Charcoal broiled	Deep-fried
Sauteed	Barbecued
Baked	Smoked
Broiled	Prepared from scratch
Roasted	Poached
Fried in butter	

REPRESENTATION OF VERBAL AND VISUAL PRESENTATION

When menus, wall placards, or other advertising contains a pictorial representation of a meal or platter, it should portray the actual contents with accuracy. Examples of *visual* misrepresentations include using mushroom pieces in a sauce when the picture depicts mushroom caps; using sliced strawberries on a shortcake when the picture depicts whole strawberries; using numerous thinly sliced meat pieces when the picture depicts a single thick slice; using five shrimp when the picture depicts six; and using a plain bun when the picture shows one with sesame seeds.

Examples of *verbal* misrepresentation include a server asking a customer whether she wants sour cream or butter with her potato but serving imitation sour cream or margarine, or a server telling a guest that pies are "baked in our kitchen" when in fact they are purchased elsewhere.

REPRESENTATION OF DIETARY
OR NUTRITIONAL CLAIMS

As of the writing of the second edition of this book, the Nutrition Labeling and Education Act (NLEA) of 1990 affects foodservice operations by requiring all nutritional and health-related claims on menus to be substantiated. Thus, if a menu item is said to be "low-fat," it must contain 3 or fewer grams of fat per portion. A list of commonly used nutrition terms and their Food and Drug Administration definitions follow.

TERM	FDA DEFINITION
Calorie-free	Contains fewer than 5 calories per portion
Cholesterol-free	Contains less than 2 mg cholesterol per portion AND 2 grams or less of saturated fat per portion
Extra lean	Contains less than 5 grams of fat, less than 2 grams of saturated fat, and less than 95 mg of cholesterol per portion AND per 100 grams
Fat-free	Contains less than 0.5 grams of fat per portion
Fresh	Not defined, except that the term must be used only to imply that food has not been processed or preserved
Good source of	Food in which a nutrient is present at 10 to 19 percent of the Daily Value
Healthy	Not defined, but should conform with other terms with which it is used
High	Food in which a nutrient is present at 20 percent or more of the Daily Value
Lean	Contains less than 10 grams of fat, less than 4 grams of saturated fat, and less than 95 mg of cholesterol per portion AND per 100 grams
Less	Contains 25 percent less of a nutrient than a *reference food* per serving; same as *reduced*
Light	Food that has either one-third fewer calories or 50 percent less fat compared to a similar food
Low-calorie	Contains 40 calories or less per portion
Low-cholesterol	Contains less than 20 mg of cholesterol per portion (for meals and main dishes, must also contain no more than 2 grams of saturated fat per 100 grams)
Low-fat	Contains 3 grams or less of fat per portion (for meals and main dishes, derives no more than 30 percent of its calories from fat)

TERM	FDA DEFINITION
Low-sodium	Contains less than 140 mg of sodium per portion
More	Food that provides at least 10 percent more of the Daily Value of a given nutrient than the *reference food*
Natural	Not defined
Percent (%) fat free	Food that meets the definition for "low-fat" and "fat-free"
Reduced	Contains 25 percent less of a nutrient than a *reference food* per serving; same as *less*
Sodium-free	Contains less than 5 mg of sodium per portion
Sugar-free	Contains less than 0.5 grams of sugar per portion

Also under the NLEA, only certain health claims are permitted.

- Dietary fiber and cancer
- Dietary fiber and heart disease
- Dietary fat and cancer
- Fruits/vegetables and cancer
- Saturated fat and heart disease
- Sodium and high blood pressure
- Calcium and osteoporosis

Health claims are not permitted for foods that exceed FDA-specified levels for fat, saturated fat, cholesterol, or sodium. Health claims and explanations of terms must place all explanations either next to the menu item being described or in a place to which the customer is directed to look (such as, "See the back of this menu for details").

 # State Laws Prohibiting Discrimination in Places of Public Accommodation

	Gender	Sexual Orientation	Physical Disability	Mental Disability
Alabama	X			
Alaska				
Arizona				
Arkansas				
California	X	X	X	
Colorado	X		X	
Connecticut	X		X	X
Delaware	X			
District of Columbia				
Florida	X		X	
Georgia				
Hawaii				
Idaho	X			
Illinois				
Indiana				
Iowa	X			
Kansas	X		X	
Kentucky	X			
Louisiana	X			
Maine	X			
Maryland				
Massachusetts	X			
Michigan				
Minnesota	X		X	
Mississippi				
Missouri				
Montana				

	Gender	Sexual Orientation	Physical Disability	Mental Disability
Nebraska	X			
Nevada				
New Hampshire	X		X	X
New Jersey*	X	X		
New Mexico	X			
New York†	X		X	
North Carolina				
North Dakota				
Ohio				
Oklahoma				
Oregon	X		X	X
Pennsylvania	X		X	
Rhode Island				
South Carolina				
South Dakota				
Tennessee				
Texas				
Utah	X			
Vermont				
Virginia				
Washington			X	
West Virginia	X		X	
Wisconsin	X		X	
Wyoming				

*By Supreme Court interpretation of Civil Rights Act, public accommodations section [*Seven Eleven Wines and Liquors v. Division of ABC*, 50 N.J., 329, 235 A.2d 12 (1967)].

†The state of New York prohibits discrimination based on marital status.

D | State-by-State Adoption of UCC 2–318

In states where no alternative or language is adopted, the courts are free to apply the appropriate alternative.

Alabama   *Section 7–2–318*
Omits words "who is in the family or household of his buyer or who is a guest in his home."

Alaska*   *Section 45.02.318*
No change.

Arizona   *Section 44–2335*
Alternative A adopted.

Arkansas   *Section 85–2–318*
Alternative A adopted.

California   *Section 2–318*
Not adopted.

Colorado   *Section 4–2–318*
No alternative chosen. Language adopted reads: "A seller's warranty whether express or implied extends to any person who may reasonably be expected to use, consume or be affected by the goods and who is injured by breach of the warranty."

Connecticut   *Section 42A–2–318*
No alternative chosen. Language adopted reads: "This section is neutral with respect to case law or statutory law extending warranties for personal injuries to other warranty."

Delaware
Not adopted.

District of Columbia   *Section 28–2–318*
No change.

Florida   *Section 672.318*
No alternative chosen. Language adopted reads: "A seller's warranty whether express or implied extends to any natural person who is in the family or household of his buyer, who is a guest in his home or who is an employee, servant *or* agent of his buyer if it is reasonable to expect that such persons may use, consume or be affected by the goods and who is injured by breach of the warranty. A seller may not exclude or limit the operation of this section."

Georgia   *Section 2–318*
No alternative chosen. Language adopted reads: "No privity is necessary to support an action for a tort; but if the tort results from the violation of a duty, itself the consequence of a contract, the right of action is confined to the parties and privies to that contract, except in cases where the party would have had a right of action for the injury done, independently of the contract, and except as proved in Code section 109A–2–318. However, the manufacturer of any personal property, sold as new property, either directly or through a dealer or any other person, shall be liable in tort, irrespective of privity, to any natural person who may use, consume or be affected by the property and who suffers injury to his person or property because the property when sold by the manufacturer was not merchantable and reasonably suited to the use intended and its condition when sold is the proximate cause of the injury sustained; a manufacturer may not exclude or limit the operation thereof."

Hawaii   *Section 490:2–318*
Alternative C adopted.

Idaho   *Section 28–2–318*
Alternative A adopted.

Illinois
Not adopted.

Indiana
Not adopted.

Iowa   *Section 554.2318*
Alternative C adopted.

Kansas   *Section 2:318*
No alternative adopted. Language adopted reads: "A seller's warranty whether express or implied extends to any natural person who may reasonably be expected to use, consume or be affected by the goods and who is injured in person by breach of the warranty."

Kentucky
Not adopted.

Louisiana
Not adopted.

Maine   *Section 2:318*
No alternative adopted. Language adopted reads: "Lack of privity between plaintiff and defendant shall be no defense in any action brought against the manufacturer, seller or supplier of goods for breach of warranty, express or implied, although the plaintiff did not purchase the goods from the defendant, if the plaintiff was a person whom the manufacturer, seller or supplier might reasonably have expected to use, consume or be affected by the goods."

Maryland   *Section 2–318*
Alternative A adopted. Language adopted reads: " . . . or any other ultimate consumer or user of the goods or persons affected thereby," added following "home".

Massachusetts   *Section 2:318*
Alternative A adopted. "Lack of privity between plaintiff and defendant shall be no defense in any action brought against the manufacturer, seller, lessor or supplier of goods for breach of warranty, express or implied, although the plaintiff did not purchase the goods from the defendant, if the plaintiff was a person whom the manufacturer, seller, lessor or suppler might reasonably have expected to use, consume or be affected by the goods. A manufacturer, seller, lessor or supplier may not exclude or limit the operation of this section: Failure to give notice shall not bar recovery under this section unless the defendant proves that he was prejudiced thereby. All actions under this section shall be commenced within three years next after the date the injury arose."

Michigan
Not adopted.

Minnesota   *Section 336.2–318*
Alternative A second sentence reads: "A seller may not exclude or limit the operation of this section."

Mississippi   *Section 75–2–318*
Alternative A adopted. Language reads: "In all causes of action for personal injuries or property damage or economic loss brought on account of negligence, strict liability or breach or warranty, including actions brought under the provisions of the Uniform Commercial Code, privity shall not be a requirement to maintain said action." Mississippi Code 11–7–20.

Missouri
Not adopted.

Montana
Not adopted.

Nebraska
Not adopted.

Nevada
Alternative A adopted.

New Hampshire   *Section 382–A: 2–318*
Not adopted. Language adopted reads: "Actions of Warranties against Manufacturers, Sellers or Suppliers of Goods. Lack of privity shall not be a defense in any action brought against the manufacturer, seller or supplier of goods to recover damages for breach of warranty, express or implied, or for negligence, even though the plaintiff did not purchase the goods from the defendant, if the plaintiff was a person whom the manufacturer, seller or supplier might reasonably have expected to use, consume or be affected by the goods. A manufacturer, seller or suppler may not exclude or limit the operation of this section."

New Jersey
Not adopted.

New Mexico
Not adopted.

New York   *Section 12:318*
Not adopted. Language adopted reads: "A seller's warranty, whether express or implied, extends to any natural person if it is reasonable to expect that such person may use, consume or be affected by the goods and who is injured in person by breach of warranty. A seller may not exclude or limit the operation of this section."

North Carolina
Not adopted.

North Dakota   *Section 41–02–35*
Alternative C adopted.

Ohio
Not adopted.

Oklahoma
Not adopted.

Oregon
Not adopted.

Pennsylvania
Alternative A adopted.

Rhode Island   *Section 6A–2–318*
Not adopted. Language adopted reads: "A seller's or a manufacturer's or a packer's warranty whether express or implied including but not limited to a warranty of merchantibility provided for in 6A–2–318 of this chapter, extends to any person who may reasonably be expected to use, consume or be affected by the goods and who is injured by breach of the warranty. A seller or a manufacturer or a packer may not exclude or limit the operation of this section."

South Carolina   *Section 36.2–318*
Not adopted. Language adopted after "natural person" reads: ". . . who may be expected to use, consume or be affected by the goods and whose person or property is damaged by breach of the warranty."

South Dakota   *Section 2–318*
Not adopted. Language adopted in first sentence of the section reads: "A seller's warranty whether express or implied extends to any person who may be reasonably expected to use, consume or be affected by the goods and who is injured by breach of warranty."

Tennessee   *Section 37–2–318*
Not adopted. Language adopted reads: "In all causes of action for personal injury or property damage brought on account of negligence, strict liability or breach of warranty, including actions under the provisions of the Uniform Commercial Code, privity shall not be a requirement to maintain said action."

Texas   *Section 2.318*
Not adopted. Language adopted reads: "This chapter does not provide whether anyone other than a buyer may take advantage of an express or an implied warranty of quality made to the buyer or whether the buyer or anyone entitled to take advantage of a warranty made to the buyer may sue a third party other than the immediate seller for deficiencies in the quality of the goods. These matters are left to the courts for their determination."

Utah   *Section 70A–2–318*
Alternative C adopted.

Vermont   *Section 2–318*
Not adopted. Words "who is in the family or household of his buyer or who is a guest in his home" omitted.

Virgin Islands
Not adopted.

Virginia   *Section 8.2–318*
Not adopted. Language adopted reads: "Lack of privity between plaintiff and defendant shall be no defense in any action brought against the manufacturer or seller of goods to recover damages for breach of warranty, express or implied, or for negligence, although the plaintiff did not purchase the goods from the defendant, if the plaintiff was a person whom the manufacturer or seller might reasonably have expected to use, consume or be affected by the goods; however, this section shall not be construed to affect any litigation pending on June twenty-nine, nineteen hundred sixty-two."

Washington
Not adopted.

West Virginia   *Section 46–2–318*
Alternative A adopted.

Wisconsin
Not adopted.

Wyoming   *Section 34–21–235*
No alternative adopted. Language adopted reads: "A seller's warranty whether expressed or implied extends to any person who may reasonably be expected to use, consume or be affected by the goods and who is injured by breach of the warranty."

State Dramshop Acts

Alabama	Selling to minors is illegal.
Alaska	Licensed servers are liable for injuries resulting from intoxication of a customer.
Arizona	Licensed servers are liable for selling alcohol to an obviously intoxicated customer.
Arkansas	Under *Carr v. Turner* (1965), common law liability is not imposed on the provider.
California	Employers have been found liable for employee conduct stemming from employer events.
Colorado	Maximum award for injury and support is $150,000 from a server.
Connecticut	Maximum award for injury and support is $50,000 from a server.
Delaware	Commercial servers are virtually immune from liability.
District of Columbia	If drinking takes place in D.C., but accident occurs in Virginia or Maryland, the plaintiff can sue under D.C. law.
Florida	A seller or provider of alcohol is not liable for injury or damage caused by served person unless that person is a known, habitual drinker or minor.
Georgia	Recently lowered BAC from 0.12 to 0.10.
Hawaii	29th state to allow automatic revocation of person's license for drunk driving.
Idaho	Servers of alcohol can be liable for injuries caused by an intoxicated person.
Illinois	Maximum recovery for injury and support is $40,000 from a server.
Indiana	Considered strictest state in holding servers liable.

Iowa	Only licensed servers of alcohol are liable for actions of intoxicated persons.
Kansas	Courts have ruled that employers owe no duty to third parties for wrongful acts of employees.
Kentucky	Licensed servers are not liable unless it is proven they knowingly served an intoxicated customer.
Louisiana	Legislature says consumption, not service, is proximate cause of any damage.
Maine	Maximum recovery for injury and support is $250,000 from a server.
Maryland	With Virginia, considered the most lenient on punishing servers.
Massachusetts	Servers of an intoxicated person may be liable if injuries arise from the person's actions.
Michigan	Recent tort reforms better protect servers from liability.
Minnesota	Courts have found that it is not the duty of a seller to "carefully scrutinize" drinkers.
Mississippi	Legislature says consumption, not service, is proximate cause of any damage.
Missouri	1985 Statute of Limitations excludes social hosts from government regulation.
Montana	Servers are not liable unless a person was obviously intoxicated at the time of purchase.
Nebraska	Repealed dram shop act in 1935.
Nevada	1990 U.S. Supreme Court ruling: First time drunk-driving offenders are not entitled to a jury trial.
New Hampshire	A server providing alcohol to intoxicated persons is liable for injuries.
New Jersey	1984 N.J. Supreme Court ruling against party host precipitated investigation of social host liability.
New Mexico	In August 1993, became the first state to qualify for special anti-drunk-driving federal grant.
New York	Selling to an intoxicated person makes the server liable for damages.
North Carolina	Maximum award for injury and support is $500,000 from servers.
North Dakota	Selling to an intoxicated person makes the server liable for damages.

Ohio	Courts have found social hosts liable when serving minors who then cause third-party injuries.
Oklahoma	Driving under the influence (DUI), reached at 0.10 faces stiffer penalties than driving while intoxicated (DWI), reached between 0.05 and 0.09.
Oregon	Commercial and social hosts are not liable unless they have served a visibility intoxicated person.
Pennsylvania	Existing cases have ruled that social hosts are not liable except where negligence is involved.
Rhode Island	Maximum award for injury and support is $200,000 from servers.
South Carolina	No common law liability for social hosts.
South Dakota	To date, social hosts have not been found liable for injuries to third parties.
Tennessee	Servers can be held liable if they are found to have served an obviously intoxicated person.
Texas	Selling and/or serving alcohol to an intoxicated person makes the server liable for damages.
Utah	Maximum award for injury and support is $300,000 from servers.
Vermont	Under common law, social hosts may be tried for negligence.
Virginia	Courts have said that liquor laws must come from the state legislature.
Washington	Courts have found companies liable for providing liquor to intoxicated employees.
West Virginia	*Walker v. Griffith* (1986) upheld legal action against a tavern owner.
Wisconsin	Third-person liability extends primarily to those selling or serving minors.
Wyoming	No person who legally sells alcohol can be held liable.

State Equal Opportunity Laws

State	Civil Rights	Gender	Marital Status	Sexual Orientation	Age	Physical Disability	Mental Disability
Alabama						X	
Alaska	X	X	X		X	X	
Arizona	X	X			X		
Arkansas		X				X	
California	X	X	X		X	X	
Colorado	X	X			X	X	
Connecticut	X	X	X		X	X	X
Delaware		X			X		
District of Columbia	X	X	X	X	X	X	
Florida	X	X	X		X	X	
Georgia		X			X	X	
Hawaii		X	X		X	X	
Idaho		X			X		
Illinois	X	X	X		X		
Indiana	X	X				X	
Iowa	X	X			X	X	X
Kansas	X	X				X	
Kentucky		X			X (40–65)		
Louisiana		X			X	X	
Maine	X	X			X	X	X
Maryland	X	X	X		X	X	X
Massachusetts	X	X			X		
Michigan	X	X	X		X		
Minnesota	X	X	X		X	X	
Mississippi							
Missouri	X	X				X	
Montana	X	X	X		X	X	X
Nebraska	X	X	X			X	

State	Civil Rights	Gender	Marital Status	Sexual Orientation	Age	Physical Disability	Mental Disability
Nevada	X	X			X	X	
New Hampshire	X	X	X		X	X	X
New Jersey	X	X	X		X	X	
New Mexico	X	X			X	X	X
New York	X	X	X		X	X	
North Carolina		X			X	X	
North Dakota		X			X		
Ohio	X	X			X	X	
Oklahoma	X	X					
Oregon		X	X		X	X	X
Pennsylvania	X	X			X	X	
Rhode Island	X	X			X	X	
South Carolina	X	X			X		
South Dakota	X	X					
Tennessee	X	X			X		
Texas		X			X (public)		
Utah	X	X			X		
Vermont		X			X	X	X
Virginia		X				X	X
Washington	X	X	X		X	X	X
West Virginia	X	X			X	X	
Wisconsin	X	X		X	X	X	
Wyoming	X	X		X	X	X	
Puerto Rico		X			X		
Virgin Islands		X			X		

NOTE: All states not adopting employment opportunity laws are covered by the Federal Civil Rights Act, Title VII (race, color, religion, national origin, gender, pregnancy, equal pay, age, disability).

# Highlights of the Americans with Disabilities Act of 1990

TITLES OF THE AMERICANS WITH DISABILITIES ACT

Title I Employment
Title II Public Accommodations
Title III Transportation
Title IV State and Local Government Operations
Title V Telecommunications

DEADLINES FOR COMPLIANCE

Title I Employers with 25 or more employees—July 26,1992
         Employers with 15 to 24 employees—July 26, 1994
Title II ADA requirements become effective—January 26, 1992
          New facilities that must be accessible—January 26, 1993
Title III Public bus, rail systems—January 26, 1992
           Privately operated bus, van systems—January 26, 1992
Title IV ADA requirements become effective—January 26, 1992
Title V ADA requirements become effective—January 26, 1993

GUIDELINES FOR COMPLYING WITH TITLE I (EMPLOYMENT) OF THE ADA

Title I of the Americans with Disabilities Act prohibits discrimination against anyone with a physical or mental impairment that substantially limits one or more major life activities, anyone with a record of such impairment, and anyone regarded as having such an impairment. Employers are obliged to make reasonable accommodations for employees unless accommodation presents an undue hardship to the employer.

Major Life Activities   This term includes, but is not limited to, the following activities: breathing, walking, caring for oneself, seeing, hearing, performing manual tasks, speaking, learning, working.

People/conditions *NOT* protected by the ADA:

- Current drug users (former and recovering drug users, and people wrongly thought to be drug users, *are* protected)
- Alcoholics who do not perform requirements of their jobs (former and recovering alcoholics *are* protected)
- Homosexuals and bisexuals, *unless* they are discriminated against because they have or are believed to have AIDS or HIV viruses
- Transvestites, pedophiles, transsexuals, people with gender disorders, exhibitionists, voyeurs, compulsive gamblers, kleptomaniacs, pyromaniacs, people with mental disorders brought about by use of psychoactive drugs (such as LSD)

Reasonable Accommodations   These include adjustments and modifications to a job and its tasks that make it possible for a qualified person with a disability to perform the essential requirements of a job.

Undue Hardship   This covers significant costs or difficulty incurred by an employer when making an accommodation.

Essential Functions   These are the critical tasks that a person, with or without accommodation, must be able to perform in order to perform in a certain position.

GUIDELINES FOR COMPLYING WITH TITLE II (PUBLIC ACCOMMODATIONS) OF THE ADA

Title II of the Americans with Disabilities Act requires business that serve the public to make readily achievable changes that will make their facilities, services, and public areas accessible to people with disabilities.

Readily Achievable   Changes that are "easily accomplishable and able to be carried out without much difficulty or expense." Small business (those with fewer than 30 employees or less than $1 million in gross receipts) that spend from $250 to $10,250 on ADA modifications to existing facilities may claim 50 percent of those expenses as a tax credit.

Following are some areas to check for accessibility in a public hospitality operation.

Parking lot:

- Designated parking spaces for people with disabilities, located near the facility's entrance
- Drop-off area near the entrance
- Barrier-free pathway from parking lot to entrance

Entrance:

- Barrier-free entrance
- Door at least 32 inches wide
- Door that can be opened easily

Interior hallways, aisles, and pathways:

- Barrier-free (steps, bumps, etc.) hallways, aisles, and pathways
- Floor or carpeting that is easily navigated by someone with a wheelchair
- Absence of protruding objects that a sight-impaired person might run into
- Elevator that is large enough to accommodate a wheelchair, and has controls that are written in Braille and are no more than 48 inches from the floor

Public rest rooms:

- Located on the same level as other public areas, or accessible to them
- Doors that are at least 32 inches wide
- Room for maneuverability of a wheelchair
- Grab bars next to at least one toilet
- At least 30 inches of clearance under sinks
- Faucets, towel racks, dispensers, and hand-dryers that are easily reached

Glossary

Abandoned property   Property that is deliberately discarded by the owner

Abandonment   When a tenant vacates the rented premises before the lease term expires, without the landlord's consent

Acceptance   In contracts, agreement by the recipient of an offer to the terms of an offer

Accession   Increase in property by production of the property or by actual addition to the property

Accessory use   Property use that is incidental to the main use for which an area is zoned

Administrative law   Laws created and often administered and enforced by federal, state, and local agencies

Adulteration   Contamination of food products as a result of unsanitary handling during preparation, packing, or storage

Adverse possession   Method of acquiring property without voluntary transfer by the owner; this method requires the possessor to occupy the property, openly and visibly, for a period of time fixed by state statutes

Affirm   In an appeal, the appellate court may *affirm* or uphold the lower court's decision rather than reverse or remand it

Affirmative defense   Response by the defense that attacks the legal right of the plaintiff to make a claim, caused by a plaintiff act or failure to act that may entitle the defense to a favorable judgment

Affixation   Attachment of property to real estate with the intention to make the property a permanent fixture

Agreement   Essential contract element that requires that all parties agree to the terms of an offer and accept the offer

Air rights   Legal rights to the air space above land

Alcoholic   One who is addicted to liquor; in legal/foodservice terms this is relevant when the server knew or should have known the patron was an alcoholic

Alien   Any person who is not a United States citizen; the person may be a *legal* or an *illegal* alien

Answer   First response by the defense to the plaintiff's charge or complaint

Antitrust   Actions taken by businesspersons that violate laws designed to prevent monopolies

Appeal   1) Request, usually by the loser in a lawsuit, for an appellate court to review a trial court decision; 2) Actual trial of an appeal

Appellate court   Intermediate review courts at the federal or state level; these courts are between the trial and supreme courts in terms of review authority

Appraisal right   Right to redeem shares of stock at fair market value

Arbitration   Method of settling disputes whereby an arbitrator decides the case; this method binds all parties to the decision

Assault   Any act or threat that creates a reasonable risk of physical harm to a person

Assignee   Person to whom another person assigns rights and interest in real property

Assignment   Transfer of interest and rights in real property

Assignor   The person who assigns rights and interest in real property to the assignee

Assumption of risk   A legal defense that is used when the plaintiff should have known about the risk and voluntarily assumed it

Attachment   1) Enforceable security interest in property; 2) Physical act of seizing property by the sheriff, as ordered by a court

Attorney-client privilege   Duty all lawyers owe clients not to disclose information given them in their role as lawyers

Bailment   Voluntary transfer of goods from one person to another for safekeeping

Bankruptcy   Federal procedure that enables a person or business to escape nonexempt debts when circumstances make a debtor incapable of meeting them

Battery   Act that physically harms a person

Bilateral contract   Contract in which both parties are expected to perform because a promise has been exchanged for a promise

Bill of lading   Document, issued to the shipper of goods by the carrier, that lists the goods accepted for transport

Blacklisting   In franchising, a *per se* antitrust violation where franchisees boycott a supplier's business for refusing to fund a promotion of the franchised product

Blue laws   State and local laws prohibiting conducting certain business operations on Sunday

Board of directors   The governing body of a corporation

Bona fide occupational qualification   In the Civil Rights Act of 1964, an exemption to the rule of equal employment opportunity may be created if the difference in treatment is justified because of the requirements of the job

Breach of contract   Nonperformance of contract terms by one of the parties

Breach of warranty   Nonperformance according to the terms of a warranty of sale

Brief   Summary of facts of the trial of a case given on appeal

Burglary   Breaking into and entering a building with the intent to steal when the business is unoccupied or very few people are present

Capacity   In contracts, the legal capacity to enter into and perform a contract, including the emotional maturity and mental competence to comply

Case law   Law created by judicial decision by the highest court to act on an issue; also called *common law*

Casualty insurance   Insurance designed to protect property from risks other than fire, such as strikes and robberies

Causation   Direct relationship between a legal harm and the alleged cause

Caveat emptor   Let the buyer beware

Caveat vendor   Let the seller beware

Certificate of occupancy   Certificate issued by local building inspectors after the building has been approved for business use

Certiorari   Permission by a court saying it will hear a case; the request to the court is called a *petition for certiorari*

Chapter 11 bankruptcy   A bankruptcy procedure for individuals, partnerships, and corporate debtors whereby they apply for reorganization of financial structure

Chapter 7 bankruptcy   Straight bankruptcy whereby all nonexempt debts are absolved

Chapter 13 bankruptcy   Form of bankruptcy that permits the debtor to pay debts according to a plan agreed to by the creditors

Child labor laws   Laws designed to protect minors from unfair and exploitative labor practices

Civil rights laws   Federal, state, and local laws that prohibit discriminatory practices against members of particular racial or minority groups

Civil unrest   Disturbances caused because of riots or politically motivated terrorist activity

Class action suit Suit brought on behalf of a group of people who have an interest in the outcome

Closely-held corporation Family-owned corporation or one owned by a few close individuals; also called a close corporation

Closing arguments Arguments presented at a trial by attorneys for both the plaintiff and the defendant, after they have rested their cases

Collateral Personal property that may be held as security for a loan or other extension of credit

Commercial zoning Allocating land use to businesses

Common law Law created by the highest court to deal with an issue; this law remains in effect until changed by statute; also called *case law*

Communication In contract law, the offer to contract that limits the offer to specific persons

Comparative negligence Defense to liability claims, which basically assigns a percentage of any judgment against the defendant to the plaintiff; also called *contributory negligence*

Compensatory damages Damage awards not intended to punish the defendant in a lawsuit, but only to compensate the plaintiff for injuries or losses

Complaint First pleading in a lawsuit, which notifies the parties involved of the purpose and facts of the case

Compromise Voluntary settlement of a dispute between the parties, without recourse to lawsuits or third parties

Concealment Failure to give information relevant to an insurance claim that might cause the carrier to deny that claim

Conciliation Another term for *mediation*, which is the action of third parties who try to settle a dispute

Confiscation Taking of private property by the government for public use

Confusion Mixing of the property of two persons

Consequential damages Damages that arise from the results of a breach of contract, but were not part of the contract terms

Consideration Money or other compensation that is given by one party to a contract to the other party in exchange for something of value

Constructive eviction Conditions that force a tenant to move, such as the presence of vermin

Contempt of court Action or refusal to act that obstructs a court's work or undermines its dignity

Contingent fee Attorney's fee based on a percentage of a favorable award or judgment

Contract   Legally enforceable agreement between two parties, in which each agrees to perform according to the terms of the agreement

Contract law   Body of laws that enforce legal contracts

Contract of sale   Contract for the purchase and transfer of goods from seller to buyer

Contributory negligence   Defense to liability suits where the plaintiff is partially responsible for the illness or injury, and the court assigns a percentage of the award or damages to the plaintiff; also called *comparative negligence*

Conversion   Use of another's property without legal justification

Copyright   Method of protecting literary and audiovisual works

Corporation   Form of business created by statute that allows a group of persons to work together for profit

Counterfeit   Not the real thing; when applied to money, the difference may be hard to detect

Counteroffer   Rejection of a contract offer that causes a new offer to be made, except in the case of a firm offer

Credit   Delay of payment through mutual agreement

Creditor   Person to whom money is owed

Criminal trespass   Presence on private property of persons who intend to commit a crime

Cross-examination   Examination, at a trial, of an opponent's witnesses or the opposing party

Damages   Physical or economic harm that requires a court to require the person who caused the harm to compensate the victim; the compensation is also called damages

Debt   Money owed by a debtor to a creditor

Debtor   Person who owes money to a creditor

Deed   Document of title to property; legal rights to property are contained in the deed

Defamation   False statement that injures a person's reputation

Defamation by computer   False information attributable to a company that owns or distributes computer printout information

Default   Failure to perform a legal duty; failure to comply with any of the procedural requirements of a trial may result in a *judgment by default*

Defendant   Person or entity against whom legal action is taken

Definiteness   In contract law, terms must be specific and firm enough for a court to enforce

Delivery   Actual transfer of property from one person to another

Demolition lien   Legal claim by which a regulatory body may demolish a building, and then seek compensation from the owner

Demurrer   Response to a complaint filed by the plaintiff in a lawsuit

Department of Justice   Federal cabinet department that enforces antitrust laws, and is the legal arm of the White House

Depositions   Sworn testimony in a trial, which is taken down by a court stenographer

Derivative shareholder action   Lawsuit to defend the rights of the corporation

Destination contract   Requires the seller to deliver the goods to a specific place

Disaffirm   In contract law, to set the contract aside or refuse to perform according to the terms

Disbarment   Process by which a lawyer's right to practice law is taken away

Discharge   1) In contract law, a contract may be discharged by performance or by mutual agreement of both parties; 2) In bankruptcy law, a person is released from debt through a discharge in bankruptcy

Disclosure Statement   In franchising, a statement that includes the financial condition of the franchiser as well as terms of the agreement; the FTC requires that the franchisor furnish prospective franchisees with this statement

Discovery   In a lawsuit, the process of gathering evidence before the trial

Dismissal   Court order that removes a lawsuit from the authority or action of the court

Disorderly conduct   Fights, noise, or other conduct that threaten the peace and safety of an area

Disparagement of goods   False, libelous statements about someone's product, business, or property; also called *slander of title*

Doctrine of unconscionable conduct   Legal doctrine that protects the victim of a contract or legal proceeding from being denied fair treatment because of unreasonable terms

Document of title   Document that gives the owner rights in the property the document covers

Dormant partner   Silent partner who does not participate in management, and also a secret partner, not known to the public

Dramshop acts   State acts that define liability for liquor sales to patrons who injure or kill third persons

Dramshop liability   Liability to third parties created by the sale of alcohol to a patron who injures another person

Economic strike Union employee strike related to hours, wages, and working condition disputes

EEOC Equal Employment Opportunity Commission

Eminent domain Government power to take private property for public use

Encumber To put up land as security to finance improvements

Equal Employment Opportunity Commission (EEOC) Federal enforcement agency for the employment section of the Civil Rights Act of 1964, the Age Discrimination Act, and the Equal Employment Opportunity Act of 1972

Equitable risk distribution According to the Uniform Commercial Code, retailers can sue growers or producers of food products for losses caused by their negligence

Equity 1) The value of interest in property; 2) Fairness

Equity of redemption Right to prevent foreclosure

Evidence Information presented at trial

Examination During a trial, questioning of a witness or party to the case by either attorney

Excess liability policy Insurance policy that protects the insured in the event the basic policy does not cover a liability claim

Exclusion In insurance, situations the carrier refuses to cover

Exculpatory clause Unenforceable, illegal contract clause whereby one party conditions the offer on the other party's waiver of legal rights related to the contract

Executed contract One not fully performed by both parties

Executory contract One not completed by one party

Express contract One in which all of the terms and conditions are fully stated

False imprisonment Personal tort that interferes with a person's right to move freely

FDA Food and Drug Administration

Featherbedding Paying employees for work not done

Federal Trade Commission (FTC) Enforcement agency with broad enforcement powers over federal trade and commerce laws, including such foodservice areas as food labeling, franchising, and advertising

Fee simple Form of ownership without any restriction on estate disposal or use

Fee simple absolute Same as *fee simple*

Fidelity insurance   Insurance that bonds an employee and protects the employer from losses arising out of employee actions, including criminal misconduct

Fiduciary   One in a position of trust

Fiduciary duty   Duty owed a client by a lawyer; the client has the right to expect that the lawyer can be trusted with personal and business information

Fire insurance   Insurance to protect premises from fire and other related hazards

Firm offer   In contracts, an offer that is good only for a specified time period

Fixtures   Anything attached to land or a building that is not readily removable

Food and Drug Administration (FDA)   Federal enforcement agency that controls the manufacture, distribution, and sale of food and drug products; this agency is the source of regulations and model standards

Foreclosure   Process that legally ends a person's rights in property; if the mortgage is not paid, a bank will usually take back the property to sell it

Foreclosure lawsuit   Lawsuit to take property back when the mortgage debt is not paid off

Foreign or natural test   Test applied by the courts in food liability cases that determines liability for food fitness based on whether or not the matter that caused the illness or injury was natural to the food item

Foreseeable risk   Risk or hazard that is visible and could be prevented

Forged check   Check that is genuine except that the signature of the actual owner is imitated

Formal contract   Tangible, negotiable instrument

Franchise   License from the owner of a trademark or service name giving another the right to sell the product and use the name in advertising

Franchise disclosure rule   FTC rule that requires franchisors to diclose certain information to a prospective franchise

Franchisee   Person who has bought into a franchise and has the right to use the products and trade names of the franchisor

Franchisor   Person or corporate entity that owns trademark product names exclusively, but licenses those rights to others to distribute the products

Fraud   Deceit; in legal terms, behavior designed to cheat another out of legal rights

Fraudulent transfers   Gift or sale for less than fair market value with intent to delay or defraud creditors from obtaining legitimate claims

Freehold estate   Estate that may be held indefinitely

FTC   Federal Trade Commission

Full partner   Same as *general partner*

Function contract   Foodservice contract that reserves a specific space and an exact time for a special event

Future interests   Rights of a future owner of property

General partner   Full partner in a partnership who has unlimited liability and management powers and shares fully in the profits

General warranty deed   Property deed that conveys the largest number of rights

Genuineness of assent   Legal phrase that means a contract must be fully negotiated by both parties

Gift   Method of acquiring or transferring real property whereby the owner voluntarily gives it away

Grading   Federal evaluations of meat products that enable buyers to select standardized products

Grandfather clause   Clause contained in local building codes that allows existing buildings to be exempt from the newer requirements

Grand larceny   Theft of property or money of a value over a certain amount set by law

Gross lease   Type of lease in which the tenant's sole responsibility is to pay rent

Ground lease   Type of lease in which the landlord rents out vacant land and the tenant agrees to erect a building, which the tenant may, in turn, mortgage

Guarantor   Person or entity who assures a creditor that the debt will be paid

Home rule   Power granted by the state to certain cities or countries to regulate certain areas

Hot cargo clause   Clause as part of a labor agreement whereby the employer agrees not to deal in another's products

House rule   Rule or set of rules foodservice operators use to maintain the class and type of an operation

Implied contract   One in which the terms and conditions are implied by the conduct of the parties

Implied warranty   Unspoken guarantee by a seller that a product is fit for sale

Improper venue   Wrongful filing of a claim in a jurisdiction unable to hear the case

Incompetent   Person who is legally defined as lacking sufficient mental capacity to understand the consequences of his or her acts, particularly regarding contracts

Informational picketing   Union representatives distributing literature outside a business

Inheritance   Method of obtaining property; a will transfers specific property to a person or party

Insanity   Recognized mental incompetence that is judicially declared and can render the person's contracts void

Insolvency   When debts exceed assets

Insurable interest   Legally recognized interest in an insured item

Intangible damage   Damage that is not physically apparent, such as harm to one's reputation

Intangible property   Property that has no physical substance, but represents rights or interest in property

Intentional tort   Legal wrong by which the wrongdoer wants to harm another

Interrogatories   Form of discovery used before a trial in which the person is given a written set of questions and responds with written answers

Interstate commerce   Trade between the states

Interstate succession   When an owner of property dies without a will

Intoxicated   Legally drunk or incapable of being in complete control of one's actions

Intrastate commerce   Trade within a state

Invasion of privacy   Use of a person's name or picture for commercial purposes, or disclosure of private facts without permission

Involuntary petition   Petition for bankruptcy filed by the creditor or creditors of the debtor

Joint venture   Form of partnership in which two or more persons agree to enter into a single business enterprise

Judgment NOV   Judgment notwithstanding the verdict

Jurisdiction   Power of a court to hear and decide a case

Jury   Panel of private citizens who are assigned to decide which side is right or wrong in a court case

Kosher   Food that is prepared in the Orthodox Jewish manner

Landlord   Person or entity who owns and leases land and/or a building

Landmark   Site or property declared by a local commission to be of historical value and in need of preservation

Larceny   Theft

Law merchant   Common law system of rules used by sellers

Lease   Contract to use land, a building, or part of a building for a specific time period

Legal duty   Legal responsibility one person owes another, and which is enforceable in court

Legality   In contract law, a contract must not violate any laws in order to be enforceable

Letter of intent   Letters that announce *intent*, such as to make reservations for a special function or space available for a function, and are not generally accepted as contracts

Liability   Enforceable responsibility one person has to another

Liability insurance   Insurance designed to protect foodservice operators from losses resulting from patron liability claims

Libel   Written defamation of a person's reputation

Lien   Legal claim against property, or interest in property, that entitles the bearer to take over the property, or part of it, in the event that the debt is not settled

Life estate   Form of ownership in which owner rights are for life only

Life tenant   Nonowner occupier of land

Limited partnership   Form of business in which the limited partner has limited liability in exchange for lack of management control

Liquidated damages   Provision included in a contract that provides for compensation in the event of a failure of one of the parties to perform

Listing contract   Real estates broker's agreement

Mailbox rule   Rule governing contracts which requires that an acceptance of an offer be properly posted and mailed

Majority   Legal age to form contracts, vote, drink, or go into business; set by state law

Malicious mischief   Intended injury to real or personal property of another

Malpractice   Misconduct or negligence by a professional in the process of handling another's business; used here to apply to lawyers

Manifestation of authority   In franchising, representation of control or authority given to patrons through advertising

Mediation   Method of settling disputes by which a third person aids in negotiating by moderating and making recommendations that are not binding

Mental distress   Intentional tort that causes emotional harm

Merger   Union of two or more companies, where one becomes a part of the other

Minor   Person who has not reached legal age, or majority, as set by state law

Misbranding   False and misleading advertising, usually on the product label

Misdemeanor   Lesser criminal offense than a felony, punishable by a fine or a brief jail term

Misrepresentation   Tort that involves false statements or concealments of facts for personal gain

Mixed zoning   Local zoning that allows both residential and commercial use

Monopoly   Illegal control over the manufacture, sale, distribution, or price of a product

Mortgage   Method of putting up land or buildings as security for a loan or to finance improvements of land

Mortgagee   Lender in a mortgage plan

Mortgagor   Person taking out a mortgage on land or a building

Motion   In a trial, a request by either party that the judge must act on or dismiss

Mutual mistake   In contract law, a mutual misunderstanding on one of the terms of a contract

National Labor Relations Board (NLRB)   Enforcement board created by the National Labor Relations Act, which enforces union and employer labor practices

Natural persons   In law, refers to persons rather than artificial entities such as corporations

Negligence theory   Liability theory that requires the injured person to prove negligence on the part of another

Negotiable instrument   Promise to pay

Negotiation   Discussion of contract terms to obtain the best deal

Net lease   Lease in which the tenant assumes incidental costs, such as utilities, insurance, or taxes

Nominal damages   Small compensation, awarded by a court to one party to a contract for its breach by the other party

Nominal partner   Partner in name, who may be liable under certain circumstances

Non compos mentis   Describes the state of mind of a person who is unable to understand the legal consequences of his or her actions or abilities to contract

Nonconforming use   Form of zoning in which a businessperson applies for permission for special use when an area has been restricted to one use, such as for residential purposes

Non-freehold estate   Restricted estates that may be held for a specific time

Nonprofit   Organization not engaged in a profit-making activity, or classified as such for tax purposes

Nontrading partnership   Partnership that provides a service rather than a product

Not-for-profit   Tax classification covering businesses not engaged in profit-making activities

Objectionable conduct   Misbehavior or conduct that is offensive to others

Objectionable appearance test   Use of observation of an individual to determine sobriety

Occupational Safety and Health Administration (OSHA)   Federal agency assigned to enforce regulations pertaining to the on-the-job health and safety of workers; OSHA is the source of laws and standards regarding worker safety

Offer   To make a proposal; the first step in a contract

Open-end mortgage   Form of mortgage whereby money may be borrowed at different times under the same agreement

Opening statement   Statement given by attorneys for both sides in a trial, which presents what it is they will attempt to prove

Operation of law   In contract law, circumstances that result in automatic termination of an offer

OSHA   Occupational Safety and Health Administration

Ostensible partner   Same as *nominal partner*

Overrule   1) A high court may overturn the decision of a lower court by overruling it; 2) A judge may refuse to acknowledge an objection during a trial by overruling it

Partial performance   Incomplete performance of contract terms

Partnership   A form of business organization in which two or more persons agree to conduct a business for profit

Partnership by estoppel   Legal principle that may make nominal or ostensible partners liable if customers act on the belief that the person is a full partner

Patent   Government grant giving an inventor the exclusive right to manufacture, use, and distribute a product for 17 years

Penalty clause   Contract clause that requires a penalty to be charged to the party who violates the contract

Per diem   Per day; during a trial, an attorney may charge for each day of the trial

Perfection   Protection of a security interest through attachment

Per se violation   Antitrust violations that cannot be justified by proper motive, absence of intent, or economic necessity

Personal property   Moveable property

Petition of certiorari   Request by a party to a lower court trial that a higher court review the lower court's decision

Petit larceny   Less serious theft

Plaintiff   Person bringing suit

Police power   Government's right to enforce laws

Policy   Insurance contract between the carrier and the insured, which provides the coverage conditions

Polygraph   Lie detector; mechanical device run by an operator, and used to determine whether or not a person is telling the truth

Possession   Simple ownership of property

Possessory   Form of landholding where the occupier has certain rights in, but does not own, real estate

Preemptory challenge   Challenge of a juror by the attorney for either side; the court must comply with these challenges, but their number is limited by law

Preferential liens   Prohibited transfer of cash or property by the debtor in a bankruptcy proceeding

Preferential payment   Prohibited payment made by a debtor to one creditor in preference to the others

Preliminary negotiations   Negotiations that precede actual contract offers; accepting a bid is an example

Premium   Payment for insurance coverage; this charge is not a set fee and may be based on a number of factors

Preservation zoning   Zoning designed to preserve historical landmarks

Pre-trial hearing   Conference held between the parties and the judge before a trial to settle any disputes, and to determine the length of the trial, the number of witnesses that may be called, and any special requirements

Price fixing   Setting the prices of products by franchisees or franchisors to harm competitors

Prima facie   "On the face of it"; fact that will be assumed to be true, unless evidence disputes it

Private corporation   Corporation established for the benefit of its owners, as distinguished from a *public* corporation

Privity of contract   Direct relationship between the parties to a contract

Prohibition   Historical period in the United States when it was illegal to make or sell alcoholic beverages

Property   In law, legally protected rights in or ownership of anything having recognized value

Proximate cause   Actual cause of an illness or injury; in liability cases, this is a requirement

Proxy   Piece of paper representing a shareholder's right to vote

Proxy fight   Corporate voting battle between existing management and those seeking to control the company, in which shareholder votes are sought

Public corporation   One owned by the general public through issues of shares

Publicly-held corporations   Private corporations that issue some stock to the public

Punitive damages   Money awarded to a person who has been harmed by another's malicious or deliberate act; distinguished from *compensatory damages*

Purchase   Method of obtaining property by ordering it from the seller and paying for it

Purchase money mortgage   Where the seller of real property finances the buyer's purchase in exchange for the mortgage

Purchase order   Simple contract form that contains a promise to pay, a description of the goods, and the place to which they are to be shipped

Quantum meruit   Compensation for work done

Quasi-contracts   Duties imposed by a court on one party to a contract in order to correct an injustice

Quiet enjoyment   Right to peaceful and undisputed use of real property

Quitclaim deed   Property deed, which provides the buyer only with those rights the seller has at the time of the sale, and includes any limitations

Raised genuine check   Same as *forged check*

Ratification   Performance of a voidable contract, obliging the other party to perform

Reaffirmation agreement   Agreement between the creditor and the debtor whereby the debtor agrees to repay debts that would otherwise be discharged in bankruptcy; the agreement must be approved by a court

Real estate   Property that is immoveable, such as land or a building

Real property   Same as *real estate*

Reasonable care   In tort law, the standard of care expected of a reasonable person by another in a particular set of circumstances

Reasonable expectations test   Test applied by courts in food liability cases that determines liability for food fitness based on whether one might reasonably expect to find the matter that caused the injury or illness in the final, processed item

Reasonable person test   Test used in liability for safety cases, used to determine whether the accused party exercised ordinary care to protect patrons against unreasonable risks of harm

Rebuttal   Disputing evidence presented by the opposition in a trial, by presenting contrary evidence

Recovery deduction   Depreciation allowance allowed by the Economic Recovery Tax Act of 1981, which accelerates the time in which purchases may be written off

Re-direct examination   In a trial, the plaintiff's re-examination of a witness, after the defendant's attorney has examined that witness, to restore his or her credibility or address any damaging evidence brought out during the examination

Reformation   Revision of a contract by a court to express the real intentions of the parties

Remand   To send back; a higher court may send a case back to a lower court with instructions as to how the latter is to look at it again

Request for admissions   Motion by either side in a trial to obtain specific facts from the other party

Rescission   Right to cancel a contract because of fraud, duress, mistake, or failure of consideration by the other party

Reservation contract   In foodservice law, a contract involving the use of an establishment's tables where a specific table or tables are reserved in advance for the patron for a specific date and time, without payment or deposit

Residential zoning   Local restriction of land use to residential purposes

Res ipsa loquitur "The thing speaks for itself"; in liability law, the use of circumstancial evidence of negligence to impose liability

Respondeat superior Legal rule that makes the employer liable for an employee's conduct when that conduct occurs in the course of employment and is intended to benefit the employer

Restitution Return of each party to a contract to the position occupied prior to the contract

Restoration In a contract between a seller of majority and a buyer who is a minor, if the seller fully performs and the minor avoids performance, the minor must restore the goods or services received to the seller; the minor is not required to compensate the seller for any damage caused

Retainer Agreement to employ an attorney

Reversal Higher court rejection of a lower court decision on a case

Reverse discrimination When a member of a group not normally the object of discrimination is discriminated against

Riot Violent, uncontrolled public disorder

Robbery Felony in which property is taken from the person of the victim or in the victim's presence by means of force, violence, or threat of the same

Rule-of-reason violation Violation of antitrust laws involving marketing activities that create monopolies; the conduct may be justified by proof of economic necessity

Sale and leaseback Real estate transaction in which the seller sells land for full value and the buyer agrees to resell the land to the seller

Sale on approval Transaction in which the goods are delivered to the buyer, who does not assume title or risk of loss until approval or acceptance of the goods

Sale on return Transaction in which the buyer assumes risks of loss and holds title unless he or she returns the goods

Secondary boycott Illegal union boycott of a neutral employer to force the latter to harm the employer with whom the union has a dispute

Secret partner In a partnership, one whose presence in the firm is not known to the public but who may manage the business

Secured creditor Creditor whose claim is legally enforceable and secured by collateral

Security interest Protected claim on property, usually secured by collateral

Separate but equal Civil rights doctrine which predated the Civil Rights Act of 1964, and which held that accommodations for white and nonwhite people could be separate as long as they were equal in quality

Service contract Contract to secure professional services

Service mark Mark used to indicate and distinguish the services of one person from another

Services Work performed by one party in return for payment

Sex discrimination Action or treatment meant to deny rights or fair treatment to a person on the basis of sex

Sexual harassment Treatment of employees designed to demean them by lewd comments, suggestion, or demands for sexual favors as a condition of employment

Sexual preference Sexual orientation

Shareholders Owners of a corporation

Shipment contract Contract specifying that the seller deliver goods to a carrier

Shortchange Action where a cashier may be cheated out of correct change by quick, deliberate action on the part of a patron

Silent partner In a partnership, a person who does not take part in managing the business

Slander Oral defamation of a person's reputation; proof of damage to the victim is usually essential

Slander of title Defamation of goods, which consists of false statements about a person's product, business, or property

Small claims courts Municipal courts set up to decide on cases not requiring an attorney and for an amount of money not higher than set by the state

Sole proprietorship Form of business organization with one owner who runs it and has sole liability for the business

Special warranty deed Property deed in which the seller guarantees he or she will not diminish the value of the land

Specific performance Court-ordered correction of a breach of contract where one party is required to perform the duties specified in the contract rather than pay damages to the other party

Spot zoning Form of zoning granted by special request where a landowner requests that a single property be exempt from regular zoning; the special zoning must benefit the surrounding community

Squatting Attempt to take property merely by occupation; this is illegal unless the rightful owner creates a lawful tenancy

Statements of intention Promises to make a future contract; not an offer

Statute of limitations Period of time set by law during which a claim can be decided or a legal wrong corrected; once this period runs out, the claimant has no recourse to the courts

Statutory law   Law formed by elected officials at the local, state, or federal level

Straight bankruptcy   Complete bankruptcy, where all nonexempt debts are discharged and remaining assets are distributed to creditors by a trustee

Strict liability   Common law liability theory that assigns liability regardless of proof of negligence or breach of warranty

Subchapter S corporation   Tax option form of incorporation that permits corporations that qualify to be taxed the same as partnerships

Sublease   Agreement whereby the tenant gives up part of the lease to another tenant

Supoena ad testificandum   Requirement that case-related documents be brought to court by the person in charge of them

Supoena duces tecum   Court order requiring a person or company to provide documents for a trial

Substantial performance   Contract performance adequate for the provider to recover payment

Subsurface rights   Rights to soil below the surface of land and other natural material on it or under it

Summary judgment   Judgment in favor of one side before a trial ends, based on a motion filed by that side

Summons   Notification of a lawsuit or legal action against a party, which directs that party to answer the complaint

Sustain   Court agreement to a point made by either the attorney for the plaintiff or the attorney for the defendant

Tangible damages   Damage that is visible and easily verifiable, such as physical injury

Tangible property   Property that has physical substance

Tenancy at sufferance   Occupying land, or squatting; this form of tenancy is not recognized by law

Tenancy at will   One created by the landlord for as long as he or she desires

Tenancy for years   One in which the duration is specified in the lease

Tenancy from period to period   One that does not specify the duration of the lease, but requires that rent be paid at a certain time

Tenant   Person who occupies and leases land and/or a building from a landlord

Tenants in partnership   Ownership in common of a partnership; none of the parties has the exclusive right to handle partnership business without consulting the other parties

Tender   1) Offer; 2) When the seller performs the requirements of a contract

Terrorism   Hostile, violent activities designed to frighten, intimidate, or harm a person or group of people

Theft of services   Using a stolen or canceled credit card to obtain services *or* intentionally avoiding payment by misrepresentation of fact or by avoidance

Tip credit   Tips added to the cash wage in order to meet the federal minimum wage scale

Title   1) Ownership right to goods; 2) Document showing ownership in goods

Tort   Civil wrong arising out of a legal duty owed by one person to another

Totally fabricated check   Check that is not drawn on a real account and has no real owner

Trade fixtures   Personal property installed on premises as part of a lease permitting use for a specific purpose

Trademark   Mark that is fixed on tangible goods to identify and protect the product registered by the owner

Trading partnership   One engaged in buying, selling, or trading goods and services for profit

Transcript   Official, typed record of a court proceeding

Trespass   Wrongful entry onto privately owned property

Trespass to land   Wrongful entry onto privately owned land

Trial court   First court to hear a case

Trial de novo   Retrial of an entire case before a reviewing court

Trustee   1) In bankruptcy, the person appointed to distribute the assets of the bankrupt; 2) A fiduciary, or person in a position of trust

Truth-in-menu   Government guidelines and laws specifying truthful advertising of products on a menu

Tying contract   Agreement required by a franchisor specifying that franchisee purchases of certain items must be made from the franchisor

Unforseeable cause   Direct cause of an injury that could not be foreseen by the defendant in a negligence suit

Unilateral contract   Agreement whereby one party promises to perform only in exchange for an act by another party

Unilateral mistake   Misunderstanding of contract terms where only one party makes the mistake

Unintentional tort   Negligence or a wrong committed by one person which injures the person or property of another

United States Department of Agriculture (USDA)   Federal cabinet department that enforces laws relating to farming and is the source of regulations in this area

Unjustified act   Malicious mischief not provoked by another's conduct

USDA   United States Department of Agriculture

Variance   Legal permit issued by local government to allow a property owner to use land for other than the specified zoned use

Venue   Area in which a case may be tried or in which the court has jurisdiction

Verdict   Decision on a court case by a jury

Voidable title   Ownership that cannot hold up in court because the "owner" has no real title

Void title   Term meaning the seller had nothing to sell

Voluntary petition   Petition by a debtor requesting that the debtor be judged a bankrupt

Walkout   Customer who uses a food service's food or services without paying

Wanton act   Malicious mischief committed in total disregard of the consequences

Warranty of seisin   Right to exclusive title and the right to convey the property

Wildcat strike   Illegal strike not called for by a union

Workers' compensation   1) Insurance required by federal law to cover workers for on-the-job injuries; 2) The federal law requiring such insurance

Writ of certiorari   Petition asking a high court to review a case

Zoning   Use of police power to regulate local land use

Index

*NOTE: *n* indicates a selection appearing in the Notes section following a chapter.